Animal Welfare

Volume 21

The Animal Welfare series has been designed to help contribute towards a culture of respect for animals and their welfare by producing academic texts addressing how best to provide for the welfare of the animal species that are managed and cared for by humans. Books in the series do not provide a detailed blue-print for the management of each species, but they do describe and discuss the major welfare concerns, often in relation to the wild progenitors of the managed animals. Welfare has been considered in relation to animals' needs, concentrating on nutrition, behaviour, reproduction and the physical and social environment. Economic effects of animal welfare provision were also considered where relevant, as were key areas where further research is required.

More information about this series at https://link.springer.com/bookseries/5675

Sophie Riley

The Commodification of Farm Animals

Co-ordinating Editor: Stephanie Torrey

Springer

Sophie Riley
Faculty of Law
University of Technology Sydney
Sydney, NSW, Australia

ISSN 1572-7408
Animal Welfare
ISBN 978-3-030-85869-8 ISBN 978-3-030-85870-4 (eBook)
https://doi.org/10.1007/978-3-030-85870-4

This Springer imprint is published by the registered company Springer Nature Switzerland AG.
The registered company address is: Gewerbestrasse 11, 6330 Cham, Switzerland

This book is dedicated to the billions of animals who are reared and slaughtered for food, fiber, and other products each year, and also, to the memory of Faith Patterson, Anne Ruprecht, and Marion Lazar.

Foreword

Throughout the world an estimated 150 billion animals are killed each year to provide food and other products for humans.[1] The public is often unaware of how their food is produced, perhaps mistakenly secure in the notion that free-range eggs lack cruelty or that organic meat means high levels of animal welfare. Yet, this is not necessarily the case. Commencing with Ruth Harrison's seminal tome, *Animal Machines*, first published in 1964, debate on institutionalized cruelty in the animal product sector has not stopped.[2] Although there are countless books and journal articles on animal cruelty and animal welfare, insufficient attention has been paid to the historical origins of law and policy that underpin present-day practices.

This is where Sophie Riley's book, *The Commodification of Farm Animals*, differs from others. In common with these books, *The Commodification of Farm Animals* examines anti-cruelty regulation and animal welfare tenets, but does so against a background of economic and commercial webs that bring together law and policy, philosophy, veterinary science, and international law. In doing this, the book extends existing scholarship to formulate a concept of commodification, based on the "commodification pathway." This is described as "a utility-driven means of animal management, which objectifies animals as goods in the marketplace, prioritizes human uses, and lacks meaningful engagement with ethical principles" (Chap. 1, Sect. 1.1). The argument is structured so that each of the early chapters deals with one steppingstone of the commodification pathway – the growth of

[1]Taylor, and Heather Fraser, "The Cow Project: Analytical and Representational Dilemmas of Dairy Farmers' Conceptions of Cruelty and Kindness" (2019) 8 (2) Animal *Studies Journal*, 133 (quoting statistics generated by the Food and Agriculture Organization), available from https://ro.uow.edu.au/asj/vol8/iss2/10

[2]Ruth Harrison, *Animal Machines,* Stuart. London (1964); F W Rogers Brambell, *Report of the Technical Committee to Inquire into the Welfare of Animals Kept under Intensive Livestock Husbandry Systems*, Her Majesty's Stationery Office, (1965, Reprinted 1967), available from https://edepot.wur.nl/134379; Karen Sayer, "Animal Machines: The Public Response to Intensification in Great Britain, c. 1960-c. 1973" (2013) *Agricultural History*, 473, available from DOI: https://doi.org/10.3098/ah.2013.87.4.473; Diana Stuart and Ryan Gunderson, "Nonhuman Animals as Fictitious Commodities: Exploitation and Consequences in Industrial Agriculture," (2020) 28 (3) *Society and Animals*, 291, 300–304, available from https://doi.org/10.1163/15685306-12341507

markets, the metaphorical shadow cast by the Enlightenment, the focus of the veterinary profession on farm animals as goods in trade, and the similar focus of international law. In the latter case, the book traces and evaluates the development of what may be loosely termed, "quarantine treaties," breaking new ground. Furthermore, Riley has undertaken original research in five key areas discussed below.

Historical Analysis

Existing analyses largely concentrate on events of the twentieth century, particularly from the time of World War II. Undoubtedly, the "productionist paradigm" of intensive agriculture became evident from the mid-twentieth century, exemplified in approaches of the Food and Agriculture Organization, which focused on how to feed populations following the ravages of World War II.[3] It was an approach that was immensely successful in harnessing scientific and technological advances to improve efficiency and increase outputs, leading to increasingly intensive production. Yet, the use of scientific and technological advances facilitated views of animals as "manufactured, tradable products."[4]

The *Commodification of Farm Animals* does not dispute the fact that beginning in the mid-twentieth century animal production (at least in Western jurisdictions) became progressively more intensified. However, the book argues that intensive production was not so much driven by technology, as by commercial and economic imperatives, which technology was able to sustain. Gunderson has made these arguments in the context of intensive production in the twentieth century, and as this book argues, comparable patterns are apparent by at least the nineteenth century.[5]

While Linda Kalof accurately concludes that the pathway toward commodification had started from the sixteenth century, arguably the most far-reaching changes occurred during the nineteenth and mid-twentieth centuries.[6] This period saw economic and commercial activities related to the growth of markets merge with scientific and technological advances, thus laying the foundations for the commodification pathway which allowed market forces to shape animal well-being in a commercial context. Chapters 2 and 3 of the *Commodification of Farm Animals*, which respectively deal with market expansion and the Enlightenment, provide background to the development of this pathway, helping to explain how trade and commerce was prioritized over animal well-being.

[3]Tim Lang, "Achieving Access to Ethical Food: Animal and Human Health Come Together" in Jacky Turner and Joyce D'Silva (eds), *Animals, Ethics and Trade*, 261, 263, Earthscan (2006).

[4]Ben Mepham, "The Ethical Matrix as a Decision-making Tool, With Specific Reference to Animal Sentience" in Jacky Turner and Joyce D'Silva (eds), *Animals, Ethics and Trade*, Earthscan (2006), 134, 141.

[5]Ryan Gunderson, "From Cattle to Capital: Exchange Value, Animal Commodification, and Barbarism" (2011) 39 (2) *Critical Sociology*, 259, 261.

[6]Linda Kalof, *Looking at Animals in Human History*, Reaktion Books, London (2007), 135.

The Role of the Veterinary Profession

The evolution and development of the veterinary profession has been the subject of much commentary and debate, and this literature provides a starting point for the *Commodification of Farm Animals*.[7] The discussion extends the inquiry by linking three developments of the nineteenth century and analyzing their impact on the growth of markets and animal well-being: government regulation, from the 1860s dealing with the cattle plagues; the importance of animal disease to trade and commerce; and the role of the veterinary profession. It is critical to these arguments that government harnessed veterinary knowledge with the aim of ensuring that trade and commerce did not introduce and spread animal diseases. However, as Bernard Rollins has indicated, this resulted in the "historical subordination of veterinary medicine to agriculture."[8] This focus led to the exigencies of the marketplace predominating, and as Riley concludes (Chap. 6), reduced animals to the status of bulk commodities.

Consequently, by the time of World War II, regimes had already marketed animals in commercially driven ways, providing post-1945 regimes with existing pathways to market animals as efficiently as any other commodity. The emphasis on disease-free shipments also became synonymous with a narrow version of what later came to be known as animal welfare, so that animal disease and by association animal well-being were regarded as trade issues, a stance that more or less continues to the present. In an innovative approach to research in this field, the book also underscores the influence of veterinarians by examining selected international veterinary and industry conference instruments (sourced from the library of the Royal College of Veterinary Surgeons) to identify what was important to stakeholders and, thus, highlight what was prioritized in the treatment of animals.

---

[7] Among others: Frederick Smith, "The Work of the British Army Veterinary Corps at the Fronts" (1919) 75 (1) *The Veterinary Journal*, 8; John Francis, "John Gamgee (1831–1894): Our Greatest Veterinarian" (1962) 118, *The British Veterinary Journal*, 430; Paul Brassley, "Animal Health and Veterinary Medicine," in Joan Thirsk (general editor) *The Agrarian History of England and Wales* Volume VII 1850–1914, Part 1, 587, 588–589, 591–592, Cambridge University Press (2000); Vanessa Carli Bones, and James W Yeates,"The Emergence of Veterinary Oaths: Social, Historical and Ethical Considerations", (2012) 2 (1) *Journal of Animal Ethics*, 20; Iain Patterson, *A Great British Veterinarian Forgotten: James Beart Simonds 1810–1904*, J A Allen & Co, London (1990); J R Fisher "Not quite a Profession: The Aspirations of Veterinary Surgeons in England in the Mid Nineteenth Century" (1993) *Institute of Historical Research*, 284; Michael Worboys, "Germ Theories of Disease and British Veterinary Medicine, 1860–1890" (1991) 35 *Medical History*, 308, 318; Wilson, Alison, "One Hundred and Fifty Years of the State Veterinary Service, 1865–2015" (2015) 72 (3) *Science in Parliament*, 8, available from http://www.scienceinparliament.org.uk/wp-content/uploads/2015/09/150-YEARS-OF-THE-STATE-VETERINARY-SERVICE-by-Dr-Alison-Wilson.pdf; James Yeates, *Veterinary Science: A Very Short Introduction*. OUP (2018).

[8] Bernard E Rollin, "Putting the Horse before the Descartes: My Life's Work on Behalf of Animals," Temple University Press, Philadelphia (2011), 22.

International Trade, Quarantine, and Animal Disease

Another neglected area of study broached by the book centers on international instruments dealing with animal health and quarantine and how these influenced farm animal commodification. Riley evaluates a selection of instruments including early treaties of "Friendship, Commerce and Navigation," before moving to treaties specifically dealing with the management of animal disease. The discussion traces the evolution of quarantine from a unilateral national initiative, assumed to belong to the national domain, to a matter that was more appropriately dealt with at the international level. For this reason, her book also touches on nascent international standards that at the end of the twentieth century were developed by the World Organisation for Animal Health (OIE).[9]

Notwithstanding these developments, as Riley highlights in Chap. 5, Sect. 5.6

> While veterinary science was critical to the health of animals, the fact that animal disease, and by implication animal wellbeing, was categorised as an issue of international trade, meant that the regime contributed greatly to the commodification of farm animals. Animal health came to be regarded less as a matter of animal wellbeing, and more as an issue of protecting national territory against diseased shipments.

She reinforces this argument by concluding that when, during the latter part of the twentieth century, animal welfare became the basis for law and policy, the health of farm animals similarly became equated with a narrow version of welfare, focusing on animal merchantability and profitability of the sector.

Use of the Phrase "Animal Welfare"

One of the most informative parts of the book is found in Chap. 6, which identifies how the phrase "animal welfare" came into common usage. It is generally accepted that the Brambell Report popularized the term from about 1965.[10] Before that time, the term was used only a handful of times, with even Ruth Harrison not using that term in *Animal Machines*. Riley has researched prior references and how they were understood, providing a timeline of society's relationship to animals:

[9] *International Agreement for the Creation at Paris of an International Office for Dealing with Contagious Diseases of Animals and Annex*, opened for signature January 25, 1924, [1925] *ATS* 15 (entered into force January 12, 1925). The organization has 182 members. The original name of the OIE was the Office International des Épizooties. However, in May 2003 the name was changed to the World Organisation for Animal Health, while keeping the historical acronym, OIE.

[10] F W Rogers Brambell, *Report of the Technical Committee to Inquire into the Welfare of Animals Kept under Intensive Livestock Husbandry Systems*, above 2.

- Veterinarian George Fleming, who in 1871 used the term with respect to the need to look after domestic animals.[11]
- Henry Salt, who in 1892 used the term with respect to maintaining the "welfare" of captive wild animals.[12]
- Major Charles Westley Hume, who in 1926 established the University of London Animal Welfare Society (ULAWS) to tackle animal cruelty, particularly in the field of scientific research.[13]

Chapter 6 also concludes that the Brambell Report inferred the existence of animal welfare legislation from law and policy dealing with anti-cruelty regulation. In this way, the regime which prevailed in the United Kingdom drew the notion of animal welfare from existing anti-cruelty regulation, which reinforced the use of animals within recognized commercial parameters. It is a pathway that still strongly forms the backbone of law and policy today.

Further Interpretation of Norbert Elias

Chapter 7 of the book, "A Sufficient Level of Repugnance," continues the groundbreaking approach of the research by using the philosophies of Norbert Elias to evaluate humanity's relationship to farm animals. Commentators, such as Tester, have linked the growth of anti-cruelty regulation to Elias' work in general,[14] but Riley's analysis specifically evaluates farm animals and also the advent of Ag-Gag laws as a decivilizing process. Given that much animal production occurs in secret, animal abuse in production systems may never see the light of day. This influences how society, particularly consumers, makes decisions and wields political pressure.[15] Consequently, as Riley pointedly notes, Norbert Elias' historical evaluations are crucial to understanding how society arrived at the point of animal commodification, and additionally, are also relevant for the future direction of regimes, because they identify weaknesses and gaps, underscoring where improvements might be made.

Riley concludes her groundbreaking book with a quote from Robert Garner noting that when human and farm animal interests clash, regulators are unlikely to

[11] George Fleming, *Animal Plagues: Their History, Nature, and Prevention*, Chapman and Hall (1871), XV. (reprint).

[12] Henry Salt, *Animals' Rights: Considered in Relation to Social Progress*, (1894), revised edition G Bell and Sons Ltd. (1922), 26, available from https://ia600901.us.archive.org/32/items/cu31924030305332/cu31924030305332.pdf

[13] S M Wickens, (compiler and editor) *Science in the Service of Animal Welfare, A Chronicle of Eighty Years of* UFAW, UFAW, Wheathampstead (2007), 6, 7.

[14] Keith Tester, *Animals & Society the Humanity of Animal Rights*, Routledge, New York (1991), 64.

[15] David Harvey and Carmen Hubbard, "Reconsidering the Political Economy of Farm Animal Welfare: An Anatomy of Market Failure" (2013) 38 *Food Policy*, 105, 110.

elevate animal interests, yet this is precisely what needs to happen if animals are to be treated less like commodities and more like the sentient and feeling beings they truly are.

I like to say it is a matter of *who* we eat, not *what* we eat because so-called food animals were once individuals who experienced rich and deep emotional lives. They were not unfeeling objects who did not care what happened to themselves, their families, or their friends. Science shows they care very much.

I hope that *The Commodification of Farm Animals* enjoys a broad global audience because the nonhumans who wind up on our plates need all the help they can get. Humane-washing glosses the horrific life experiences of these individuals because the abuse to which they are interminably exposed – hidden behind the walls of animal processing plants – is incomprehensible and far too often written off when people say something like, "We're doing the best we can and that's all we can do." The "best" is not close to what these animals truly need. Indeed, what they want, need, and deserve is to be respected, be treated with dignity, and be removed from human production chains.

Boulder, CO
June 2021

Marc Bekoff

Animal Welfare Series Preface

Animal welfare is attracting increasing interest worldwide, and the knowledge and resources are available to, at least potentially, provide better management systems for farm animals, as well as companion, zoo, laboratory, and performance animals. The key requirements for adequate food, water, a suitable environment, companionship, and health are important for animals kept for all of these purposes.

The attention given to animal welfare in recent years derives largely from the fact that the relentless pursuit of financial reward and efficiency, to satisfy market demands, has led to the development of intensive animal management systems that challenge the conscience of many consumers, particularly in the farm and laboratory animal sectors. Livestock are the world's biggest land users, and the farmed animal population is increasing rapidly to meet the needs of an expanding human population. This results in a tendency to allocate fewer resources to each animal and to value individual animals less than the group. In these circumstances, the importance of each individual's welfare is diminished.

Increased attention to welfare issues is just as evident for zoo, companion, sport, and wild animals. Of growing importance is the ethical management of breeding programs since genetic manipulation is now technically advanced. There is less public tolerance of the breeding of extreme animals if it comes at the expense of animal welfare (e.g., brachycephalic dogs). The quest for producing novel genotypes has fascinated breeders for centuries. Dog and cat breeders have produced a variety of deformities that have adverse effects on their welfare, but nowadays the breeders are just as active in the laboratory, where the mouse is genetically manipulated with equally profound effects.

In developing countries, human survival is still a daily uncertainty for many, so that provision for animal welfare has to be balanced against human welfare. Animal welfare is usually a priority only if it supports the output of the animal, be it food, work, clothing, sport or companionship. However, in many situations the welfare of animals is synonymous with the welfare of the humans that look after them, because happy, healthy animals will be able to assist humans best in their struggle for survival. In principle, the welfare needs of both humans and animals can be provided for, in both developing and developed countries, if resources are properly husbanded. In reality, the inequitable division of the world's riches creates physical and psychological poverty for humans and animals alike in many parts of the world.

The intimate connection between animals and humans that was once so essential for good animal welfare is rare nowadays, having been superseded by technologically efficient production systems where animals on farms and in laboratories are tended by increasingly few humans in the drive to enhance labor efficiency. With today's busy lifestyles, companion animals too may suffer from reduced contact with humans, although their value in providing companionship, particularly for certain groups such as the elderly, is beginning to be recognized. Animal consumers also rarely have any contact with the animals that are kept for their benefit.

In this estranged, efficient world, people struggle to find the moral imperatives to determine the level of welfare that they should afford to animals within their charge. A few people, and in particular many companion animal owners, strive for what they believe to be the highest levels of welfare provision, while others, deliberately or through ignorance, keep animals in impoverished conditions in which their health and wellbeing can be extremely poor. Today's multiple moral codes for animal care and use are derived from a broad range of cultural influences, including media reports of animal abuse, guidelines on ethical consumption and campaigning and lobbying groups.

This series has been designed to contribute towards a culture of respect for animals and their welfare by producing learned treatises about the provision for the welfare of the animal species that are managed and cared for by humans. The early species-focused books were not detailed management blue-prints; rather they described and considered the major welfare concerns, often with reference to the behavior of the wild progenitors of the managed animals. Welfare was specifically focused on animals' needs, concentrating on nutrition, behavior, reproduction, and the physical and social environment. Economic effects of animal welfare provision were also considered where relevant, as were key areas where further research is required.

In this volume, the series again departs from the single vertebrate species model to a detailed analysis of how animals have become commodities in our current social landscape. Starting with a historical assessment, Sophie Riley analyses the origins of this dehumanizing development, concluding that, although there was major activity in this field in the mid to late twentieth century, it actually began in the sixteenth century. The commercialization of animals, particularly farm animals, developed alongside the growth in other tradable commodities, dragging with it a veterinary profession that eschewed its responsibilities to animals to render services in support of a global manufacturing industry. Controlling animal disease had become particularly important because it had the potential to hinder animal growth and production, and in particular trade because of the risk of spreading infectious diseases. Sophie Riley charts the development of the term animal welfare, which was of little importance prior to 1965, the time of the legendary Brambell Report on calf production in the UK. She discovers that the term has come to be embedded within a commercial framework, so that animals' feelings only matter if they can be accommodated within this framework. Sophie Riley concludes her seminal work by charting how animal production is becoming hidden from society, who would find it distasteful and disturbing. From Ag-gag laws to secret animal production locations, this decivilizing process is perhaps the final stage in the transition of animals to mere commodities. However, society has a habit of rejecting secret government-sponsored cruelty to any animals, human or not, as the last century was testament to. Sophie Riley's book will help to shape a future in which commodification is revoked, be it by proper provision of resources for farm animals, replacement of farm animal production on farms to cellular growth in laboratories or rejection of animal consumption in favor of plants.

Former Professor of Animal Welfare and Director, Centre for Animal Welfare and Ethics, University of Queensland, Brisbane, QLD, Australia
Clive Phillips

Trouw Nutrition and Department of Animal Biosciences, University of Guelph, Guelph, ON, Canada
Stephanie Torrey

Preface

This book provides a historical evaluation of how and why farm animals became commodified (focusing on the United Kingdom), arguing that the reasons are found in commerce and economics, feasibly more than, technology.

The idea for this book grew out of material researched and then discarded for my PhD. That degree, which was completed some 15 years ago, focused on border controls and the rules of the World Trade Organization in the management of invasive alien species (e.g., weeds, pathogens, and diseases). The material that was abandoned related to treaties and international agreements, negotiated up to the mid-twentieth century, which dealt with border controls and quarantine in the international trade of farm animals. Given the lengthy time that has elapsed from that research to the present, it is an understatement to say that this book has had a prolonged incubation.

This is partly explainable by the fact that, although the plight of farm animals was concerning, I only thought about in a formal way when I began to teach animal law and policy at the University of Technology Sydney. One of the most enduring memories from that time was dismay at how disconnected the policy underpinnings of the regime were from the practical application of the law. Although ostensibly based on sentience and animal welfare, regulation clearly depended less on these concepts and more on the use that humans had for the animal in question. Consequently, and notwithstanding equivalence in terms of sentience, the law allows animals raised for food to be treated in ways that would never be tolerated, say, for companion animals.

Every now and again thoughts of the research I had cast aside bubbled to the surface. Many questions came to my mind: why did these agreements omit any mention of animal well-being or welfare; why was animal health treated as a trade issue; and was this a pattern replicated elsewhere? In reality, farm animals were treated as commodities in processes that extended beyond their legal classification as personal property. After all, companion animals are also classified as personal property, yet they are treated very differently. From time to time, I researched further, concentrating on animal diseases and treaties for animal protection. Preliminary inquiries confirmed the emergence of a pattern, but one which did not provide clear-cut answers, instead raising further issues. Foremost among these was the role of the veterinary profession. Where did the profession stand with respect to animal well-being in the context of trade, and did the profession have input into quarantine treaties?

A persistent theme that emerged was the fact that intensive production of farm animals appeared to follow a settled pathway from at least the nineteenth century. Yet, conventional wisdom specified that this trajectory did not appear until after World War II, when the need to feed growing populations merged with advances in technology, providing both the need and means for increased animal production. This laid the responsibility for intensive farming at the door of the kind of technology that developed from the mid-twentieth century. Yet, technology had already played a part in animal production, long before World War II. In the nineteenth century, the industrial revolution saw the introduction of railways, steamships, and refrigeration, which allowed the animal product sector to expand. Thus, technology had already provided the means to intensify production prior to 1945.

To my mind, the real issue was what drove the need for, and use of, technology. The answer lay in consumer demand and the need to maintain supply. This not only saw farm animals treated as commodities, but also increasingly subjected them to regulation dominated by the vagaries of the marketplace. Their commodification was thus an economic and commercial issue, which relied on technology for its fulfillment.

By 2014 my thoughts had sufficiently crystallized to write a halfway decent application for study leave, to research archival material at the Royal College of Veterinary Surgeons in London. That research proved to be very significant, answering many questions and shaping the structure of the book. However, upon my return to Sydney, thoughts of writing up the material and preparing a book proposal were set aside for 2 years while I completed an administrative role as the Director of Students at the University of Technology Sydney. Once that role was completed, the research recommenced and the writing began. As with much of academic life recently, the project was interrupted by the COVID-19 pandemic, but the lockdowns also provided time to focus and complete the project.

The book is not intended to provide the definitive explanation of how and why farm animals became commodified, but rather to provide a snapshot of how this occurred in one jurisdiction – the United Kingdom. The introduction to the book explains that the United Kingdom was selected for study, not because it is particularly blameworthy, but because it was the epicenter of the meat markets for the period under discussion and the laws of the United Kingdom became influential in other parts of the world.

I hope that the reader enjoys this book and that it provides material for thought and further research.

Sydney, Australia Sophie Riley

Vote for This Book from the Community

Dr. Sophie Riley's book adds notably to the texts already available on animal welfare law. Starting with an historical analysis and describing the economic and commercial forces that are at work, Dr. Riley demonstrates the dangers of unbridled commodification of farm animals. She invokes international law to analyze and evaluate the development of "quarantine treaties." There must be re-engagement with ethical principles to bring the human mind into focus upon our empathy with animal species. Animals are not just things. Fortunately, an increasing numbers of citizens are coming to realize this and to demand change. Dr. Riley's book explains why the change is happening and should accelerate.

The Honorable Michael Kirby, AC CMG
Sydney, Australia

Acknowledgments

This book would not have been possible without the assistance of many people and institutions, whom I would like to thank and acknowledge. First, my sincere thanks to the Library of the Royal College of Veterinary Surgeons and to Clare Boulton in particular, for advice and assistance in locating source documents on international veterinary conferences. Second, much appreciation goes to the librarians at the University of Technology Sydney, especially Helen Chan, Jackie Edwards, Carl Hoschke, and Hossain Salahuddin for arranging inter-library loans and sourcing journal articles and other material not easily available. Third, heartfelt thanks to the many organizations who provide their material freely on the Internet: Sara Belingheri of the Wellcome Trust (UK); Phil Browning, archivist at the RSPCA (UK); the HathiTrust Organisation; the digitized treaty base of the United Nations; the Australasian Legal Information Institute; the British Statutes Project; Internet Archive Organization; Google Books; the website of the RSPCA (NSW); the website of Voiceless, the Animal Protection Institute (Australia); the website of Compassion in World Farming; and the website of the Universities Federation for Animal Welfare. Apologies in advance, if I have unintentionally omitted anyone.

A special acknowledgment to Richard Perren, whose encyclopedic work on farm animals and agricultural markets formed the starting point for Chap. 2 of the book and weaves its way through other chapters.

In addition, thanks to Bibhuti Sharma, my editor at Springer, for his patience as the COVID-19 pandemic slowed down the process of writing. And, last but not least, gratitude and thanks go to my husband, Frank Riley, who read drafts of each of the chapters to ensure that they were readable, that the arguments flowed, and that infinitives remained unsplit.

Contents

List of Abbreviations

EU	European Union
FAO	Food and Agriculture Organization of the United Nations
MFN	Most favored nation
OIE	World Organisation for Animal Health
RSPCA	Royal Society for the Prevention of Cruelty to Animals
UFAW	Universities Federation for Animal Welfare
UK	United Kingdom
ULAWS	University of London Animal Welfare Society
USA	United States of America
WTO	World Trade Organization

List of Figures

List of Tables

1 Introduction: For What Is the Animal But the Profits Thereof?

Abstract

Intensive farming, or factory farming as it sometimes called, currently involves between 50 and 60 billion animals per year. Its use increased following the ravages of World War II, a fact often explained by advances in science and technology, which were critical to feeding the world's population from the mid-twentieth century. This book, however, argues that intensive production was not so much driven by technology as by commercial and economic imperatives, which technology was able to sustain. These imperatives had shaped the animal product sector from at least the nineteenth century, along a "commodification pathway" that may be described as a utility-driven means of animal management, which objectifies animals as goods in the marketplace, prioritises human uses and lacks meaningful engagement with ethical principles. In effect, the pathway leads to the regulation of animals as bulk commodities, paying insufficient attention to individual wellbeing. When animals become commodified, there are few limits on their use for meat and other products, so that the value of animals is determined by the marketplace, leading to animal exploitation and regard for them as a form of capital.

Keywords

Animals as capital · Commodification of animals

S. Riley, *The Commodification of Farm Animals*, Animal Welfare 21,
https://doi.org/10.1007/978-3-030-85870-4_1

1.1 Animals and Profit

The title for the introduction to this book is a wordplay on a statement made by Edward Coke (1552–1634) in 1628, "for what is the land but the profits thereof".[1] Although not all dealings with land or animals are motivated by profit-seeking, transactions with both are based on property rights that "conceptually separate"[2] land and animals from the benefits of ownership. Farm animals in particular are exploited to produce meat, fibre and other products, in circumstances of total reliance on humans.[3] Yet, neither this vulnerability nor the fact that they are living beings in their own right has persuaded humans that they should not use animals in profit-driven ways.[4] Instead, as this book argues, farm animals have become increasingly commodified.

The word "animal" is used to describe non-human animals, and the term "farm animal(s)" refers to animals "kept or raised in captivity", consistent with the definition developed by the Food and Agriculture Organization (FAO) of the United Nations.[5] However, the FAO definition is very broad, ranging from cattle, sheep, pigs, poultry and camelids to insects such as silkworms and bees, while the scope of this book is more limited, focussing on sheep and cattle, with occasional references to pig and poultry production.[6] As explained in more detail in Sect. 1.4, one of the main reasons for this limitation ensues from the fact that sheep and cattle were the primary focus of regulation throughout the nineteenth century in the UK. This not only draws attention to animals who were important to law and policy throughout the formative years of regulatory regimes overseeing the animal product sector but also keeps the material manageable. At the same time, the term farm animal is not

---

[1] Edward Coke, *The First Part of the Institutes of the Lawes of England*, Society of Stationers, London (1628), L 1 C 1 Sect 1 of Fee Simple 4 b, available from https://play.google.com/books/reader?id=NOgyAAAAIAAJ&hl=en&pg=GBS.PA5

[2] Robyn Bartel and Nicole Graham, "Property and Place Attachment: A Legal Geographical Analysis of Biodiversity Law Reform in New South Wales", (2016) 54 (3) *Geographical Research*, 267, 270, 272, https://doi.org/10.1111/1745-5871.12151

[3] Dinesh Joseph Wadiwel, "Like One Who is Bringing his Own Hide to Market", (2016) 21 (2) *Angelaki Journal of the Theoretical Humanities*, 65, 69, https://doi.org/10.1080/0969725X.2016.1182725

[4] F Bailey Norwood and Jayson L Lusk, *Compassion, by the Pound: The Economics of Farm Animal Welfare*, Oxford University Press, New York (2011) 66; Amanda Whitfort, "Justice and the Vulnerable: Extending the Duty to Prevent Serious Crimes Against Children to The Protection of Agricultural and Research Animals", (2018) 39 *Adelaide Law Review*, 125, 128–132.

[5] *Constitution of the Food and Agriculture Organization of the United* Nations, opened for signature 16 October, 1945, [1945] *ATS* No 9, entered into force on 16 October, 1945; The FAO has 194 member nations, one organization and two associate members, available from http://www.austlii.edu.au/au/other/dfat/treaties/1945/9.html; Food and Agriculture Organization of the United Nations, *Livestock Statistics – Concepts, Definitions and Classifications*, FAO (2020), Sections II.1 II.2, available from http://www.fao.org/economic/the-statistics-division-ess/methodology/methodology-systems/livestock-statistics-concepts-definitions-and-classifications/en/

[6] Food and Agriculture Organization of the United Nations, *Livestock Statistics – Concepts, Definitions and Classifications*, above, 5.

universally favoured, with critics pointing to its connotations of financial gain.[7] Nevertheless, because the term identifies a subset of animals who are the focus of this book, it is used for that reason.

Worldwide, the manner in which farm animals are raised and sold presents significant regulatory challenges for animal wellbeing, a point exacerbated by the fact that intensive farming, or factory farming as it sometimes called, involves between 50 and 60 billion animals per year.[8] One of the challenges stems from the need to feed the world's population, bearing in mind that not all agricultural land can be used to grow crops.[9] Indeed, the demand for food has been used to explain the prevalence of intensified agricultural production following the end of World War II.[10] However, some 59% of the "world's crop calories are wasted", with 25% being squandered by consumers and retailers, 9% used on biofuels and 25–30% used as animal feed.[11] These statistics call into question whether better use could be made of existing crop yields, eliminating or substantially reducing the need for intensive animal production.

Intensive animal production also attracts criticism because of its entrenched institutionalised cruelty, a fact acknowledged from the 1960s with the publication of Ruth Harrison's *Animal Machines*.[12] Although advances in animal husbandry and veterinary science have resulted in animals being healthier today than they were in previous centuries, pressures of production also mean that today's animals lead poorer quality lives "than those of their ancestors".[13] One of the main reasons for

---

[7] F Bailey Norwood and Jayson L Lusk, *Compassion, by the Pound: The Economics of Farm Animal Welfare*, above 4, 66.

[8] Compassion in World Farming, indicates over fifty billion per year, *Ending Factory Farming*, available from, https://www.ciwf.org.uk/factory-farming/; Gary L Francione, "The Abolition of Animal Exploitation" in Gary L Francione and Robert Garner (eds), *The Animal Rights Debate: Abolition or Regulation?*, 1, 2, Columbia University Press, New York (2010), estimates 56 billion animals per year; Dinesh Joseph Wadiwel, "Like One Who is Bringing his Own Hide to Market", above 3, 65, estimates over 60 billion per year.

[9] Gordon Gatward, *"Livestock Ethics"*, Chalcombe Publications, Lincoln (2001), 10.

[10] Tim Lang, "Achieving Access to Ethical Food: Animal and Human Health Come Together" in Jacky Turner and Joyce D'Silva (eds), *Animals, Ethics and Trade*, 261, 263, Earthscan, UK (2006).

[11] Emily S Cassidy, Paul C West, James S Gerber and Jonathan A Foley, "Redefining Agricultural Yields: from Tonnes to People Nourished per Hectare" (2013) 8 *Environmental Research Letters*, 1, 4–5, https://doi.org/10.1088/1748-9326/8/3/034015; additional discussion, Compassion in World Farming, web site, *Ending Factory Farming*, available from, https://www.ciwf.org.uk/factory-farming/

[12] Ruth Harrison, *Animal Machines*, first published 1964 by Vincent Stuart Publishers, 2013 Edition J Harrison and J Wilson, CABI, Wallingford, Oxfordshire (2013); Donald M Broom, "Ruth Harrison's Later Writings and Animal Welfare Work", in Ruth Harrison, *Animal Machines*, 21, 21, Vincent Stuart Publishers, 2013 Edition J Harrison and J Wilson, CABI, Wallingford, Oxfordshire (2013).

[13] Peter Sandøe, Stine B Christiansen and Björn Forkman, "Animal Welfare: What is the Role of Science?" in Jacky Turner and Joyce D'Silva (eds), *Animals, Ethics and Trade*, 41, 46, Earthscan, UK (2006); David Harvey and Carmen Hubbard, "Reconsidering the Political Economy of Farm Animal Welfare: An Anatomy of Market Failure", (2013) 38 *Food Policy*, 105, 106.

this development derives from the fact that the view of farm animals as manufactured and tradeable goods, apparent from at least the nineteenth century, has strengthened with the passage of time.[14]

This is a curious development, if it is kept in mind that the earliest anti-cruelty laws, initiated in the United Kingdom (UK) in 1822, proscribed cruelty to farm animals, specifically cattle, horses and sheep.[15] Yet almost 100 years later, farm animals are arguably far less protected than other animals, such as companion animals. One proffered explanation is that humans have long used farm animals in accordance with "exploitation principle[s]",[16] so that intensified production became the natural goal of the sector.[17] In reality, while advances in technology from the mid-twentieth century are thought to have initiated intensified animal production, such advances were not unique to that century, with innovations in refrigeration and transportation already having stimulated intensive production from the nineteenth century.[18] As Gunderson argues, one needs to look behind technology to understand that the sector was propelled by economic incentives where "capital's blind drive for self-expansion and self-accumulation" shaped the animal product sector.[19] Accordingly, market expansion and intensive animal farming were not so much driven by technology, as by commercial and economic imperatives, which technology was able to sustain.

For these reasons, this book argues that the building blocks of intensive production can be traced to events of the nineteenth and early twentieth centuries, when authorities laid the foundations for regulating animal production but failed to deal meaningfully with animal wellbeing, let alone threshold issues regarding the legitimacy of animal use. This period saw the instigation of anti-cruelty legislation, growing societal interest in preventing cruelty to animals, increasing importance of veterinary science and the negotiation of international instruments to manage trade in the sector. Throughout these initiatives, science and technology provided the

[14]Gordon Gatward, "*Livestock Ethics*", above 9, 9; Ben Mepham, "The Ethical Matrix as a Decision-making Tool, With Specific Reference to Animal Sentience" in Jacky Turner and Joyce D'Silva (eds), *Animals, Ethics and Trade*, 134, 141, Earthscan, UK (2006).

[15]1822, *An Act to Prevent the Cruel and Improper Treatment of Cattle 1822*, (3 Geo IV c 71), *The Statutes of the United Kingdom of Great Britain and Ireland 3 George IV. 1822*, his Majesty's statute and law printers London, sold by Butterworths and son, 403, available from https://archive.org/details/statutesunitedk10britgoog/page/n436

[16]Steven M Wise, "Entitling Non-Human Animals to Fundamental Legal Rights on the Basis of Practical Autonomy", in Jacky Turner and Joyce D'Silva (eds), *Animals, Ethics and Trade*, 87, 92, Earthscan, UK (2006).

[17]Compassion in World Farming, web site, *Ending Factory Farming*, available from, https://www.ciwf.org.uk/factory-farming/

[18]John P Huttman, "British Meat Imports in the Free Trade Era", (1978) 52 (2) *Agricultural History*, 247, 259.

[19]Ryan Gunderson, "From Cattle to Capital: Exchange Value, Animal Commodification, and Barbarism", (2011) 39 (2) *Critical Sociology*, 259, 259–260. In addition, an entire volume has been devoted to the subject, Brett Clark and Tamar Diana Wilson (eds), *The Capitalist Commodification of Animals*, Volume 35, Emerald Publishing Limited, Bingley UK (2021).

means for satisfying market demand and fulfilling ever-growing commercial objectives. However, commercial and economic imperatives paved the way, not only recruiting technology and science but also exerting political influence over law and policy, so that farm animals came to be regulated as commodities in the marketplace.[20]

The notion of a commodity is central to this book and is discussed in more detail in Sect. 1.2, as well as Chap. 6. It forms the basis of the "commodification pathway", which may be described as a utility-driven means of animal management, which objectifies animals as goods in the marketplace, prioritises human uses and lacks meaningful engagement with ethical principles. In effect, the pathway leads to the regulation of animals as bulk commodities, paying insufficient attention to individual wellbeing. As such, the concept can extend beyond farm animals to situations such as puppy farms, where animals are arrogated to human uses and managed in accordance with substandard industry practices.[21]

In the context of farm animals, the pathway evolved in stages, each step viewing animals as goods in trade. The origins of the first stage stem from expanding urbanisation during the eighteenth and nineteenth centuries, leading to escalating demand for animal products.[22] Technology was important to this trend because advances in refrigeration and transportation allowed producers to meet this demand. The second stage derives from the way society managed its relationship to farm animals. Although the nineteenth century saw the beginnings of a strong anti-cruelty movement, market expansion occurred against the backdrop of high levels of acceptable violence against animals. This is not to say that animal wellbeing had been ignored throughout history. It had been a matter of concern from ancient times, and for many religions, including Islam, Buddhism and Hinduism.[23] In addition, while domination over animals was reinforced by Judaeo-Christian beliefs,[24] this

[20] Chris Otter, "The Vital City: Public Analysis, Dairies and Slaughterhouses in Nineteenth-century Britain, (2006) 13 *Cultural Geographies*, 517, 517; Gergely Baics and Mikkel Thelle, "Introduction: Meat and the Nineteenth-Century City" (2017) *Urban History*, 1, 1, https://doi.org/10.1017/S0963926817000414; Richard Perren, *Taste, Trade and Technology: the Development of the International Meat Industry Since 1840* (First published 2006 Ashgate) Routledge (2017), Abingdon, Oxfordshire 50; Karen Raber, "From Sheep to Meat, From Pets to People", in Matthew Senior, Matthew (ed), *A Cultural History of Animals in the Age of Enlightenment*, Berg, Oxford (2011) 73, 73.

[21] Kimberley Smith, *Governing Animals Animal Welfare and the Liberal State*, Oxford University Press, New York (2012), 92.

[22] Robert C Allen, "Why was the Industrial Revolution British?", (2009) 4 *Oxonomics*, 50, 52; Mark B Tauger, *Agriculture in World History*, Routledge, Abingdon, Oxfordshire (2011), 103.

[23] E Szűcs, R Geers, E N Sossidou and D M Broom, "Animal Welfare in Different Human Cultures, Traditions and Religious Faiths", (2012) 25 (11) *Asian-Australian Journal of Animal Sciences*, 1499, 1501–1503, https://doi.org/https://doi.org/10.5713/ajas.2012.r.02

[24] Andreas-Holger Maehle, "Cruelty and Kindness to the 'Brute Creation': Stability and Change in the Ethics of the Man-Animal Relationship 1600–1850", in Aubrey Manning and James Serpell, (eds) *Animals and Human Society Changing Perspectives*, Routledge, Abingdon, Oxfordshire (1994), 81, 82, 86–7.

tenet had been accepted from antiquity.[25] Similarly, sporadic regulation such as *An Act Against Plowing by the Tayle, and Pulling the Wooll Off Living Sheep 1635* (Ireland)[26] and the *Massachusetts Body of Liberties 1641*[27] prohibited specific acts of cruelty against animals, but these prohibitions were not widespread. In the UK, it was not until the passage of the 1822 Cattle Act that the concept of anti-cruelty gained momentum.[28] As already noted, this legislation specifically targeted farm animals, yet by the late nineteenth century, when market expansion, both domestically and internationally, was in full swing, the treatment of farm animals was of secondary importance to profitability of the sector. The third stage involves advances in veterinary science that concentrated on diseases, such as sheep pox and cattle plague, which were important to trade. This focus was also reflected in treaties of the nineteenth and twentieth centuries, which regulated international trade and which in common with national regimes used veterinary knowledge to shape the notion of animal wellbeing to suit market demands.

These stages, all steppingstones in the commodification pathway, created settled frameworks before World War II, which governed animals in ways that increased their production value. Accordingly, farm animals were regarded as "eating machines", readily converted to "eggs, oven-ready chickens, or beefsteaks as quickly as possible".[29] By the time of World War II, farm animal production had already moved away from a system based on animal stewardship, towards one based on intensive production.[30] This transformation raised issues concerning the extent to which the market should be regulated, how it should be regulated and who should regulate it, issues that to this day have not been satisfactorily addressed.[31]

In truth, society finds it difficult to reconcile their relationship with animals because of the many and contradictory ways humans relate to them: people eat animals, use them, keep them as companion animals, admire them, experiment on them and understand that they feel pain and suffering.[32] Some, such as Nussbaum,

---

[25] Andrew Linzey, *Creatures of the Same God: Explorations in Animal Theology*, Lantern Books, New York (2009), 10–12.

[26] Richard Ryder, *Animal Revolution*, Basil Blackwell, Oxford (1989), 53–54; detailed discussion Piers Beirne, *Confronting Animal Abuse: Law, Criminology and Human-Animal Relationships*, Rowman and Littlefield, Maryland (2009), Chap. 1, 21–67.

[27] *Massachusetts Body of Liberties 1641*, paragraphs 92 and 93, available from https://history.hanover.edu/texts/masslib.html

[28] Linda Kalof, *Looking at Animals in Human History*, Reaktion Books, London (2007), 51–52, 89.

[29] F M L Thompson, "The Second Agricultural Revolution, 1815–1880", (1968) 21 (1) *The Economic History Review*, 62, 65.

[30] Harriet Friedman and Philip McMichael, "Agriculture and the State System, the Rise and Decline of National Agricultures, 1870 to the Present", (1989) XXIX-2 *Sociologia Ruralis*, 93, 106.

[31] Roger Horowitz, Jeffrey M. Pilcher, and Sydney Watts, "Meat for the Multitudes: Market Culture in Paris, New York City, and Mexico City over the Long Nineteenth Century", (2004) 109 (4) *American Historical Review*, 1055, 1057–1058.

[32] F Bailey Norwood and Jayson L Lusk, *Compassion, by the Pound: The Economics of Farm Animal Welfare*, above 4, 31.

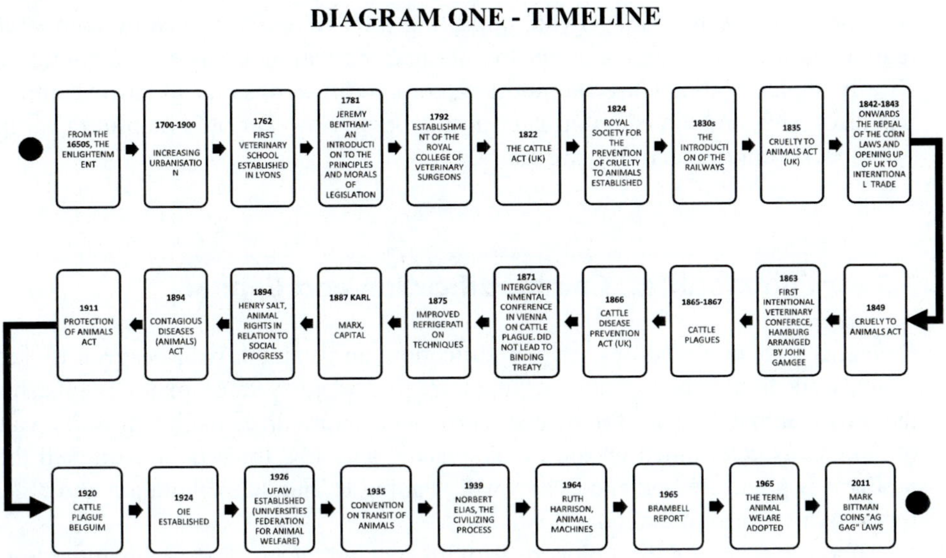

Fig. 1.1 Timeline

argue that animals are entitled to dignity, to "flourish" as their species was intended to.[33] This last point extends the debate far beyond superficial nods to animal sentience and turns a spotlight on the legitimacy of human "domination of the animal world", as well as society's complicity in failing to deal with institutionalised cruelty in production systems.[34] As the discussion in this book develops, the arguments increasingly link commodification to state-sanctioned violence against farm animals, so that by the last chapter this theme culminates in arguments that articulate how economic biases support violence becoming entrenched in law and policy.

The book however does not provide detailed economic analyses, instead identifying and examining "relational webs" which emerged during the nineteenth and early twentieth centuries.[35] A timeline, labelled in Fig. 1.1, sets out the main events that occurred during the period under discussion. In particular, the material evaluates how and why stakeholders gained and exercised power during this time. This form of historical analysis allows conceptualisation of a frame of reference,[36] which helps society understand why policies of the nineteenth and early twentieth

[33] Martha C Nussbaum, "Beyond 'Compassion and Humanity': Justice for Nonhuman Animals", in Cass R Sunstein and Martha C Nussbaum (eds) *Animal Rights, Current Debates and New Directions*, 299, 305–306, Oxford University Press New York (2004).

[34] Abigail Woods, "From Cruelty to Welfare: The Emergence of Farm Animal Welfare in Britain, 1964–71", (2011) 36 (1) *Endeavour*, 14, 16.

[35] Kristin Hoganson, "Meat in the Middle: Converging Borderlands in the U.S. Midwest, 1865–1900", (2012) 98 (4) *The Journal of American History*, 1025, 1026.

[36] Andrew Linzey, *Creatures of the Same God, Explorations in Animal Theology*, above 25, 5.

centuries have been so long-lived, outlasting their initial relevance by compelling regimes in the twenty-first century to continue regulating animals as commodities. Clearly, topics dealing with commodities, commodification and capital are central to this discussion and are taken up later in this book; however, it is helpful to furnish some background material at this point.

1.2 Commodities, Commodification and Capital

References to commodities, commodification and capital with respect to farm animals are found throughout the literature and identify three major points: how the use of animals' bodies for meat and products commodifies them,[37] how the value of animals is determined by the marketplace[38] and how trade in animals and their products is predicated on endless growth, leading to animal exploitation and regard for them as a form of capital.[39]

Although each of these points deals with specific facets of animal commodification, the foundation rests on broader questions pertaining to humanity's relationship with farm animals. In particular, these questions underscore the links between capitalism and commodification, probing how both concepts lead to substandard animal wellbeing. Although commodification is often seen as the opposite of treating animals as sentient beings, the latter does not stop the use of animals, with the dividing line being sufficiently fine for Gary Francione to observe that veganism, which proscribes animal use, is critical to ensuring that animals are not looked upon as commodities.[40] Although Francione argues this point based on ethical reasons, in practical terms, the very fact that society puts a price on animals and their products reconstructs both into little more than "consumer goods", governed by the vagaries of the marketplace.[41] The dominance of economic priorities is not new, with Linda Kalof describing commodification in the seventeenth century as a pathway to "untold misery" and suffering by animals.[42]

[37] Carol J Adams, *The Sexual Politics of Meat*, Continuum, New York (2010 anniversary edition), 51.

[38] Kimberley Smith, *Governing Animals Animal Welfare and the Liberal State*, above 21, 87.

[39] Christian Stache, "Conceptualising Animal Exploitation in Capitalism: Getting Terminology Straight", (2020) 44(3) *Capital and Class*, 401, 417, available from https://journals.sagepub.com/doi/abs/10.1177/0309816819884697

[40] F Bailey Norwood and Jayson L Lusk, *Compassion, by the Pound: The Economics of Farm Animal Welfare*, above 4, 31; Gary L Francione, "The Abolition of Animal Exploitation", above 8, 62.

[41] Virginia De John Anderson, *Creatures of Empire, How Domestic Animals Transformed Early America*, Oxford University Press, Oxford (2004), 68; Kimberley Smith, *Governing Animals Animal Welfare and the Liberal State*, above 21, 87.

[42] Linda Kalof, *Looking at Animals in Human History*, above 28, 135.

Commodification effectively objectifies animals and reduces them to a means of generating products and profits for humans.[43] However, even beyond these goals, producers also regard animals as "commodities of exchange for profit", in other words as a form of capital.[44] Carol J. Adams describes the process as derogating animals to the status of "absent referents", so that the animal and its death are forgotten and separated from the goods and profit created out of its body.[45] This process is particularly significant for the animal product sector because the measure of a robust economy includes rates of consumption, thereby absorbing farm animals "into the world of commodities", treating them more as a means of accumulating wealth and generating capital, rather than producing food.[46]

For some, these developments represent a problem that can be pinned down to the way capitalism operates, although others question whether a different system would lead to a better outcome for animals, without impacting livelihoods in the sector.[47] A major difficulty derives from the fact that while it is theoretically possible for the government to restrict the way society uses animals, the government is already complicit in the economic biases that shape the sector.[48] Law and policy ostensibly protect animals from cruelty, yet invariably exceptions permit routine husbandry practices and intensive farming, subordinating animal wellbeing to commercial imperatives.[49] This situation exists because animals are objects rather than subjects of the law. Consequently, the law allows owners to create and unlock surplus value in their animals by way of intensive production and the tenets of the marketplace, every stage being sanctioned and facilitated by the state.[50]

---

[43] Josephine Donovan, "Aestheticizing Animal Cruelty", (2011) 38 (4) *College Literature*, 202, 203, 211;

Charles Thorpe and Brynna Jacobson, "Abstract Life, Abstract Labor, Abstract Mind" in Brett Clark and Tamar Diana Wilson (eds), *The Capitalist Commodification of Animals*, 59, 97, Volume 35, Emerald Publishing Limited, Bingley UK (2021).

[44] Dinesh Joseph Wadiwel, "Like One Who is Bringing his Own Hide to Market", above 3, 69; John Bellamy Foster and Brett Clark, "Marx and Alienated Speciesism", (2018) 70 (7) *Monthly Review*, 1, 14, https://doi.org/10.14452/MR-070-07-2018-11_1

[45] Carol J Adams, *The Sexual Politics of Meat*, above 37, 51.

[46] Matthew Hilton, *Consumerism in Twentieth-Century Britain: The Search for a Historical Movement*, Cambridge University Press, Cambridge (2003), 4; Ryan Gunderson, "From Cattle to Capital: Exchange Value, Animal Commodification, and Barbarism", above 19, 261; Brett Clark and Tamar Diana Wilson, 'The Capitalist Commodification of Animals: A Brief Introduction", in Brett Clark and Tamar Diana Wilson (eds), *The Capitalist Commodification of Animals*, 1, 1, Volume 35, Emerald Publishing Limited, Bingley UK (2021).

[47] Steven McMullen, *Animals and the Economy*, Palgrave Macmillan (2016), 4; John Bellamy Foster and Brett Clark, "Marx and Alienated Speciesism", above 44, 12.

[48] Steven McMullen, *Animals and the Economy*, above 47, 4.

[49] Pamela Fiber-Ostrow and Jarret S Lovell, "Behind a Veil of Secrecy: Animal Abuse, Factory Farms, and Ag-Gag Legislation", (2016) 19 (2) *Contemporary Justice Review*, 230, 230, https://doi.org/10.1080/10282580.2016.1168257; Dinesh Joseph Wadiwel, "Like One Who is Bringing his Own Hide to Market", above 3, 74–75.

[50] Karl Marx, "*Capital, A Critique of Political Economy*", Volume II Book One: The Process of Circulation of Capital, Progress Publishers, Moscow, USSR, first published in German in 1885,

1.3 Structure of the Book

The book is arranged so that Chaps. 2, 3, 4 and 5 each deal with a specific steppingstone of the commodification pathway, while Chap. 6 deals with the ethical consequences of that pathway, and Chap. 7 looks to the future.

The first steppingstone, discussed in Chap. 2, is market expansion, which saw supply and demand increase throughout the nineteenth century, extending the animal product sector both domestically and internationally. The UK was at the centre of these developments, as its population transformed from a predominantly rural society to one based on the manufacture and consumption of industrialised goods and services.[51] Animals and their products, once raised for personal or community use, came to be regarded as a commodity, manufactured for the mass market and for profit. Accordingly, consumerism created a cycle of supply and demand reconstructing agriculture into a commercial undertaking and laying the foundations for economies of scale and perceptions of farm animals as industrial goods.

Chapter 3 deals with the Enlightenment and its influence on the human-farm animal relationship. The period was one of scientific and social progresses where rationality and reason became the analytical lenses for scrutinising all levels of society. However, until the end of the eighteenth century, the focus on logic, reason and cognition underscored differences between humans and animals, reinforcing religious views that animals were not an appropriate subject for moral concern.[52] This cast a metaphorical shadow over attempts to include animals within humanity's social and moral spheres, until the issue was re-framed as one of sentience rather than reason.[53] Accordingly, from the nineteenth century, the ethical basis of human-animal relationships centred on utilitarianism,[54] which in a legal context resulted in the passage of a range of anti-cruelty legislation. However, in somewhat of a paradox, the focus on sentience stymied deeper evaluation of the legitimacy of using animals, especially in intensive agricultural systems. Against this backdrop, animal pain and suffering acquired nuanced meanings, depending less on animal sentience and more on the uses humans had for the animal. Consequently, by the twentieth century, the protection afforded to farm animals eroded to a level that was

---

English edition first published in 1956, 122, available from https://www.marxists.org/archive/marx/works/download/pdf/Capital-Volume-II.pdf

[51] R M Hartwell, *The Industrial Revolution and Economic Growth*, Methuen and Co Ltd., London (1971), 158.

[52] John Passmore, "The Treatment of Animals", (1975) 36 (2) *Journal of the History of Ideas*, 195, 202.

[53] Gary Steiner, *Anthropocentrism and its Discontents*, University of Pittsburgh, Pittsburgh (2005), 153.

[54] Gary L Francione, "Animal Welfare and the Moral Value of Nonhuman Animals", (2010) 6 (1) *Law, Culture and the Humanities*, 24, 26.

lower than protection afforded to other domestic animals, such as companion animals.[55]

Chapter 4 discusses the regulatory consequences of animal disease, predominantly during the latter part of the nineteenth century, concentrating on the UK, for reasons set out more fully in Sect. 1.4. During this time, the veterinary profession came under increasing scrutiny, particularly for its early failures to control cattle plague, which led to calls for improved training and stricter licensing conditions.[56] By the time the plagues of the 1860s had died down, veterinarians were positioned as gatekeepers, acquiring formal roles as government inspectors and certifiers of animal health. This rise in the status of the profession also saw the initiation of professional gatherings, designed to share knowledge and ideas. Nevertheless, these developments occurred against the backdrop of growing trade and commerce in the sector, leading to animal disease being regarded as a trade issue. Consequently, animal health became subsumed into the practicalities of market transactions and the merchantability of goods, thereby providing a further steppingstone towards commodification.

Chapter 5 evaluates how the evolution of international instruments dealing with quarantine, trade in animals and disease prevention shaped the notion of animal wellbeing in an international context. This chapter proceeds from Chap. 4 because as the trade in farm animals and their products became more globalised, it required safeguards against the transmission of disease across international boundaries.[57] Accordingly, international markets followed the pathway already set at the national level, harnessing advances from veterinary science to support trade.[58] Trading partners negotiated treaties requiring inspections and certification of shipments as part of quarantine regulation. As with national jurisdictions, this resulted in animal disease being seen as a trade issue but this time equated with the "global good" deriving from international trade.[59] The focus on disease-free shipments contributed

---

[55] For example, *European Convention for the Protection of Pet Animals 1987*; the Convention was adopted on 13 November 1987, *ETS* No 125 (entered into force 1 May 1992) and had 24 ratifications as of July 2019, available from https://rm.coe.int/168007a67d . Preamble recognises the special relationship pets have with humans; Elizabeth Ann Overcash, "Unwarranted Discrepancies in the Advancement of Animal Law: the Growing Disparity in Protection Between Companion Animals and Agricultural Animals", (2012) 90 *North Carolina Review*, 837, 864–872.

[56] Veterinary Surgeons Act 1881, 44 & 45 Victoria c 62, Preamble, The Public General Statutes Passed in the Forty Fourth and Forty Fifth Years of the Reign of Her Majesty Queen Victoria, Printed by G.E. Eyre and W. Spottiswoode, printers to the Queen and W. Clowes and Sons, printers to the Council of Law Reporting, London, (1881), 371, available from https://archive.org/details/publicgeneralst05walegoog/page/n385; discussion, Paul Brassley, "Animal Health and Veterinary Medicine", in Joan Thirsk (general editor) *The Agrarian History of England and Wales* Volume VII 1850–1914, Part 1, 587, 591, Cambridge University Press, Cambridge (2000).

[57] R M Hartwell, *the Industrial Revolution and Economic Growth*, above 51, 207–210.

[58] Ibid, 218.

[59] Mark Harrison, "A Global Perspective: Reframing the History of Health, Medicine, and Disease", (2015) 89 (4) *Bulletin of the History of Medicine*, 639, 642.

greatly to commodification because animals were seen in terms of bulk shipments, which international trade promoted, rather than seen as individual living beings.

Chapter 6 draws together the threads of arguments presented in the preceding four chapters, addressing why anti-cruelty regulation of the nineteenth century, and animal welfare regulation of the twentieth century, did not lead to greater ethical treatment of farm animals. Given that from the nineteenth century, market growth in the sector was accompanied by anti-cruelty legislation, an increasing number of animal protection organisations as well as general critiques of the sector, the question "Whither Ethics?" is an important one.[60] The heart of the regulatory dilemma is centred on how to support production and trade while simultaneously dealing with cruelty and animal disease. By the end of the nineteenth century, resolution of this predicament had been set in favour of commerce and trade, exacerbated by the duplicity of national regulation, which dealt with only the worst cases of cruelty and which was reinforced by treaties at the international level, which rarely concerned themselves with animal wellbeing.

These developments ensured that farm animals were primarily treated as articles of trade and their wellbeing subject to the whims of the marketplace.[61] Within these structures, the individual animal became invisible, subsumed into the merchantability of the whole, receiving only what little protection the market could spare. Chapter 6 concludes by evaluating animal commodification and exploitation against the theoretical framework of Karl Marx's (1818–1883) *Capital*, particularly his explanation of different forms of value.[62]

Chapter 7, the last substantive chapter of the book, looks to the future by examining animal commodification in the context of Norbert Elias's civilising process.[63] Elias argues that Western European society, from the Middle Ages to the turn of the twentieth century, has steadily become less violent as a result of increasing disgust and repugnance at violence.[64] Elias's work can be extrapolated to farm animals because violence is an inherent part of animal production. A critical feature of Elias's theory is that the state holds a monopoly over violence, which it should use to pacify society and thus diminish individual and group violence.[65] In addition, farm animals are completely reliant on humans, linking violence against them with society's prevailing stage of civilisation. Although Elias's work is historical in nature, the civilising process is relevant to understanding current regimes because it identifies gaps and weaknesses, highlighting areas in need of reform.

---

[60] Kimberley Smith, *Governing Animals Animal Welfare and the Liberal State*, above 21, 28.

[61] F Bailey Norwood and Jayson L Lusk, *Compassion, by the Pound: The Economics of Farm Animal Welfare*, above 4, 258.

[62] Karl Marx, "*Capital, A Critique of Political Economy*", above 50.

[63] Norbert Elias, *The Civilizing Process, Sociogenetic and Psychogenetic Investigations,* Blackwell Publishing, UK (Revised Edition 2000).

[64] Stephen Mennell, "Civilizing Processes", (2006) 23 (2–3) *Theory, Culture and Society*, 429, 429.

[65] Stephen Quilley, "The Land Ethic as an Ecological Civilizing Process: Aldo Leopold, Norbert Elias, and Environmental Philosophy", (2009) 31 (2) *Environmental Ethics*, 115, 115.

At the same time, the civilising process is subject to de-civilising forces that shape and sway the progress of civilisation, including the content and implementation of law and policy. The political power wielded by the animal product sector has emerged as just such a de-civilising force, evident by the anomaly of regulatory capture and the introduction of "Ag-Gag" laws, which limit the ability of whistle blowers to inform regulators and the public of substandard production practices.[66] These events have afforded a substantial degree of ascendancy to market forces, which will continue to exert de-civilising pressures if not counterbalanced.

1.4 Scope and Limits of the Book

The scope of this book is limited by the jurisdictions and periods selected for discussion. With respect to the former, Chaps. 2 and 3 focus on the UK, and Chaps. 4 and 5 provide a more international perspective, while Chaps. 6 and 7 draw examples as appropriate. The time span concentrates on the nineteenth to mid-twentieth centuries, although the discussion also touches on earlier and later events where relevant. In particular, Chaps. 6 and 7 extend beyond the mid-twentieth century to identify how the commodification pathway unfurled after the adoption of animal welfare policies.

The UK has been selected as the focus of this book for a number of reasons. First, the UK Parliament enacted some of the earliest anti-cruelty laws, and they became influential in common law countries around the world, including Australia, New Zealand and North America.[67] Second, by enacting these laws, the UK indicated that societal concern at animal cruelty was a valid reason for government intervention, creating the (unrealised) potential for strong regulation of the sector.[68] Third, during the time under discussion, the UK was the centre of the global trade in animals and their products.[69] Although increasing demand for animal products was not unique to the UK,[70] their markets drove the demand because they were a nation of meat eaters who regarded animal products as a mark of good health and status.[71]

[66] Discussion in Chap. 7, Sect, 7.4.1 of this book.

[67] David S Favre and Vivien Tsang, "The Development of Anti-Cruelty Laws During the 1800s" (1993) Spring (1) *Detroit College of Law Review*, 1, 2.

[68] David Harvey and Carmen Hubbard, "Reconsidering the Political Economy of Farm Animal Welfare: An Anatomy of Market Failure", above 13, 107.

[69] Felicity Barnes and David M. Higgin, "Brand Image, Cultural Association and Marketing: 'New Zealand' Butter and Lamb Exports to Britain, C. 1920–1938", (2017): *Business History*, https://doi.org/10.1080/00076791.2017.1344223, 1, 3–4.

[70] Roger Horowitz, Jeffrey M. Pilcher, and Sydney Watts, "Meat for the Multitudes: Market Culture in Paris, New York City, and Mexico City over the Long Nineteenth Century", above 31, 1057.

[71] Patrick Karl O'Brien, "Path Dependency, or Why Britain Became an Industrialized and Urbanized Economy Long before France", (1996) 49 (2) *The Economic History Review*, 231, 239; Chris Otter, "The Vital City: Public Analysis, Dairies and Slaughterhouses in Nineteenth-century Britain, above 20, 526.

Indeed at the turn of the twentieth century, the "British market continued to absorb over 60% of the world meat trade".[72]

With respect to the time frame of this book, the nineteenth to mid-twentieth centuries were periods of significant change. Notwithstanding the fact that commodification of farm animals had started from at least the sixteenth century, the most profound changes occurred during the nineteenth to mid-twentieth centuries.[73] This period saw the growth of markets, scientific and technological advances as well as the establishment of animal protection organisations. The tussle for dominance that ensued led to a form of "corporate capitalism" which irretrievably fractured the bonds of animal stewardship, so that raising animals came to be regarded as a means of raising capital, commodifying animals and acting as a stumbling block to meaningful reform.[74]

Finally, two additional points warrant special mention. Although Chaps. 4 and 5 critique the role of the veterinary profession, this is not intended to critique the work of veterinarians. Anyone who has had the privilege of sharing their life with a companion animal, or indeed any other kind of animal, will know first-hand the amazing work the profession does. Rather, the critique is aimed at government and industry who then as now harness the expertise and dedication of veterinarians for overtly economic gain.

Last but not least, the book deliberately avoids delving into the animal welfare/animal rights debate, because strictly speaking it is not necessary for the study that has been undertaken. Nevertheless, the book is highly critical of the way anti-cruelty regulation and subsequently animal welfare regulation has evolved in a commercial context, indicating that something beyond the utilitarian underpinnings of animal welfare is required to improve the plight of farm animals.

Bibliography

Adams CJ (2010) The sexual politics of meat, anniversary edn. Continuum, New York

Allen RC (2009) Why was the industrial revolution British? Oxonomics 4:50

Anderson VJ (2004) Creatures of empire, how domestic animals transformed early America. Oxford University Press, Oxford

Baics G, Thelle M (2017) Introduction: meat and the nineteenth-century city. Urban Hist:1. https://doi.org/10.1017/S0963926817000414

Barnes F, Higgin DM (2017) Brand image, cultural association and marketing: 'New Zealand' butter and lamb exports to Britain, C. 1920–1938: Bus Hist:1. https://doi.org/10.1080/00076791.2017.1344223

[72] I R Phimister, "Meat and Monopolies: Beef Cattle in Southern Rhodesia, 1890–1938", (1978) 19 (3) *The Journal of African History*, 391, 392.

[73] Linda Kalof, *Looking at Animals in Human History*, above 28, 135; Ryan Gunderson, "From Cattle to Capital: Exchange Value, Animal Commodification, and Barbarism", above 19, 262.

[74] John Bellamy Foster and Brett Clark, "Marx and Alienated Speciesism", above 44, 12.

Bartel R, Graham N (2016) Property and place attachment: a legal geographical analysis of biodiversity law reform in New South Wales. Geogr Res 54(3):267. https://doi.org/10.1111/1745-5871.12151

Beirne P (2009) Confronting animal abuse: law, criminology and human-animal relationships. Rowman and Littlefield, Maryland

Brassley P (2000) Animal health and veterinary medicine. In: Thirsk J (general ed) The Agrarian history of England and Wales, volume VII 1850–1914, part 1. Cambridge University Press, Cambridge, p 587

Broom DM (2013) Ruth Harrison's later writings and animal welfare work. In: Harrison R (ed) Animal machines. Vincent Stuart Publishers, p 21 (J Harrison J, Wilson J, CABI, Wallingford)

Cassidy ES, West PC, Gerber JS, Foley J (2013) Redefining agricultural yields: from tonnes to people nourished per hectare. Environ Res Lett 8(1):4–5. https://doi.org/10.1088/1748-9326/8/3/034015

Clark B, Wilson TD (2021) The capitalist commodification of animals: a brief introduction. In: Clark B, Wilson TD (eds) The capitalist commodification of animals, vol 1. Emerald Publishing Limited, Bingley UK, p 35

Donovan J (2011) Aestheticizing animal cruelty. Coll Lit 38(4):202

Elias N (2000) The civilizing process, Sociogenetic and psychogenetic investigations, Revised edn. Blackwell Publishing, UK

Favre DS, Tsang V (1993) The development of anti-cruelty laws during the 1800s. Detroit College of Law Review 1(Spring):1

Fiber-Ostrow P, Lovell JS (2016) Behind a veil of secrecy: animal abuse, factory farms, and ag-gag legislation. Contemporary Justice Review 19(2):230. https://doi.org/10.1080/10282580.2016.1168257

Foster JB, Clark B (2018) Marx and alienated speciesism. Mon Rev 70(7):1. https://doi.org/10.14452/MR-070-07-2018-11_1

Francione GL (2010a) The abolition of animal exploitation. In: Francione GL, Garner R (eds) The animal rights debate: abolition or regulation? Columbia University Press, New York, p 1

Francione GL (2010b) Animal welfare and the moral value of nonhuman animals. Law, Culture and the Humanities 6(1):24

Friedman H, McMichael P (1989) Agriculture and the state system, the rise and decline of National Agricultures, 1870 to the present. Sociol Rural XXIX(2):93

Gatward G (2001) Livestock ethics. Chalcombe Publications, Lincoln

Gunderson R (2011) From cattle to capital: exchange value, animal commodification, and barbarism. Crit Sociol 39(2):259

Harrison M (2015) A global perspective: reframing the history of health, medicine, and disease. Bull Hist Med 89(4):639

Harrison R (2013) *Animal Machines* first published 1964 Vincent Stuart Publishers (Edition J. Harrison and J. Wilson, CABI, Wallingford)

Hartwell RM (1971) The industrial revolution and economic growth. Methuen and Co Ltd., London

Harvey D, Hubbard C (2013) Reconsidering the political economy of farm animal welfare: an anatomy of market failure. Food Policy 38:105

Hilton M (2003) Consumerism in twentieth-century Britain: the search for a historical movement. Cambridge University Press, Cambridge

Hoganson K (2012) Meat in the middle: converging borderlands in the U.S. Midwest, 1865–1900. J Am Hist 98(4):1025

Horowitz R, Pilcher JM, Watts S (2004) Meat for the multitudes: market culture in Paris, new York City, and Mexico City over the long nineteenth century. Am Hist Rev 109(4):1055

Huttman JP (1978) British meat imports in the free trade era. Agric Hist 52(2):247

Kalof L (2007) Looking at animals in human history. Reaktion Books, London

Lang T (2006) Achieving access to ethical food: animal and human health come together. In: Turner J, D'Silva J (eds) Animals, ethics and trade. Earthscan, UK, p 261

Linzey A (2009) Creatures of the same god, explorations in animal theology. Lantern Books, Brooklyn

Maehle A-H (1994) Cruelty and kindness to the 'brute creation': stability and change in the ethics of the man-animal relationship 1600–1850. In: Manning A, Serpell J (eds) Animals and human society changing perspectives. Routledge, Abingdon, p 81

Marx K 1956 *Capital, a critique of political economy* Volume II Book One: The Process of Circulation of Capital, Progress Publishers, Moscow, USSR, first published in German in 1885 (English edition). https://www.marxists.org/archive/marx/works/download/pdf/Capital-Volume-II.pdf

McMullen S (2016) Animals and the economy. In: Palgrave Macmillan

Mennel S (2006) Civilizing processes. Theory Cult Soc 23(2–3):429

Mepham B (2006) The ethical matrix as a decision-making tool, with specific reference to animal sentience. In: Turner J, D'Silva J (eds) Animals, ethics and trade. Earthscan, UK, p 134

Norwood FB, Lusk JL (2011) Compassion, by the pound: the economics of farm animal welfare. Oxford University Press, New York

Nussbaum MC (2004) Beyond 'compassion and humanity': justice for nonhuman animals. In: Sunstein CR, Nussbaum MC (eds) Animal rights, current debates and new directions. Oxford University Press, New York, p 299

O'Brien PK (1996) Path dependency, or why Britain became an industrialized and urbanized economy long before France. Econ Hist Rev 49(2):231

Otter C (2006) The Vital City: public analysis, dairies and slaughterhouses in nineteenth-century Britain. Cult Geogr 13:517

Overcash EA (2012) Unwarranted discrepancies in the advancement of animal law: the growing disparity in protection between companion animals and agricultural animals. North Carolina Rev 90:837

Passmore J (1975) The treatment of animals. J Hist Ideas 36(2):195

Perren R (2017) Taste, trade and technology: the development of the international meat industry since 1840 (first published 2006 Ashgate). Routledge, Abingdon

Phimister IR (1978) Meat and monopolies: beef cattle in southern Rhodesia, 1890–1938. J Afr Hist 19(3):391

Quilley S (2009) The land ethic as an ecological civilizing process: Aldo Leopold, Norbert Elias, and environmental philosophy. Environ Ethics 31(2):115

Raber K (2011) From sheep to meat, from pets to people. In: Senior M (ed) A cultural history of animals in the age of enlightenment, 73. Berg, Oxford

Ryder R (1989) Animal revolution. Basil Blackwell, Oxford

Sandøe P, Christiansen SB, Forkman B (2006) Animal welfare: what is the role of science? In: Turner J, D'Silva J (eds) Animals, ethics and trade. Earthscan, UK, p 41

Smith K (2012) Governing animals animal welfare and the Liberal state. Oxford University Press, New York

Stache C (2020) Conceptualising animal exploitation in capitalism: getting terminology straight. Capital and Class 44(3):401. https://journals.sagepub.com/doi/abs/10.1177/0309816819884697

Steiner G (2005) Anthropocentrism and its discontents. University of Pittsburgh, Pittsburgh

Szűcs E, Geers R, Sossidou EN, Broom DM (2012) Animal welfare in different human cultures, traditions and religious faiths. Asian-Australian Journal of Animal Sciences 25(11):1499. https://doi.org/10.5713/ajas.2012.r.02

Tauger MB (2011) Agriculture in world history. Routledge, Abingdon

Thompson FML (1968) The second agricultural revolution, 1815–1880. Econ Hist Rev 21(1):62

Thorpe C, Jacobson B (2021) Abstract life, abstract labor, abstract mind. In: Clark B, Wilson TD (eds) The capitalist commodification of animals, vol 35. Emerald Publishing Limited, Bingley UK, p 59

Wadiwel DJ (2016) Like one who is bringing his own Hide to market. Angelaki Journal of the Theoretical Humanities 21(2):65. https://doi.org/10.1080/0969725X.2016.1182725

Whitfort A (2018) Justice and the vulnerable: extending the duty to prevent serious crimes against children to the protection of agricultural and research animals. Adelaide Law Review 39:125

Wise SM (2006) Entitling non-human animals to fundamental legal rights on the basis of practical autonomy. In: Turner J, D'Silva J (eds) Animals, ethics and trade. Earthscan, UK, p 87

Woods A (2011) From cruelty to welfare: the emergence of farm animal welfare in Britain, 1964–71. Endeavour 36(1):14

Websites, Factsheets

Compassion in World Farming, *Ending Factory Farming*, available from, https://www.ciwf.org.uk/factory-farming/

Food and Agriculture Organization of the United Nations, *Livestock Statistics—Concepts, Definitions and Classifications*, FAO (2020). http://www.fao.org/economic/the-statistics-division-ess/methodology/methodology-systems/livestock-statistics-concepts-definitions-and-classifications/en/

Legislation, Regulation

1635, *Act Against Plowing by the Tayle, and Pulling the Wooll off Living Sheep 1635* (Ireland)

1641, *The Massachusetts Body of Liberties 1641*. https://history.hanover.edu/texts/masslib.html

1682, Edward Coke, *The First Part of the Institutes of the Lawes of England*, Societe of Stationers, London (1628). https://play.google.com/books/reader?id=NOgyAAAAIAAJ&hl=en&pg=GBS.PA5

1822, *An Act to Prevent the Cruel and Improper Treatment of Cattle 1822*, (3 Geo IV c 71), *The Statutes of the United Kingdom of Great Britain and Ireland 3 George IV. 1822*, his Majesty's statute and law printers London, sold by Butterworths and son, 403. https://archive.org/details/statutesunitedk10britgoog/page/n436

1881, *Veterinary Surgeons Act 1881, 44 & 45 Victoria c 62*, The Public General Statutes Passed in the Forty Fourth and Forty Fifth Years of the Reign of Her Majesty Queen Victoria, Printed by G.E. Eyre and W. Spottiswoode, printers to the Queen and W. Clowes and Sons, printers to the Council of Law Reporting, London, (1881), 371. https://archive.org/details/publicgeneralst05walegoog/page/n385

International Instruments

1945, *Constitution of the Food and Agriculture Organization of the United* Nations, [1945] ATS No 9. http://www.austlii.edu.au/au/other/dfat/treaties/1945/9.html

1987, *European Convention for the Protection of Pet Animals 1987*, *ETS* No 125. https://rm.coe.int/168007a67d

2 "Meating" the Demand: Markets and Commodification

Abstract

In the UK, supply and demand for animal products increased throughout the nineteenth century, a consequence of the combined impacts of population growth, urbanisation, improved living standards, intensified farm production and increasing international trade. Improved techniques and refinements in agricultural production, such as crop rotations, increased capital expenditure and selective breeding, gained momentum during the eighteenth and nineteenth centuries, innovations categorised as an Agricultural Revolution. These advances initiated production methods better suited to growing market conditions, so that farm animals, once produced for personal or community use, came to be regarded as another commodity, manufactured for the mass market and profit. During this time, consumerism gained a foothold, so that market growth not only met the demand for animal products but also fuelled it. This created a cycle of supply and demand that transformed pastoral agriculture into a commercial undertaking, laying the foundations for economies of scale and increasing regard for farm animals as industrial goods. What is more, these events gradually transformed societies in the UK from agriculturists to producers and consumers of industrialised goods and services, a trend that also included animals and their products.

Keywords

Agricultural Revolution · Commodification of animals · Consumerism urbanisation

S. Riley, *The Commodification of Farm Animals*, Animal Welfare 21,
https://doi.org/10.1007/978-3-030-85870-4_2

2.1 Introduction

In the UK, at the beginning of the nineteenth century, meat was considered a luxury, predominantly available to the wealthy.[1] This led to low demand, which was matched by low output, primarily sourced from small farm holdings.[2] These limitations constrained the size of the market, although over the course of the nineteenth century, supply and demand increased, expanding the sector, both domestically and internationally. Globally, production and consumption of meat rose from approximately 10 kg per person per annum in the early 1800s to 40 kg per person per annum in the early 1900s.[3] In the UK, consumption was higher, with the annual per capita rate reaching 41.3 kg in 1880 and 59.5 kg by 1909–1913.[4]

The popularity of meat stemmed from a confluence of population growth, urbanisation, improved living standards, intensified farm production and increasing international trade.[5] The UK was at the epicentre of these developments, gradually transforming its societies from agriculturists to producers and consumers of industrialised goods and services.[6] Livestock and animal products, which had once been produced for personal or community use, came to be regarded as another commodity, manufactured for the mass market and for profit.

This chapter evaluates market growth, one facet of the pathway by which farm animals became commodified. The analysis is conducted against the backdrop of consumerism and commercialisation of animal products, which occurred across the nineteenth century. The term "consumerism" does not have a settled meaning.[7] It has been used to describe the growth of consumption per se, to categorise activism by consumers and to critique undue consumption of material goods.[8] As used in this discussion, consumerism focusses on the latter meaning, to describe a form of market growth that not only met the demand for animal products but also fuelled it. This created a cycle of supply and demand that transformed pastoral agriculture into a commercial undertaking, laying the foundations for economies of scale and standpoints that increasingly regarded farm animals as industrial goods.

---

[1] John Burnett, *Plenty and Want, a Social History of Food in England from 1815 to the Present Day*, Routledge, London (1989, 3rd Edition), 11.

[2] Gareth Shaw, "Changes in Consumer Demand and Food Supply in Nineteenth-Century British Cities", (1985) 11 (3) *Journal of Historical Geography*, 280.

[3] Ibid.

[4] Richard Perren, "The North American Beef and Cattle Trade, with Great Britain, 1870–1914", (1971) 24 (3) *Economic History Review*, 430.

[5] Gareth Shaw, "Changes in Consumer Demand and Food Supply in Nineteenth-Century British Cities", above 2, 281.

[6] R M Hartwell, *The Industrial Revolution and Economic Growth*, Methuen and Co Ltd. (1971), 158.

[7] Robert Schwalger. "Evolution and Applications of the Term Consumerism: Theme and Variations", (1994) 28 (2) *The Journal of Consumer Affairs*, 347.

[8] Matthew Hilton, *Consumerism in Twentieth-Century Britain: The Search for a Historical Movement*, Cambridge University Press, Cambridge (2003), 4.

2.2 Population Growth and Urbanisation

In Western Europe, populations grew dramatically in the eighteenth and nineteenth centuries, as illustrated by the UK, where numbers increased from approximately 5 million in 1700 to 8.6 million in 1800[9] and to 41 million by 1911.[10] This growth was distributed comparatively evenly across the country, until 1850 when industrialisation concentrated population build-up in urban areas.[11] Although urbanisation briefly weakened in the 1860s, it again gained momentum in the 1870s and 1880s, becoming part of general trend that saw rural and semi-rural dweller decline from 80% of the population at the beginning of the nineteenth century to 20% by 1911.[12]

The search for higher wages was an important element of urbanisation and was exacerbated by the fact that agricultural workers in areas transitioning to industrialisation were usually among the most poorly paid.[13] Labourers were also mired in society's negative attitudes towards the poor, which arguably persist till the present, and certainly persisted until the end of the eighteenth century. Stereotypical views regarded the impoverished as idle and a burden on society, capped by a somewhat contradictory conclusion that they nevertheless loved luxuries.[14] This aspiration was seen as detrimental to social order, leading to beliefs that low-paid workers should continue to receive low wages, as befitted their position in society.[15] Although these views stranded labourers in a cycle of poverty, the perspectives were consistent with the then-favoured labour theory of value, which held that the price of commodities should be low enough for manufacturers to make a profit, even if this meant wages were stagnant.[16]

From the later part of the eighteenth century, however, perceptions altered and society accepted that higher wages could not only help alleviate poverty but could also support a market for products and services, forged by an increasingly industrialised society.[17] Higher incomes meant greater discretionary spending, including for purchasing animal products, such as meat.[18] Meat eating was also

[9] James C Riley, "A Widening Market in Consumer Goods", in *Early Modern Europe, an Oxford History* (Euan Cameron ed) (OUP) Oxford: New York, 1999, 233.

[10] Richard Perren, *Taste, Trade and Technology: the Development of the International Meat Industry Since 1840*, (2006 Ashgate) Routledge, Abingdon, Oxfordshire (2017), 7, 20.

[11] R M Hartwell, *The Industrial Revolution and Economic Growth*, above 6, 126.

[12] John Burnett, *Plenty and Want, a Social History of Food in England from 1815 to the Present Day*, above 1, 3.

[13] Gregory Clark, "Farm Wages and Living Standards in the Industrial Revolution: England, 1670–1869", (2001) 53 (3) *The Economic History Review*, 477, 498.

[14] R M Hartwell, *The Industrial Revolution and Economic Growth*, above 6, 73.

[15] Ibid, 74.

[16] M Modrzejewska, (2017). "Adam Smith's Concept of Value of Labour. Anglo-American Perspective till Mid- Century", (2017) 8 (25) Horizons of Politics, 65, 68.

[17] R M Hartwell, *The Industrial Revolution and Economic Growth*, above 6, 74.

[18] John Burnett, *Plenty and Want, a Social History of Food in England from 1815 to the Present Day*, above 1, 3.

seen as an indicator of higher living standards, although to what extent dietary patterns improved during the eighteenth and nineteenth centuries and whether they in fact raised living standards remains unsettled. The standard of living would not have been high by today's benchmarks, since exploitation of child labour was pervasive and work conditions were inadequate and insecure.[19] However, in a general sense, living standards improved, even if this occurred slowly. Farmers in 1850 certainly enjoyed a better life-style than those of the mid-eighteenth century, and wages for both agricultural labourers and urban workers increased, even though urban workers enjoyed higher raises than agricultural workers.[20]

With respect to dietary patterns, statistics extrapolated from Smithfield, the oldest of London's markets, signify that more animals were being slaughtered, suggesting that meat eating was increasing.[21] In 1790–1794, for example, there was a 32% increase in the sale of cattle compared to 1732–1736.[22] These types of statistics, however, may underestimate the extent of meat eating, because they do not take into account the size of animal carcasses; nor do they include the consumption of fish or the trade in delicatessen and preserved meats.[23] Intake of animal products is thus likely to have been higher than signified by statistics derived from the trade at Smithfield.

At the same time, the popularity of meat did not develop uniformly. There were always sections of society who followed vegetarian diets,[24] and the agricultural depression of the 1830s and 1840s temporarily curtailed the propensity towards greater meat consumption.[25] Dietary patterns also tended to be divided along gender lines. By the 1890s, most men ate at least one main meal of meat or fish per day, yet many women did not, largely existing on eating bakery goods and drinking tea.[26] Notwithstanding these variations, the long-term demand for animal products generally increased throughout the nineteenth century and particularly from the 1850s onward.[27] Although fresh meat was still considered a luxury, it had become one of the few indulgences available to workers, whether that meat was pork, beef or

[19] R M Hartwell, *The Industrial Revolution and Economic Growth*, above 6, 129, 340, 344.

[20] E J T Collins, "The Rationality of 'Surplus' Agricultural Labour: Mechanization in English Agriculture in the Nineteenth Century", (1987) 35 (1) *The Agricultural History Review*, 36, 45.

[21] R M Hartwell, *The Industrial Revolution and Economic Growth*, above 6, 334.

[22] G E Fussell, "Animal Husbandry in Eighteenth Century England: Part 1", (1937) 11 (2) *Agricultural History*, 96, 10, 102.

[23] R M Hartwell, *The Industrial Revolution and Economic Growth*, above 6, 334, 340.

[24] Terry Newholm, Sandra Newholm and Deirdre Shaw, "A History for Consumption Ethics", (2015) 57 (2), *Business History*, 290, 298.

[25] John Burnett, *Plenty and Want, a Social History of Food in England from 1815 to the Present Day*, above 1, 18.

[26] D J Oddy, "Working-Class Diets in Late Nineteenth-Century Britain", (1970) 23 (2) *The Economic History Review*, 314, 321.

[27] Robert C Allen, "Why was the Industrial Revolution British?", (2009) 4 *Oxonomics*, 50, 50.

mutton.[28] Meat came to be regarded as essential to good health, a point underscored by the fact that it also formed an important component in the diet of the military.[29] These developments strengthened the market in animal products and created challenges with regard to maintaining supply.[30]

While population increases favoured urban areas, city communities lacked sufficient land and time to grow their own food, becoming dependent on rural and peri-urban areas.[31] This dependency also occurred simultaneously with reductions in the proportion of agricultural workers who were available to produce food for urban populations.[32] Until the 1840s, the pressure that this placed on supply was aggravated by long-standing restrictions on the importation of food products into the UK.[33] In the absence of external sources, it was, therefore, critical that producers find ways of increasing domestic productivity. The ensuing transformations became characterised as the Agricultural Revolution.[34]

2.3 Agricultural Revolution and Intensified Production

Improved techniques and refinements are common features of agricultural systems, but in the UK, they gained momentum during the eighteenth and nineteenth centuries.[35] Innovations included the introduction of crop rotations, such as the Norfolk four-course system, increased capital expenditure to improve drainage and provide new equipment and selective breeding to produce livestock better suited to market conditions.[36] These transformations were categorised as an Agricultural Revolution, a concept not easily definable by set times or events, because changes and innovations were adopted in a piecemeal manner.[37]

In addition, agricultural transformation also occurred in tandem with social, technological and commercial shifts, which were as important as agricultural innovation.[38] A case in point is the enclosure movement, which was a legal process

[28] Susanne Freidberg, "Moral Economies and the Cold Chain", (2015) 88 (239) *Historical Research*, 125, 127.

[29] Ibid.

[30] D J Oddy, "Working-Class Diets in Late Nineteenth-Century Britain", above 26, 320.

[31] Chris Otter, "The Vital City: Public Analysis, Dairies and Slaughterhouses in Nineteenth-century Britain, (2006) 13 *Cultural Geographies*, 517, 520.

[32] Robert C Alle "Tracking the agricultural revolution in England", (1999) LII (2) *Economic History Review*, 209, 217.

[33] R M Hartwell, *The Industrial Revolution and Economic Growth*, above 6, 125.

[34] Ibid.

[35] Robert C Alle "Tracking the Agricultural Revolution in England", above 32, 209, 217.

[36] R M Hartwell, *The Industrial Revolution and Economic Growth*, above 6, 125.

[37] C Peter Timmer, "The Turnip, the New Husbandry, and the English Agricultural Revolution", (1969) 83 (3) *The Quarterly Journal of Economics*, 375, 392; James C Riley, "A Widening Market in Consumer Goods", above 9, 233, 242, 243.

[38] F M L Thompson, "The Second Agricultural Revolution, 1815–1880", (1968) 21 (1) *The Economic History Review*, 62, 63.

that gradually replaced shared or communal uses of land by transferring property rights to individuals.[39] Other changes included improvements in transportation, increased international trade and the growth of the financial sector, with each contributing to changes that during the nineteenth century manifested in tandem with agricultural modernisation. For these reasons, the term Agricultural Revolution, as used in this chapter, describes the agrarian, technological, social and commercial changes of the eighteenth and nineteenth centuries, which became increasingly, if not universally adopted. By the early nineteenth century, these transformations had led to greater productivity, a notion that encompasses the attainment of greater yields through using "less labour per unit of output".[40] From a practical perspective, therefore, productivity refers to the use of land, labour and equipment in a more efficient manner, to intensify production.

Prior to the Agricultural Revolution, outputs in arable products had become limited by depleted soils and inflexible adherence to a two-crop rotation system, where two fields were planted, while one was left fallow to re-establish soil fertility.[41] This meant that at any one time, approximately one-third to one quarter of arable land was not under cultivation.[42] In England, the adoption of new rotation systems, such as the Norfolk four-course system, obviated the need for a fallow season. Instead, farmers grew crops such as barley, clover, turnips, wheat and pasture grasses in continuous rotation.[43] Turnips and clover replenished nitrogen and nutrients to the soil, while pasture grasses supplied fodder for animals.[44]

The Norfolk rotation system is often attributed to Viscount Charles Townshend (1674–1738), who implemented it in the eighteenth century, earning him the moniker, "Turnip Townshend".[45] However, rather than introducing the system, he popularised it.[46] The use of nitrogen-setting plants had been practised by Flemish farmers since the Middle Ages, even if they did not understand the science that underpinned their methods.[47] In addition, the use of clover had already become widespread in the Netherlands by the sixteenth century and is likely to have been

---

[39] Leigh Shaw-Taylor, "Parliamentary Enclosure and the Emergence of an English Agricultural Proletariat", (2001) 61 (3) *The Journal of Economic History*, 640, 642.

[40] Patrick Karl O'Brien, "Path Dependency, or Why Britain Became an Industrialized and Urbanized Economy Long before France", (1996) 49 (2) *The Economic History Review*, 231, 239.

[41] J D Chambers and G E Mingay, *The Agricultural Revolution 1750–1880*, B T Batsford Ltd. (1966), 49.

[42] Ibid.

[43] Lluis Argemí, "Agriculture, Agronomy, and Political Economy: Some Missing Links", (2002) 34 (2) *History of Political Economy*, 449, 452, 454.

[44] Ibid.

[45] C Peter Timmer, "The Turnip, the New Husbandry, and the English Agricultural Revolution", above 37, 380.

[46] Ibid.

[47] Lluis Argemí, "Agriculture, Agronomy, and Political Economy: Some Missing Links", above 43, 452, 454.

introduced from there to England in the early seventeenth century.[48] The Norfolk system provided efficiencies, but its uptake depended on soil type and individual farm needs. It was well suited to sandy soils, where nitrogen-setting plants could improve yields, but it was not as successful on harder clay soils, which were predominantly used to sow pasture grasses for animal production.[49] Nevertheless, in areas where the soil was suitable, it allowed the land to be used more efficiently and, as will be discussed later, supported the raising of increased numbers of livestock.

The use of technology was another feature of the Agricultural Revolution, although its initial impact was not as extensive as might be assumed. Innovations such as the mechanical seed drill, invented by Jethro Tull (1674–1741), were important, but until the last quarter of the nineteenth century, mechanisation happened slowly.[50] Patent data up to the 1840s reveals several patents for threshing machines granted in the 1790s, but otherwise, innovations predominantly targeted improving tools, such as seed drill and scythes that were not mechanised in the true sense.[51] After the 1840s, patents largely related to improved drainage, irrigation, new fertilizers and manure systems.[52] Undoubtedly, these innovations contributed to enhanced productivity, but changes in agricultural production were not dependent on technology in the sense of a "mechanical revolution".[53] Indeed, the rural workforce was sufficiently robust to avoid wholesale reliance on machinery.[54] Instead, changes were more dependent on a combination of judicious use of technology, improved farming techniques and better economic management, although these occurred at different stages and to varying degrees.[55] A significant influence on these transformations stemmed from the enclosure movement.

In the UK, the movement gathered momentum during the Middle Ages, culminating in a large number of enclosures towards the end of the eighteenth and the beginning of the nineteenth centuries.[56] Villages throughout the UK traditionally enjoyed access to a variety of common areas such as open fields, common meadows and common waste. The latter two would be available for communal grazing

---

[48] J D Chambers and G E Mingay, *The Agricultural Revolution 1750–1880*, above 41, 8.

[49] C Peter Timmer, "The Turnip, the New Husbandry, and the English Agricultural Revolution", above 37, 392.

[50] E J T Collins, "The Rationality of 'Surplus' Agricultural Labour: Mechanization in English Agriculture in the 10 Nineteenth Century", above 20, 36.

[51] Richard J Sullivan, "The Timing and Pattern of Technological Development in English Agriculture", (1985) 45 (2) *The Journal of Economic History*, 305, 307.

[52] Ibid, 308.

[53] James B Ang, Rajabrata Banerjee, and Jakob B Madsen, "Innovation and Productivity Advances in British Agriculture: 1620–1850", (2013) 80 (1) *Southern Economic Journal*, 162, 182.

[54] Ibid.

[55] Lluis Argemí, "Agriculture, Agronomy, and Political Economy: Some Missing Links", above 43, 450.

[56] J D Chambers and G E Mingay, *The Agricultural Revolution 1750–1880*, above 41, 77.

throughout the year, while open fields would be used by individuals to grow crops or hay and would be open to communal grazing following the harvest.[57] Critiques of enclosure point to loss of access to common land, as well as the decline of small farm holdings, which by the turn of the twentieth century had reduced to one-eighth of the total area under cultivation.[58]

Changes, however, were gradual, and small farms persisted, especially near industrialised areas where they provided urban populations with a range of foodstuffs.[59] Even as late as 1911, over 40% of farmers in the UK operated in districts administered as towns and cities.[60] This indicated that enclosure had not eliminated small farm holdings, especially those that marketed their produce to nearby urban areas.[61] Milk was a popular product because it afforded farmers year-round income, unlike livestock where investment was tied up for several years until animals could be sold. Moreover, peri-urban farms afforded advantages due to their close proximity to markets, allowing farmers to avoid costly distribution and retail expenses along the supply chain.[62]

For other farmers, enclosure presented opportunities by providing greater acreage to grow turnips, root crops and clover. This facilitated raising crops and livestock together or "mixed farming" as it was known.[63] The transition to mixed farming heralded a new age in animal production for two reasons: first, more farmers raised animals, and second, animals were raised for market rather than for personal use. In consequence, ideals of productivity and economic efficiency that had been applied at the first instance to arable products also came to be applied to animals.

2.4 The Beginnings of Commodification

Traditionally, growing grains, especially wheat and oats, had been more profitable than raising livestock.[64] Wheat was a significant food crop, as were oats in the northern part of the UK, while elsewhere oats were important as fodder for horses and oxen.[65] Along with cereal cultivation, it was also customary for arable farmers to

[57] Leigh Shaw-Taylor, "Parliamentary Enclosure and the Emergence of an English Agricultural Proletariat", above 39, 640, 642.

[58] Michael Winstanley, "Industrialization and the Small Farm: Family and Household Economy in Nineteenth Century Lancashire", (1996) 152 *The Past and Present Society*, 157–176; Leigh Shaw-Taylor, "Parliamentary Enclosure and the Emergence of an English Agricultural Proletariat", above 55, 640.

[59] Michael Winstanley, "Industrialization and the Small Farm: Family and Household Economy in Nineteenth Century Lancashire", above 58, 160, 162.

[60] Ibid, 171.

[61] Ibid. 171.

[62] Ibid, 173.

[63] J D Chambers and G E Mingay, *The Agricultural Revolution 1750–1880*, above 41, 4, 92, 93.

[64] Ibid, 16.

[65] Ibid.

keep a small number of animals because sheep and cattle provided meat, wool, manure, labour, hides and tallow. Cereal and grain production peaked in approximately 1855, and thenceforth the demand for meat and dairy grew substantially, to the extent that pastoral farming became more profitable than arable farming.[66] Given the viability of livestock production, farmers soon turned to using grain crops to feed their animals, and, in this way, arable farmers became involved in animal production.[67] As part of this trend, animals were increasingly raised for profit rather than personal use, a pattern that was already in place before the mid-nineteenth century and flowed naturally, although at a quicker pace, from new rotation systems and the enclosure movement.[68]

The Norfolk rotation system provided continuous fodder and grazing crops, which enabled farmers to breed livestock throughout the year.[69] In addition, root crops and legumes, which formed the basis of the system, supplied animals with nutritious feed. This reduced fattening time, improved meat and milk outputs as well as enhanced livestock fertility.[70] Livestock also generated manure, which was used to improve the soil, increasing outputs of arable products and the number of animals that could be kept.[71] Moreover, livestock farmers frequently conducted dairying and pig-keeping together, because pigs ate milk by-products, enabling farmers to produce dairy foods side by side with pork products.[72]

The enclosure movement complemented the Norfolk rotation system by providing land to stock more animals as well as discouraging the practice of grazing mixed livestock on common areas. The latter invariably led to random breeding, which was not well-suited to commercial initiatives aimed at increasing productivity by producing animals tailored for market purposes.[73] Fattening times were a case in point. Even with nutritious feeding, it still took some 4 or 5 years before livestock could be traded.[74] Accordingly, the use of selective breeding to reduce these time-lags became a key objective of the sector.

Robert Bakewell (1725–1795), an eighteenth-century farmer from Leicestershire, was one of the first to re-establish experimental breeding methods and trial them in producing new breeds.[75] He had a keen interest in innovation, having already

---

[66] Ibid, 109–110, 181, 199.

[67] Ibid, 183.

[68] Patrick Karl O'Brien, "Path Dependency, or Why Britain Became an Industrialized and Urbanized Economy Long before France", above 40, 239.

[69] James B Ang, Rajabrata Banerjee, and Jakob B Madsen, "Innovation and Productivity Advances in British Agriculture: 1620–1850", above 53, 175.

[70] R M Hartwell, *The Industrial Revolution and Economic Growth*, above 6, 358.

[71] J D Chambers and G E Mingay, *The Agricultural Revolution 1750–1880*, above 41, 54.

[72] Ibid, 16.

[73] James B Ang, Rajabrata Banerjee, and Jakob B Madsen, "Innovation and Productivity Advances in British Agriculture: 1620–1850", above 53, 175.

[74] J D Chambers and G E Mingay, *The Agricultural Revolution 1750–1880*, above 41, 54, 55.

[75] John R Walton, "The Diffusion of Improved Sheep Breeds in Eighteenth- and Nineteenth-Century Oxfordshire", (1983) 9 (2) *Journal of Historical Geography*, 175.

established complex irrigation systems on his property, as well as conducting tests on the effectiveness of different types of manures. He bred sheep to be short, thickset and rounded, in the hope that they would mature early, be fleshy and profitable. Bakewell named the sheep the New Leicester, and as expected, they put on weight quickly and produced acceptable wool. The economic success of the breed, however, was mixed.[76] Between 1800 and 1850, the breed achieved a 20% reduction in production costs, but the sheep were also less fertile than their progenitors, and the meat they produced was very fatty, so that it could not be sold at a premium.[77] As a separate breed, the New Leicester sheep did not survive the test of time, but the breed is still important as the progenitors of sheep, such as the Devon Longwool and South Devon, that are reared to this day.[78]

Bakewell also bred longhorn cattle, which fattened more quickly than other cattle but which, like the New Leicester sheep, were susceptible to low fertility and additionally, were prone to low milk yields.[79] Although the breeds produced by Bakewell do not exist today, his methods paved the way for producers who followed, including John Ellman (1753–1852) who produced the Southdown sheep and the Colling brothers (Charles Colling 1751–1836, Robert Colling 1749–1820) and Thomas Bates (1775–1849) who bred shorthorn cattle.[80] Southdown sheep and shorthorn cattle have endured to the present day.

The use of selective breeding methods was not an eighteenth-century innovation, it having been practised for millennia. Archaeological and DNA evidence suggest that domestic cattle are descended from the extinct auroch and are the result of hybridisation of ancient taurine cattle and zebu (South-East Asian cattle).[81] However, what distinguished the breeding practices of the eighteenth century was their clear commercial focus – they were market-driven and geared towards economic efficiencies. In one respect, these efficiencies were too successful because they created the challenge of feeding additional animals. The Norfolk rotation system increased fodder and pasture, yet this was not sufficient. If sheep and cattle were to be fattened entirely on pasture, it would have required sweeping prairies, far beyond the capabilities of UK farmlands.[82] Farmers, therefore, turned to principles of "high farming", using grain crops to feed animals.[83]

[76] R M Hartwell, *The Industrial Revolution and Economic Growth*, above 6, 358.

[77] Ibid, 358.

[78] M L Ryder, "The History of Sheep Breeds in Britain", (1964) 12 (1) *The Agricultural History Review*, 1, 11, 12.

[79] J D Chambers and G E Mingay, *The Agricultural Revolution 1750–1880*, above 41, 68.

[80] R M Hartwell, *The Industrial Revolution and Economic Growth*, above 6, 333.

[81] Paolo Ajmone-Marsan, José Fernando Garcia, Johannes A Lenstra, and the Globaldiv Consortium, "On the Origin of Cattle: How Aurochs Became Cattle and Colonized the World", (2010) 19 *Evolutionary Anthropology*, 148.

[82] Richard Jeffries, *Landscape and Labour*, (Collection of Essays edited by John Pearson) Moonraker Press, Wiltshire (1979), 103, 104.

[83] J D Chambers and G E Mingay, *The Agricultural Revolution 1750–1880*, above 41, 4, 92, 93, 183.

High farming commenced approximately at the beginning of Queen Victoria's reign in 1837 and continued until the agricultural depression of the late nineteenth century.[84] In one sense, it was an extension of existing crop rotations and mixed farming, but it also incorporated the adoption of new knowledge and technology.[85] An important facet of high farming was "high feeding", a technique that used fodder and grains to produce stall-fed livestock.[86] The following extract, penned by Hippolyte Adolphe Taine (1828–1893), a Frenchman who visited England in the 1860s, describes an experimental farm based on high feeding technology:

> we stopped at a model farm...bullocks, pigs, sheep, each in a well-aired, well cleaned stall...in which the floor is a grating; beasts being fattened remain there for six weeks without moving. A narrow-gauge railway [takes]...food to the animals; they eat chopped turnips, crushed beans and 'oil cakes'. Farming on these terms is a complicated industry based on theory and experiment, constantly being perfected, and equipped with cleverly designed tools.[87]

The type of farm described by Taine was not widely adopted in the 1860s. However, operators of similar large-scale farms that emerged after enclosure demonstrated that restructured land-holdings could be competitive and could improve productivity by cutting costs.[88] By the 1850s yard and stall feeding of livestock, colloquially known as "pavestone farming", was becoming more common.[89] This development was also assisted by the railways, which transported manufactured feed quickly throughout the UK.[90] Importantly, manufactured feeding allowed farmers to keep livestock on poor soils and to increase stocking densities. Pasture required between three and four acres per cow, while pavestone farming reduced this to between one and two acres per cow.[91] Farmers therefore adopted high feeding because it was cost-effective and facilitated raising a greater number of animals more quickly.[92]

Yet, even with these efficiencies, the drive to increase productivity had limits. To start with, manufactured feed had to be purchased, which added another layer of economic complexity to animal production.[93] In addition, this feed was expensive,

[84] Ibid, 170.

[85] Ibid, 171, 184.

[86] Ibid.

[87] J D Chambers and G E Mingay, *The Agricultural Revolution 1750–1880*, above 41, 174, quoting *Taine's Notes on England* (translated by E Hyams, 1957) 132.

[88] Ibid, 181.

[89] Michael Winstanley, "Industrialization and the Small Farm: Family and Household Economy in Nineteenth Century Lancashire", above 58, 179.

[90] Ibid.

[91] Ibid.

[92] E L Jones, "The Changing Basis of English Agricultural Prosperity, 1857–73", (1962) 10 (2) *The Agricultural History Review*, 102, 112.

[93] Harriet Friedman and Philip McMichael, "Agriculture and the State System, the Rise and Decline of National Agricultures, 1870 to the Present", (1989) XXIX-2 *Sociologia Ruralis*, 93, 107.

and to intensify production further, farmers needed substantial capital outlays for drainage and construction. Discussion, therefore, not only focussed on how to source cheaper feed but also how to make farming more efficient by transforming it into a "commercial operation, organised along factory lines".[94] Richard Jeffries (1848–1887), an essay writer on rural life in the nineteenth century, noted that the UK needed to double its cattle production and called on entrepreneurs to find ways of accomplishing this while also improving returns to justify capital expenditure.[95] Perhaps somewhat surprisingly, mechanical technology did not lie at the core of these ideals, primarily because animal production in the late nineteenth century did not afford many openings for mechanisation.[96] Pastoral farmers may have made use of hay-making machines and steam-driven butter churns, but other innovations such as milking machines were not widespread until the first quarter of the twentieth century.[97]

Rather, the solution was seen in terms of commercialisation, where animal production was conducted by large corporations or joint-stock operations whose enterprises occupied vast farmlands, which were vertically integrated, employing butchers, packers and carriers who delivered animal products direct to retailers in major cities such as London.[98] By 1876, Jeffries noted that agriculture in the UK was getting closer to this ideal, concluding that production systems were akin to "methods of practice… employed in manufacture".[99]

The vision of agriculture as a manufacturing industry with animals fed turnips, grain and oil cakes in stalls has been described as the "apotheosis of high farming".[100] Accordingly, towards the end of the nineteenth century, cows in urban dairies were unlikely ever to feel grass under their feet, or eat pasture, as they were invariably kept indoors and fed on grains and by-products from breweries and distilleries.[101] Laying hens were similarly confined, a practice that became so common, that by the mid-twentieth century, it was unusual for these hens to be given access to fresh air and sunshine; instead, they were confined and fed a controlled diet, which invariably included a large amount of antibiotics.[102] Supply chains were the linchpin in these developments, because they needed to maintain a

[94] E J T Collins, "Rural and Agricultural Change", in Joan Thirsk (general ed) *The Agrarian History of England and Wales* Volume VII 1850–1914, Part 1, 72, 115, Cambridge University Press, Cambridge (2000).

[95] Richard Jeffries, *Landscape and Labour*, above 82, 105.

[96] E J T Collins, "Rural and Agricultural Change", above 94, 72, 129.

[97] Ibid.

[98] Richard Jeffries, *Landscape and Labour*, above 82, 105, 103, 105, 113.

[99] Ibid, 113.

[100] E J T Collins, "Rural and Agricultural Change", above 94, 72, 115.

[101] Chris Otter, "The Vital City: Public Analysis, Dairies and Slaughterhouses in Nineteenth-century Britain, above 31, 524.

[102] Susanne Freidberg, "Moral Economies and the Cold Chain", above 28, 136, 137.

continuous flow of stock. This not only depended on adequate production but also on adequate transportation and distribution networks.

2.5 Transportation and Distribution at the Domestic Level

As the per capita consumption of meat and animal products grew, maintaining adequate supply had become increasingly problematic. This was exacerbated by England's protectionist policies, which prior to the 1700s restricted the importation of livestock and farm produce, including from Scotland and Ireland.[103] Limitations changed slowly, commencing with the Act of the Union 1707 that joined England and Scotland, creating the UK.[104] English markets were now open to Scottish livestock, but not to Irish pastoral products and livestock, which were prohibited until the end of the eighteenth century, when pressure on food supplies forced the UK government to re-think its position.[105]

Opening up England's markets to produce from distant areas in Scotland and Ireland highlighted transportation challenges. Rural populations could readily obtain fresh produce, including meat, but prior to the railways, urban communities and distant markets depended on droving or transporting live animals by sea.[106] Droving was very much a traditional way of taking animals to market, and the routes could cover lengthy distances.[107] Cattle raising, for example, was Scotland's main commodity, and throughout the eighteenth century, it was not uncommon for pastoralists to drive as many as 320,000 heads of cattle a year from the Highlands to London.[108] Turkeys and geese from Norfolk and Suffolk were also driven to London, in droves each containing up to 1000 birds.[109] The birds would travel on foot or in carts and

---

[103] *An Additional Act against the Importation of Foreign Cattle*, Charles II, 1667 and 1668, in *Statutes of the Realm: Volume 5, 1628–80*, John Raithby, ed., (1819), 641, ss II, IV, available from *British History Online* http://www.british-history.ac.uk/statutes-realm/vol5/pp641-642; *An Act Prohibiting the Importation of Cattle from Ireland*, Charles II, 1680. in *Statutes of the Realm: Volume 5, 1628–80*, John Raithby, ed., (1819), 941, ss II, III, available from *British History Online* http://www.british-history.ac.uk/statutes-realm/vol5/pp941-942

[104] *An Act for the Union of the Two Kingdoms of England and Scotland*, 1706 Chap. 11 6 Ann, available from the National Archives, on behalf of HM Government, from, http://www.legislation.gov.uk/aep/Ann/6/11

[105] J D Chambers and G E Mingay, *The Agricultural Revolution 1750–1880*, above 41, 43; *An Act Prohibiting the Importation of Cattle from Ireland*, Charles II, 1680, ss II, III; For food products, *Importation Act*, 1774, George III c 7, *Statutes at Large*, Volume 43, J Bentham (1801) 3, available from https://play.google.com/books/reader?id=w6M3AAAAMAAJ&hl=en_US&pg=GBS.PA3 (subscription required); for food and livestock, *Importation Act*, 1774, George III c 8, *Statutes at Large*, Volume 43, J Bentham (1801) 157, available from https://play.google.com/books/reader?id=w6M3AAAAMAAJ&hl=en_US&pg=GBS.PA158

[106] R M Hartwell, *The Industrial Revolution and Economic Growth*, above 6, 183.

[107] J D Chambers and G E Mingay, *The Agricultural Revolution 1750–1880*, above 41, 24, 29, 30.

[108] Laurie Winn Carlson, *Cattle, An Informal Social History*, Ivan R. Dee, Chicago (2001) 72, 73.

[109] J D Chambers and G E Mingay, *The Agricultural Revolution 1750–1880*, above 41, 24.

along the way eat stubbles that remained after the harvest.[110] However, road transportation was expensive and slow, because drovers needed to secure access to routes and animals needed to be fed and watered.[111] As towns grew in size and the enclosure movement took hold, there were fewer pastured areas for livestock to graze, and local people became less tolerant of the noise and pollution generated by herds travelling through their towns.[112]

In the 1830s, droving declined with the introduction of the railways, an innovation that revolutionised the supply chain by providing relatively rapid and cheap distribution networks.[113] This was particularly significant for perishable animal products, because transportation times were sufficiently fast for country-killed meat to arrive in good condition to many urban areas.[114] London was by far the largest destination, and its markets received produce from all parts of the UK.[115] The railways also transported livestock that were sold directly to butchers who killed and prepared animals in the many small slaughterhouses.[116] Although railways transformed domestic markets and led to the demise of droving, they did not have the same immediate impact on coastal shipping.[117]

The coastal shipping trade was substantial for the times, even though it ebbed and flowed with the vagaries of legislative restrictions.[118] During the reign of Charles II, several acts prohibited the importation of livestock, noting that imported sheep and cattle had reduced livestock prices in England and Wales, which also led to lower rents, reducing the value of farmlands.[119] The Irish trade was regarded as particularly troublesome, so in 1680 Charles II legislated for permanent prohibitions against the importation of sheep, cattle, butter and cheese from Ireland.[120] These prohibitions

---

[110]Ibid.

[111]Kristin Hoganson, "Meat in the Middle: Converging Borderlands in the U.S. Midwest, 1865–1900", (2012) 98 (4) *The Journal of American History*, 1025, 1038.

[112]Richard Perren, *The Meat Trade in Britain 1840–1914*; Routledge & Kegan Paul, London (1978), 16.

[113]John Burnett, *Plenty and Want, a Social History of Food in England from 1815 to the Present Day*, above 1, 8.

[114]Ibid.

[115]Richard Perren, "The Meat and Livestock Trade in Britain, 1850–70", (1975) Volume XXVIII, (3) *The Economic History Review, Second Series*, 385.

[116]Chris Otter, "Civilizing Slaughter: The Development of the British Public Abattoir, 1850–1910", in *Meat, Modernity and the Rise of the Slaughterhouse* (Paula Young Lee ed), University of New Hampshire Press, New Hampshire (2008) 89, 90.

[117]J D Chambers and G E Mingay, *The Agricultural Revolution 1750–1880*, above 41, 171.

[118]Laurie Winn Carlson, *Cattle, An Informal Social History*, above 108, 72.

[119]*An Act for the Encouragement of Trade*, Charles II, 1663 in *Statutes of the Realm: Volume 5, 1628–80* (John Raithby ed) (1819), 449, s X, available from *British History Online* http://www.british-history.ac.uk/statutes-realm/vol5/pp449-452; *An Additional Act against the Importation of Foreign Cattle*, Charles II, 1667 and 1668, recital; *An Act Prohibiting the Importation of Cattle from Ireland*, Charles II, 1680. s VII.

[120]*An Act Prohibiting the Importation of Cattle from Ireland*, Charles II, 1680, Recital, ss III, VI.

were not lifted until almost a century later, occurring progressively in 1774 and 1776.[121]

After 1776, the Irish-UK trade intensified. In 1798, 72,200 cattle were exported to the UK, and by 1818 this figure had risen to 87,771.[122] Coastal shipping between Ireland and England remained busy well into the nineteenth century, and by the 1850s, shipments extended to the English port of Bristol and the Scottish port of Glasgow.[123] England and Scotland also enjoyed a popular coastal trade, which lasted until the 1870s.[124] However, as with droving, carriage by sea could be challenging. Bad weather and storms resulted in stock losses and delays, which increased the costs of transportation.[125] Moreover, even with the advent of steam ships, sea travel was still a lengthy process and realistically only suitable where markets were located close to a port.[126] Notwithstanding these challenges, coastal shipping persisted and for many years existed side by side with railway transportation.

By the 1850s railways and coastal steamers formed part of a supply chain that made it as economical to transport goods long distances, as it had been to transport them 30 or 50 kilometres.[127] Indeed, innovations in transportation facilitated the distribution of animals and their products across ever-increasing distances, becoming a significant driver of international trade.[128]

2.6 Trade, Transportation and Distribution at the International Level

International trade formed part of a global economy, which started emerging after 1500 and became more pronounced from the 1820s.[129] At first, the trade centred on commodities of high value, such as silk, gold and silver, but by the sixteenth and seventeenth centuries, England was exporting substantial quantities of wool and textiles.[130] In addition, towards the end of the seventeenth century, England had established a flourishing trade with the Americas and West Indies, exporting

[121] For food products, *Importation Act,* 1774, George III c 7; for food and livestock, *Importation Act,* 1774, George III c 8.

[122] Laurie Winn Carlson, *Cattle, An Informal Social History*, above 108, 72.

[123] Richard Perren, *The Meat Trade in Britain 1840–1914*, above 112, 17.

[124] Ibid, 23–25.

[125] Richard Perren, *The Meat Trade in Britain 1840–1914*, above 112, 23, 25.

[126] Richard Perren, "The Meat and Livestock Trade in Britain, 1850–70" above 115, 386.

[127] J D Chambers and G E Mingay, *The Agricultural Revolution 1750–1880*, above 41, 171.

[128] Chris Otter, "The Vital City: Public Analysis, Dairies and Slaughterhouses in Nineteenth-century Britain, above 31, 517.

[129] Robert C Allen, "Why was the Industrial Revolution British?" above 27, 50.

[130] Ibid.

preserved meat and livestock.[131] By the mid-nineteenth century, the focus on exports had altered, as the UK began to import an increased amount of food products.[132]

The reasons for this shift were partly based on the need to feed a growing UK population but were also a reflection of changes in government policy that grappled with the best way of harnessing international trade for the national good. At first, policies emphasised safeguarding local industries and/or maximising revenue collected at the border. The latter provided a consistent flow of capital, essential for prosperity and maintaining the military and navy.[133] Up to the mid-nineteenth century, UK regulation imposed protectionist and mercantilist measures, curtailing imports by way of tariffs and outright prohibitions.[134] These policies were enshrined in legislation, such as the seventeenth-century Acts of Trade and Navigation (Navigation Acts), the Customs Consolidation Act of 1787, and the Corn Laws of 1815.

The Navigation Acts were designed to uphold the balance of trade in favour of England, as well as weaken the competitive edge of the Dutch, who at the time held a commanding position in world trade.[135] The Acts therefore stipulated that commerce with England and its colonies could only be conducted using English ships, a requirement reinforced by prohibitions against foreign ships transporting goods between English ports.[136] These regulations were unpopular and contentious but persisted until 1849.[137] Protectionist policies also underpinned the Customs Consolidation Act of 1787, which replaced individual taxes on enumerated articles with a

[131] Virginia De John Anderson, *Creatures of Empire, How Domestic Animals Transformed Early America*, Oxford University Press, Oxford: New York (2004), 151.

[132] Kevin H O'Rourke and Jeffrey G Williamson, "When Did Globalisation Begin?", (2002) 6 (1) *European Review of Economic History*, 23, 25, 27.

[133] Istvan Hont, *Jealousy of Trade: International Competition and the Nation-State in Historical Perspective*, Belknap Press of Harvard University Press, Harvard (2010), 186.

[134] C Harley, "Trade: Discovery, Mercantilism and Technology", in *The Cambridge Economic History of Modern Britain, Volume 1: Industrialisation, 1700–1860* (Roderick Floud ed) Cambridge University Press, Cambridge (2004), 175, 187.

[135] This included *An Act for increase of Shipping, and Encouragement of the Navigation of this Nation* 1651, in *Acts and Ordinances of the Interregnum, 1642–1660* (C H Firth and R S Rait, ed), London (1911), 559–562. *British History Online* https://www.british-history.ac.uk/no-series/acts-ordinances-interregnum/pp559-562; C Harley, "Trade: Discovery, Mercantilism and Technology", above 134, 187, 188.

[136] For example, *An Act for increase of Shipping, and Encouragement of the Navigation of this Nation* 1651; Jonathan Barth, "Reconstructing Mercantilism: Consensus and Conflict in British Imperial Economy in the Seventeenth and Eighteenth Centuries", (2016) 73 (2) *William and Mary Quarterly*, 257, 272, referring to Acts of Trade and Navigation, 1651, 1660, 1662, 1663, 1670 and 1673.

[137] *An Act to amend the Laws in force for the Encouragement of British Shipping and Navigation* 1849, s 1, *Public General Statutes*, Twelfth and Thirteenth Year, Queen Victoria, 37, 1849, Printed by George Edward Eyre and William Spottiswoode and published by W Benning and Co, London (1849), 179, available from https://books.google.co.uk/books?id=RKJKAAAAMAAJ; Discussion, Jonathan Barth, "Reconstructing Mercantilism: Consensus and Conflict in British Imperial Economy in the Seventeenth and Eighteenth Centuries", above 136, 287.

single rate of duty.[138] Although the regime was simpler, the tax was still sufficiently high to create trade barriers.[139] These policies culminated with the introduction of the Corn Laws in 1815, which permitted imports of flour and grain only when domestic prices reached a certain level.[140] The legislation favoured UK grain growers by keeping the price of domestic grains artificially high.

However, even as these laws were being promulgated and implemented, political and economic philosophies had begun to entertain principles of free trade. These ideals can be traced to the laissez-faire movement of the eighteenth century, as well as growing misgivings regarding the value of tariffs "as a collective good".[141] Accordingly, by the nineteenth century, economists such as Alexander Baring (1774–1848) and Sir Robert Peel (1788–1850) began to support repeal of the Corn Laws.[142] Their viewpoints also coincided with chronic meat shortages in Western Europe, which affected the UK more than other countries because the UK population had increased almost fourfold since 1800.[143] European governments largely attributed meat shortages to insufficient supply, so interventions were designed to remedy this deficiency.[144] In France, for example, the sale of horsemeat for human consumption was legalised by 1866.[145] In the UK, the sale of horsemeat was already legal, but it never became a popular food, with its use mostly restricted to food for companion animals.[146] Instead, the UK government relaxed border controls, allowing the importation of live animals and meat. From the 1840s, incremental

---

[138] *An Act for repealing the several Duties of* Customs *and Excise, and granting other Duties in lieu thereof, and for applying the said Duties, together with the other Duties composing the Public Revenue; for permitting the Importation of certain Goods, Wares and Merchandize, the Produce or Manufacture of the European Dominions of the French King, into this Kingdom; and for applying certain unclaimed Monies, remaining in the Exchequer for the Payment of Annuities on Lives, to the Reduction of the National Debt*, 1787, 27 George III, c. 13, *Statutes at Large*, 36, Danby Pickering, Printed by John Archdeacon (1787), 23, available from, https://play.google.com/books/reader?id=XrIuAAAAIAAJ&hl=en_GB&pg=GBS.PA23; B R Leftwich, "The Later History and Administration of the Customs Revenue in England (1671–1814)", (1930) 13, *Transactions of the Royal Historical Society*, 187, 201.

[139] E J T Collins, "Food Supplies and Food Policy", in Joan Thirsk (general editor) *The Agrarian History of England and Wales* Volume VII 1850–1914, Part 1, 33, 35, Cambridge University Press, Cambridge (2000).

[140] *An Act to Amend the Law now in Force for Regulating the Importation of Corn* (1815) Public General Act, 55 George III, c 26, ss II, III and IV, *Statutes of the United Kingdom, Great Britain and Ireland*, 55 George III 1815, J Butterworth and Son, London (1815) 93, available from https://books.google.co.uk/books?id=fhY3AQAAMAAJ

[141] C P Kindleberger, "The rise of Free Trade in Western Europe 1820–1875", (1975) 35 (1) *The Journal of Economic* History, 20, 23, 24.

[142] Ibid, 27.

[143] Richard Perren, *Taste, Trade and Technology: the Development of the International Meat Industry Since 1840*, above 10, 2, 7, 20.

[144] Ibid, 2, 7.

[145] Susanne Freidberg, "Moral Economies and the Cold Chain", above 28, 127.

[146] Ibid.

tariff reform phased out restrictions and prohibitions, concluding with the repeal of the Corn Laws in 1849.[147]

The decision to lower trade barriers and open UK markets to international trade was controversial, as evinced by Peel's speech to the House of Commons in May 1842.[148] Peel argued in favour of lifting the prohibition on importation of cattle and replacing it with a tariff of one shilling per animal. He was motivated by the need to alleviate livestock shortages that had led to high meat prices, although he did acknowledge that scarcities were compounded by sheep and cattle diseases, as well as lack of capital investment in animal production.[149] Peel was also acutely aware of opposition from home producers but countered their resistance by pointing to the fact that within the UK, England already enjoyed a thriving livestock trade with Scotland and Ireland.[150] Furthermore, Peel regarded the distance between international markets and the UK as a deterrent that would keep import numbers low.[151] His comments relied on the impracticability of moving animals across lengthy distances and he did not foresee improvements in transportation that would occur from the 1870s. Peel also conveniently side-stepped the potential for livestock to introduce epizootics into the UK, a possibility that became more likely after 1846 when the tariff on livestock was eliminated, encouraging greater importation of animals.[152]

Indeed, once foreign sheep and cattle started passing through UK ports, it did not take long for livestock diseases to follow. Sheep pox was detected in 1847, suspected of having arrived in a shipment from Germany.[153] At that time, the UK border controls did not incorporate systems that could identify diseased animals, a situation which Parliament addressed when it passed the *Contagious Disorders (Sheep) Act* 1848 (Sheep Pox Act).[154] In accordance with that act, the Privy Council made

[147]Bethanie Afton and Michael Turner, "The Statistical Basis of Agricultural Performance in England and Wales 1850–1914, the Impact of Foreign Trade", in Joan Thirsk (general editor) *The Agrarian History of England and Wales* Volume VII 1850–1914, Part 1II, 2106, 2106, Cambridge University Press, Cambridge (2000).

[148]Customs Acts—The Tariff, House of Commons Debate, 10 May 1842, Sir Robert Peel, HC Deb 10 May, (1842) 63 *Hansard*, cc351–412, available from https://api.parliament.uk/historic-hansard/commons/1842/may/10/customs-acts-the-tariff

[149]Ibid, 370, 372.

[150]Ibid, 374, 379.

[151]Ibid, particularly 374 and 378.

[152]Paul Brassley, "Animal Health and Veterinary Medicine", in Joan Thirsk (general editor) *The Agrarian History of England and Wales* Volume VII 1850–1914, Part 1, 587, 589. Cambridge University Press, Cambridge (2000).

[153]Ibid, 589.

[154]*An Act to prohibit the Importation of Sheep, Cattle, or other Animals, for the Purpose of Preventing the Introduction of* Contagious *or Infectious Disorder*, 1848, Public General Act, 11 & 12 Victoria, c. 105, ss I, II, III, A Collection of the Public General Statute Passed in the Eleventh and Twelfth Year of the Reign of Her Majesty Queen Victoria 1847–8 Printed by George Edward Eyre and William Spottiswoode and published by W Benning and Co, London (1848),

enabling orders (regulations) on 4 September 1848.[155] Among other things, the regulations allowed veterinary surgeons to examine animals, and customs officers were authorised to seize and destroy infected livestock.[156] The Sheep Pox Act was followed by other legislation, such as the *Contagious Diseases Animals Act* 1878, which similarly provided for the destruction of infected animals.[157] The need to safeguard UK herds from disease had to some extent wound back philosophes of free trade. However, these restrictions did not completely halt livestock imports; instead, they led to altered trading patterns.

Between the years 1865 and 1869, most imports into the UK were sourced from Europe. The UK-Europe trade, however, had considerably declined by 1889, due to the introduction of steamships that made North American markets increasingly important.[158] Although steamships could still be delayed, they were more reliable than sailing ships and reduced voyage times made it feasible to procure products from increasingly distant regions.[159] This facilitated substituting Central and Eastern Europe trade, with trade from the Americas.[160]

The trans-Atlantic livestock trade commenced in 1868 with a shipment of cattle from the United States of America (USA) to London; this was followed by consignments of sheep and pigs, although they were never as commercially important as cattle.[161] Live exports from the USA reached their peak in the early 1900s when approximately 500,000 heads of livestock were transported each year, with the trade persisting until the outbreak of the World War I.[162] The trade in meat followed a similar pattern, initially focussing on imports from Europe but then shifting to North America and ultimately to markets south of the equator. By the turn of the

882, available from, https://play.google.com/books/reader?id=WntDAAAAcAAJ&hl=en_GB&pg=GBS.PA882

[155] Privy Council, Orders in Council, 4 September 1848, *The London Gazette*, 22 September 1848, Issue 20,899, 3477, available from https://www.thegazette.co.uk/London/issue/20899/page/3477; *The Edinburgh Gazette,* 3 November 1848, Issue 5800, 549, available from https://www.thegazette.co.uk/Edinburgh/issue/5789/page/480

[156] Privy Council, Orders in Council, 4 September 1848.

[157] *An Act for Making Better Provision Respecting Contagious and Infectious Diseases of Cattle and Other Animals and for Other Purposes*, (*Contagious Diseases Animals Act* 1878), 41 & 42 Victoria c 74, ss 21, 30; The Public General Statutes passed in the 41st and 42nd Years of the Reign of Her Majesty Queen Victoria 1878, George Edward Eyre and William Spottiswoode, London (1878), 581, available from https://archive.org/details/statutesatlarge15unkngoog/page/n592

[158] Richard Perren, *The Meat Trade in Britain 1840–1914*, above 112, 114, 157.

[159] Luigi Pascali, "The Wind of Change: Maritime Technology, Trade, and Economic Development", (2017) 107 (9) *American Economic Review*, 2821, 2823, 2827, 2838.

[160] Richard Perren, "Filth and Profit, Disease and Health: Public and Private Impediments to Slaughterhouse Reform in Victorian Britain", in *Meat, Modernity and the Rise of the Slaughterhouse* (Paula Young Lee ed), University of New Hampshire Press, New Hampshire (2008) 127, 132; Richard Perren, *The Meat Trade in Britain 1840–1914*, above 112, 114, 126.

[161] Richard Perren, *The Meat Trade in Britain 1840–1914*, above 112, 114, 115.

[162] Ibid, 160.

twentieth century, the US-UK trade had all but disappeared, by which time the European-UK trade had also been reduced to a small supply of bacon and pork products from Denmark and the Netherlands.[163] Refrigeration was critical to these altered trading regimes; it filled a gap in the market, because alternative preservation methods, such as drying, canning and pickling, were not favoured by consumers.[164]

However, early refrigeration techniques also restricted the range of preferred markets. Chilling kept produce at a temperature just above freezing, making it suitable for shorter journeys, whereas perishable products sent by way of longer trade routes needed to be frozen. The difficulty was that freezing allowed the formation of ice crystals, which damaged the taste and texture of meat, favouring North American markets because their proximity to the UK allowed shipments to be chilled rather than frozen.[165] This also meant that the USA captured the market by proffering a superior product, and when the first shipments of chilled beef reached the UK in 1875, they proved very popular, leading to swift expansion of US firms.[166]

By the latter part of the nineteenth century, refrigeration techniques had improved, and freezing no longer formed ice crystals.[167] This opened British markets to produce from Australia, New Zealand and South America, all of whom generated surpluses, gladly received in the UK.[168] In a short while, it became a common practice to ship better quality meat from the Southern Hemisphere to the UK, so that by the outbreak of World War I, UK ports had become the main destination for international trade in animal products.[169] In monetary terms, these imports were substantial, increasing from 3 million pounds in 1854 to 16 million pounds in 1900.[170] Yet, in a broader sense, neither the livestock trade nor the trade in meat was economically significant compared to UK international trade as a whole.[171] Rather, the significance of the trade lay in less obvious ways, namely, in growing consumerism and commercialisation, which came to characterise the sector, both internationally and domestically.

[163] John P Huttman, "British Meat Imports in the Free Trade Era", (1978) 52 (2) *Agricultural History*, 247, 259.

[164] Ibid.

[165] Richard Perren, *The Meat Trade in Britain 1840–1914*, above 112, 125.

[166] Ibid, 114, 126, 169.

[167] E J T Collins, "Food Supplies and Food Policy", above 139, 39.

[168] Richard Perren, *The Meat Trade in Britain 1840–1914*, above 112, 70.

[169] Bethanie Afton and Michael Turner, "The Statistical Basis of Agricultural Performance in England and Wales 1850–1914, the Impact of Foreign Trade", above 147, 2110.

[170] John P Huttman, "British Meat Imports in the Free Trade Era", above 163, 247.

[171] Ibid, 247, 254.

2.7 Consumerism and Commercialisation

As already discussed, social processes, such as enclosure and urbanisation, combined to engender the demand for animal products. Appetite for these products remained comparatively steady, notwithstanding the economic ups and downs of the second half of the nineteenth century.[172] Consumer demand was also reflected in the price of meat, which maintained its market position, even when prices of other agricultural commodities fell, particularly during the agricultural depression of the 1890s.[173]

The decision by farmers to raise more animals in response to market conditions fashioned an "entrepreneurial path" that focussed on the production of surpluses for profit and in the process, increasingly commodified animals.[174] Robert Bakewell had acted in such a manner when he produced the New Leicester sheep, having in mind the mass market – the many urban workers who could afford reasonable quality produce.[175] Commercial imperatives were also evident in high farming, where intensive feeding reduced fattening times, allowing pastoralists to sell their sheep and cattle earlier than they would otherwise have done. Other analogous changes included government intervention that facilitated speedier movement of animals along supply chains, as well as modification of market exchanges by stakeholders and emerging transformations in the composition of the rural workforce.

A telling illustration derives from bottlenecks in the supply chain. The advent of railroads, steamships and refrigerated cargo holds had increased the volume of trade, but inadequate warehousing and transportation at landing ports created congestion, hindering the movement of livestock and animal products.[176] To deal with this deficiency, regulators constructed new storage and handling facilities, as well as improved rail networks from wharves to inland areas.[177] The aim was to move animals and animal products quickly and efficiently, as would happen with any other commodity. This not only provided a solution to the problem of bottlenecks but also facilitated a greater volume of trade. Yet, increasing the amount of trade also increased the risk of introducing epizootics, a liability that necessitated organisational changes in how animals were slaughtered and marketed.

After 1848, infected animals were destroyed, while the *Contagious Diseases Animals Act* 1878 charged the Privy Council with making regulations for landing healthy animals.[178] In accordance with this delegation, the Privy Council published

[172] J D Chambers and G E Mingay, *The Agricultural Revolution 1750–1880*, above 41, 209.

[173] Richard Perren, *Agriculture in Depression, 1870–1940*, Cambridge University Press, Cambridge (1995), 9.

[174] Mark B Tauger, *Agriculture in World History*, Routledge, Abingdon, Oxfordshire (2011), 86.

[175] R M Hartwell, *The Industrial Revolution and Economic Growth*, above 6, 326.

[176] Richard Perren, *The Meat Trade in Britain 1840–1914*, above 112, 42.

[177] Chris Otter, "Civilizing Slaughter: The Development of the British Public Abattoir, 1850–1910", above 116, 96.

[178] *Contagious Disorders (Sheep), etc. Act* 1848 c107, s III; *Contagious Diseases Animals Act* 1878, ss 29, 36.

the Foreign Animals Order on 10 December 1878.[179] Among other things, the Order prevented animals from gaining entry if they originated from countries where livestock were suspected of carrying epizootics.[180] In other cases, however, foreign animals were permitted to land, provided they were offloaded at a designated "Foreign Animals Wharf" and were slaughtered within 10 days upon arrival.[181]

These requirements necessitated slaughter facilities dockside, as well as the means of inspecting a copious number of animals.[182] In 1899 at the height of the livestock trade, some 4800 animals were passing through Liverpool each week and approximately 4000 a week through London.[183] Facilities had already been built in London at Deptford that could deal with this volume of trade, but a special wharf had to be constructed at Birkenhead, opposite to Liverpool on the River Mersey.[184] These solutions addressed logistical challenges in the most economically efficient way. Yet, they also augmented centres of power whose interests were, and to this day remain, predominantly financial and commercial. Transformations in the role of intermediaries exemplify how these interests influenced the supply chain.

Up to the mid-nineteenth century, meat sales were largely undertaken by "producer-retailers…[and] large-scale butchers who did their own slaughtering".[185] Intermediaries principally comprised cattle salesmen, moneylenders and retailers. At Smithfield, cattle salesmen acted as carriers and wholesalers, transporting animals from farms to markets and then selling them to butchers who were the main retailers.[186] To facilitate payment, butchers deposited the purchase price with moneylenders who also acted as commission agents, retaining a fee per head of livestock and paying the remainder into farmers' bank accounts.[187] As the demand for meat grew, this comparatively uncomplicated structure altered, particularly with the leverage provided by international trade.

In the USA, animal production became a vast commercial undertaking, comprising three distinct industries dealing with wholesaling, slaughter of livestock and retailing.[188] Wholesaling was especially influential because US firms amalgamated,

[179] Privy Council, The Foreign Animals Order, 10 December, 1878, *The London Gazette*, Issue 24,655, page 7073, available from https://www.thegazette.co.uk/London/issue/24655/page/7073

[180] Ibid, Part I (5).

[181] Privy Council, The Foreign Animals Order, 10 December, 1878, Part II (6) and (7).

[182] Richard Perren, "Filth and Profit, Disease and Health: Public and Private Impediments to Slaughterhouse Reform in Victorian Britain", above 160, 143.

[183] Ibid, 144.

[184] Ibid.

[185] Gareth Shaw, "Changes in Consumer Demand and Food Supply in Nineteenth-Century British Cities", above 2, 288.

[186] Richard Perren, *The Meat Trade in Britain 1840–1914*, above 112, 37.

[187] Ibid, 37, 39.

[188] Roger Horowitz, Jeffrey M. Pilcher, and Sydney Watts, "Meat for the Multitudes: Market Culture in Paris, New York City, and Mexico City over the Long Nineteenth Century", (2004) 109 (4) *American Historical Review*, 1055, 1060.

operating as virtual monopolies.[189] This created a commercially based power centre whose domination was felt in both the USA and the UK.[190] Subsidiaries of US corporations obtained outlets at Smithfield, enabling them to substitute their own personnel for moneylenders and commission agents that had formerly operated onsite.[191] At the same time, the sheer size of US businesses required significant capital expenditure, elevating the role of financiers and capital markets, allowing them to become extremely influential.[192] The upshot was that stakeholders with a financial interest in animal production gained ascendency, a development that occurred in tandem with fragmentation of the supply chain as the rural workforce became more diffuse.

Throughout the nineteenth century, the proportion of agricultural workers as a percentage of the total workforce declined: from 40 to 45% in 1780, to 34% in 1801, to 24% in 1832, to 21% in 1851[193] and to 8% in 1914.[194] Two points should be kept in mind: first, these figures represent agricultural labourers as a percentage of the total work force, and second they do not take into account support workers. Although the percentage of agricultural workers declined, this belies the fact that agriculture remained a leading English industry until the mid-nineteenth century. During this time, rural employment numbers per se continued to increase, although in the long term, urban regions grew at the expense of country areas, as labourers sought employment and improved wages in the cities.[195] By the time these labour shifts occurred, farms were no longer self-sufficient, and fewer workers were directly involved in agriculture, taking with them specialised labour and knowledge.[196]

Moreover, as the proportion of on-site workers declined, the number of outside workers increased. Studies in England and Scotland reveal that in 1841, 27 outside workers supported every 100 farmers or agricultural workers, a proportion that by 1881 had climbed to 47 support staff per 100 and by 1911 had risen to 67 support staff per 100.[197] Although the type of support worker has varied over time, it has included blacksmiths, wheelwrights, saddlers, wholesalers, retailers, office staff, veterinarians and laboratory workers.[198] The increasing number of off-site workers

---

[189] Richard Perren, *The Meat Trade in Britain 1840–1914*, above 112, 165.

[190] C Knick, "Steers Afloat: The North Atlantic Meat Trade, Liner Predominance, and Freight Rates, 1870–1913", (2008) 68 (4) *Economic History Association*, 1028, 1036.

[191] Richard Perren, *The Meat Trade in Britain 1840–1914*, above 112, 165.

[192] Kenneth Pomeranz and Steven Topik, *The World that Trade Created*, M E Sharpe, New York (2006), 46.

[193] R M Hartwell, *The Industrial Revolution and Economic Growth*, above 6, 125.

[194] Patrick Karl O'Brien, "Path Dependency, or Why Britain Became an Industrialized and Urbanized Economy Long before France", above 40, 231.

[195] R M Hartwell, *The Industrial Revolution and Economic Growth*, above 6, 125.

[196] Richard Jeffries, *Landscape and Labour*, above 82, 99, 100.

[197] E J T Collins, "Introduction" in Joan Thirsk (general editor) *The Agrarian History of England and Wales* Volume VII 1850–1914, Part 1, 1, 8, Cambridge University Press, Cambridge (2000).

[198] Ibid.

reveals a tendency towards outsourcing functions and responsibilities that were once the purview of farm owners and/or agricultural labourers.[199] This resulted in specific stakeholders occupying a progressively smaller and disconnected part of the supply chain. Responsibilities came to be seen in fragmented and narrow terms, viewing animals as a commodity to be efficiently transported and marketed; a specimen to be tested for disease and/or an agricultural product to be butchered and sold as cuts of meat.

Fragmentation of the supply chain was also exacerbated by uneven development in the sector, which occurred as stakeholders attempted to resolve economic challenges on a case-by-case basis, as they adapted production to the exigencies of the market.[200] Thus, livestock was bred to fatten quickly and for the same reasons was also subjected to the use of concentrated feed stuff. Refrigeration allowed the transportation of animals and their products over wider distances, while governments harnessed the knowledge of veterinary practice to prevent the introduction and spread of epizootics. These adaptations achieved a form of equilibrium but one that was attained by implementing innovations and adjustments that displaced existing processes as the commercial need arose.[201] Market balance, therefore, did not occur because of concurrent development in industries and sectors that provided reciprocal benefits but rather because economic efficiencies facilitated supply.

As Ryan Gunderson argues, although technological innovations are often the first to be noticed in animal production, financial and commercial issues invariably underpin technology, because the latter responds to economic needs.[202] Gunderson's comments were largely made in the context of factory farming and twentieth-century markets, yet analogous patterns are evident in the late nineteenth century, because, in both cases, animals were used as a commodity to generate profits.[203] These patterns are illustrated by the logistics of the meat trade in the nineteenth to early twentieth centuries.

Chilled beef has a short shelf life, so supplies to retailers needed to be consistent. In the decade or so prior to World War I, domestic provisions in the UK were supplemented by 9000 tonnes of carcasses arriving in England every Monday morning.[204] The extensive scale of this trade was dependent on mass production, rapid turnover, reduced outlays and minimal shipping costs.[205] By the same token,

---

[199] Ibid.

[200] J R T Hughes, "Foreign Trade and Balanced Growth: The Historical Framework", (1959) 49 (2) *The American Economic Review*, 330, 334.

[201] Ibid, 335.

[202] Ryan Gunderson, "From Cattle to Capital: Exchange Value, Animal Commodification, and Barbarism", (2011) 39 (2) *Critical Sociology*, 259, 259.

[203] Ibid, 261.

[204] I R Phimister, "Meat and Monopolies: Beef Cattle in Southern Rhodesia, 1890–1938", (1978) 19 (3) *The Journal of African History*, 391, 392.

[205] Ibid, 394.

the considerable range of these operations also needed large capital investments, which meant that by the end of the nineteenth century the industry came to be concentrated in the hands of a small number of companies who integrated their operations vertically by owning ranches and production and distribution networks.[206] These enterprises embodied an intense form of profit farming, which in the UK had progressively displaced "use values", where animals were raised for personal or community use, irrespective of their commercial value.[207]

The early stages of profit farming were largely stimulated by external causes, such as urbanisation and the accelerated demand for meat, which have already been discussed. However, "internal causes", which originated as farms stopped being self-sufficient, were also important.[208] When farmers started purchasing staples, such as bread and ale, this necessitated greater attention to money matters, as well as the retention of financial records to keep track of expenditure.[209] In addition, the very thought of expenditure engendered notions of profit and loss, introducing accounting and business principles into an undertaking that hitherto had not traditionally been run as a commercial undertaking.[210] As with other businesses, expenditure needed to be justified, which in a farming context usually occurred because outlays intensified production.[211]

The need to justify expenditure was aggravated by increased competition wrought by transportation innovations. By the 1880s, railways and steamships transported animal products from increasingly distant markets, so that UK producers were not only competing with each other but also with European, American and Canadian producers.[212] As a result, production became increasingly tailored to the prescriptions of global markets, particularly with respect to cost, quality and quantity.[213] Within these frameworks, animals quickly became deconstructed from living entities to corporeal vendibles.[214]

Commodifying animals and their products by commercial exchange is not unique to the nineteenth century, the trade having existed since antiquity. In Asia, for example, trade with China in the second century BC included live birds, as well as

---

[206] Ibid.

[207] Ryan Gunderson, "From Cattle to Capital: Exchange Value, Animal Commodification, and Barbarism", above 202, 261, quoting E Mandel, *Marxist Economic Theory*, London: Merlin (1968).

[208] Richard Jeffries, *Landscape and Labour*, above 82, 101.

[209] Ibid.

[210] Ibid, 100, 101.

[211] Ibid, 107, 108, 117, 118.

[212] Robert M Schwartz, "Rail Transport, Agrarian Crisis, and the Restructuring of Agriculture: France and Great Britain Confront Globalization, 1860–1900", (2010) 34 (2) *Social Science History*, 229, 231, 234.

[213] Ibid, 231, 234.

[214] Virginia De John Anderson, *Creatures of Empire, How Domestic Animals Transformed Early America*, above 131, 68.

tribute in the form of animals such as rhinoceros and elephants.[215] Moreover, it is not easy to pinpoint a date when raising livestock in the UK transformed from use values to profit farming, although in England there were signs of this transformation from the sixteenth century.[216] What is clear, though, is that developments of the nineteenth century were critical, because increased supply fuelled demand.[217] This created a continuous and ever-expanding cycle, which saw animal production taking on the mantle of manufacturing systems.[218]

Prior to these transformations, farming had operated as an "extractive industry", although one where the product continually renewed itself.[219] Supply was limited by the physiological traits of the animal as well as the carrying capacity of the land. Manufacturing systems, however, know few permanent restraints, and limits to production are seen as temporary challenges to be overcome. Thus, in animal production, selective breeding that achieves efficient production is encouraged with as much enthusiasm as improvements in warehousing, transportation and refrigeration. Moreover, the slaughter of animals shifted from private slaughterhouses to public abattoirs, linking slaughter to distribution systems that expedited the movement of carcasses, permitting more animals to be killed more quickly.[220]

These initiatives reinforced commercial expectations of animals as a marketable commodity. Indeed, early experiments with New Leicester sheep and the introduction of concentrated feed were not only economic responses to limitations of supply but were also harbingers of the agro-industrial enterprises of the twentieth century. The latter underscored mechanisation, feed lots, farrowing crates and selective breeding of commercially desirable traits in food animals, echoing nineteenth-century techniques designed to produce a marketable product in the most economical and profitable way.[221] As Tauger concludes:

> Agriculture in the nineteenth century shook off its main medieval vestiges and rapidly acquired all the characteristics of the leading industrial sectors: mass production, world

[215] Richard L Smith, *Premodern Trade in World History*, Routledge, Abingdon, Oxfordshire (2009) 117.

[216] Ryan Gunderson, "From Cattle to Capital: Exchange Value, Animal Commodification, and Barbarism", above 202, 262.

[217] Patrick Karl O'Brien, "Path Dependency, or Why Britain Became an Industrialized and Urbanized Economy Long before France", above 40, 225.

[218] F M L Thompson, "The Second Agricultural Revolution, 1815–1880", above 38, 64.

[219] Ibid.

[220] Chris Otter, "The Vital City: Public Analysis, Dairies and Slaughterhouses in Nineteenth-century Britain, above 31, 529.

[221] Generally, Katrina Sharman, "Farm Animals and Welfare Law: An Unappy Union" in *Animal Law in Australasia* (Peter Sankoff and Steven White ed), 35, Federation Press, Annandale, NSW (2009).

markets, extensive wide-ranging trade and increasing concentration in trade and processing.[222]

2.8 Conclusion

The nineteenth century saw the widespread commercialisation of animals, encouraging economies of scale and maximisation of profits, so that by the time of World War II livestock production had altered from a system of hands-on farming to one concentrated in "continuous production systems".[223] The emphasis lay on supply, engendering a commercial bias, which as discussed in later chapters created a disconnect between the treatment of animals in food production and emerging philosophies of animal wellbeing. The next chapter commences the discussion on how this disconnect evolved, by analysing anti-cruelty regulation of the nineteenth century and identifying how farm animals came to be excluded from increasing protection afforded to other domesticated animals.

Bibliography

Afton B, Turner M (2000) The statistical basis of agricultural performance in England and Wales 1850–1914, the impact of foreign trade. In: Joan T (general ed) The Agrarian history of England and Wales volume VII 1850–1914, Part 1II. Cambridge University Press, Cambridge, p 2106

Ajmone-Marsan P, Fernando GJ, Lenstra JA, the Globaldiv Consortium (2010) On the origin of cattle: how aurochs became cattle and colonized the world. Evol Anthropol 19:148

Alle RC (1999) Tracking the agricultural revolution in England. Econ Hist Rev LII(2):209

Ang JB, Banerjee R, Madsen JB (2013) Innovation and productivity advances in British agriculture: 1620–1850. South Econ J 80(1):162

Argemí L (2002) Agriculture, agronomy, and political economy: some missing links. Hist Polit Econ 34(2):449

Barth J (2016) Reconstructing mercantilism: consensus and conflict in British Imperial economy in the seventeenth and eighteenth centuries. William Mary Q 73(2):257

Brassley P (2000) Animal health and veterinary medicine. In: Thirsk J (general ed) The Agrarian history of England and Wales volume VII 1850–1914, Part 1. Cambridge University Press, Cambridge, p 587

Burnett J (1989) Plenty and want, a social history of food in England from 1815 to the present day, 3rd edn. Routledge, London

Carlson LW (2001) Cattle, an informal social history. Ivan R. Dee, Chicago

Clark G (2001) Farm wages and living standards in the industrial revolution: England, 1670-1869. Econ Hist Rev 53(3):477

[222] Mark B Tauger, *Agriculture in World History*, above 174, 104.

[223] Harriet Friedman and Philip McMichael, "Agriculture and the State System, the Rise and Decline of National Agricultures, 1870 to the Present", above 93, 106.

Collins EJT (1987) The rationality of 'surplus' agricultural labour: mechanization in English agriculture in the nineteenth century. Agric Hist Rev 35(1):36
Collins EJT (2000a) Introduction. In: Thirsk J (general ed) The Agrarian history of England and Wales volume VII 1850–1914, Part 1, 1. Cambridge University Press, Cambridge, p 8
Collins EJT (2000b) Rural and agricultural change. In: Thirsk J (general ed) The Agrarian history of England and Wales volume VII 1850–1914, Part 1. Cambridge University Press, Cambridge, p 72
Collins EJT (2000c) Food supplies and food policy. In: Thirsk J (general ed) The Agrarian history of England and Wales volume VII 1850–1914. Cambridge University Press, Cambridge, p 33
De John Anderson V (2004) Creatures of empire, how domestic animals transformed early America. Oxford University Press, Oxford
Freidberg S (2015) Moral economies and the cold chain. Hist Res 88(239):125
Friedman H, McMichael P (1989) Agriculture and the state system, the rise and decline of National Agricultures, 1870 to the present. Sociol Rural XXIX-2:93–106
Fussell GE (1937) Animal husbandry in eighteenth century England: part 1. Agric Hist 11(2):96
Gunderson R (2011) From cattle to capital: exchange value, animal commodification, and barbarism. Crit Sociol 39(2):259
Harley C (2004) Trade: discovery, mercantilism and technology. In: Floud R (ed) The Cambridge economic history of modern britain, volume 1: industrialisation, 1700–1860. Cambridge University Press, Cambridge, p 175
Harrison M (2015) A global perspective: reframing the history of health, medicine, and disease. Bull Hist Med 89(4):639
Hilton M (2003) Consumerism in twentieth-century britain: the search for a historical movement. Cambridge University Press, Cambridge
Hoganson K (2012) Meat in the middle: converging borderlands in the U.S. Midwest, 1865–1900. J Am Hist 98(4):1025
Hont I (2010) Jealousy of trade: international competition and the nation-state in historical perspective. Belknap Press of Harvard University Press, Harvard
Horowitz R, Pilcher JM, Watts S (2004) Meat for the multitudes: market culture in Paris, New York City, and Mexico City over the long nineteenth century. Am Hist Rev 109(4):1055
Hughes JRT (1959) Foreign trade and balanced growth: the historical framework. Am Econ Rev 49(2):330
Huttman JP (1978) British meat imports in the free trade era. Agric Hist 52(2):247
Jeffries R (1979) Landscape and labour (collection of essays edited by John Pearson). Moonraker Press, Wiltshire
Jones EL (1962) The changing basis of English agricultural prosperity, 1857–73. Agric Hist Rev 10(2):102
Kindleberger CP (1975) The rise of free trade in Western Europe, 1820-1875. J Econ Hist 35(1):20
Knick C (2008) Steers afloat: the North Atlantic meat trade, liner predominance, and freight rates, 1870-1913. Econ Hist Assoc 68(4):1028–1036
Leftwich BR (1930) The later history and administration of the customs revenue in England (1671-1814). Trans R Hist Soc 13:187
Modrzejewska M (2017) Adam Smith's concept of value of labour. Anglo-American perspective till mid-19th century. Horiz Polit 8(25):65
Newholm T, Newholm S, Shaw D (2015) A history for consumption ethics. Bus Hist 57(2):290
O'Brien PK (1996) Path dependency, or why Britain became an industrialized and urbanized economy long before France. Econ Hist Rev 49(2):23
O'Rourke KH, Williamson JG (2002) When did globalisation begin? Eur Rev Econ Hist 6(1):23
Oddy DJ (1970) Working-class diets in late nineteenth-century Britain. Econ Hist Rev 23(2):314

Otter C (2006) The Vital City: public analysis, dairies and slaughterhouses in nineteenth-century Britain. Cult Geogr 13:517

Otter C (2008) Civilizing slaughter: the development of the British public Abattoir, 1850–1910. In: Lee PY (ed) Meat, modernity and the rise of the slaughterhouse. University of New Hampshire Press, New Hampshire, p 89

Pascali L (2017) The wind of change: maritime technology, trade, and economic development. Am Econ Rev 107(9):2821

Perren R (1971) The north American beef and cattle trade, with Great Britain, 1870-1914. Econ Hist Rev 24(3):430

Perren R (1975) The meat and livestock trade in Britain, 1850-70. Econ Hist Rev, Second Series XXVIII(3):385

Perren R (1978) The meat trade in Britain 1840–1914. Routledge & Kegan Paul, London

Perren R (1995) Agriculture in depression, 1870–1940. Cambridge University Press, Cambridge

Perren R (2017) Taste, trade and technology: the development of the international meat industry since 1840, (2006 Ashgate). Routledge, Abingdon

Perren R (2008) Filth and profit, disease and health: public and private impediments to slaughterhouse reform in Victorian Britain. In: Lee PY (ed) Meat, modernity and the rise of the slaughterhouse. University of New Hampshire Press, New Hampshire, p 127

Phimister IR (1978) Meat and monopolies: beef cattle in southern Rhodesia, 1890-1938. J Afr Hist 19(3):391–392

Pomeranz K, Topik S (2006) The world that trade created. M E Sharpe, New York

Riley JC (1999) A widening market in consumer goods. In: Cameron E (ed) Early modern Europe, an Oxford history. OUP, Oxford

Rollin B (2008) Animal ethics and the law. Mich Law Rev First Impress 106:143

Ryder ML (1964) The history of sheep breeds in Britain. Agric Hist Rev 12(1):1

Schwartz RM (2010) Rail transport, agrarian crisis, and the restructuring of agriculture: France and Great Britain conffront globalization, 1860–1900. Soc Sci Hist 34(2):229

Sharman K (2009) Farm animals and welfare law: an unappy union. In: Sankoff P, White S (eds) Animal law in Australasia. Federation Press, Annandale, p 35

Shaw G (1985) Changes in consumer demand and food supply in nineteenth-century British cities. J Hist Geogr 11(3):280–290

Shaw-Taylor L (2001) Parliamentary enclosure and the emergence of an English agricultural proletariat. J Econ Hist 61(3):640–642

Smith RL (2009) Premodern trade in world history. Routledge, Abingdon

Sullivan RJ (1985) The timing and pattern of technological development in English agriculture. J Econ Hist 45(2):305–308

Swalger R (1994) Evolution and applications of the term consumerism: theme and variations. J Consum Aff 28(2):347

Tauger MB (2011) Agriculture in world history. Routledge, Abingdon, p 87

Thomsen FL (1943) The impact of war on marketing farm products. J Farm Econ 25(1):120

Thompson FML (1968) The second agricultural revolution, 1815-1880. Econ Hist Rev 21(1):62

Timmer CP (1969) The turnip, the new husbandry, and the English agricultural revolution. Q J Econ 83(3):375

Walton JR (1983) The diffusion of improved sheep breeds in eighteenth- and nineteenth-century Oxfordshire. J Hist Geogr 9(2):175

Winstanley M (1996) Industrialization and the small farm: family and household economy in nineteenth century Lancashire. Past Present Soc 152:157

Legislation, Regulations, Gazettes and Hansard

1651, *An Act for increase of Shipping, and Encouragement of the Navigation of this Nation* 1651, in *Acts and Ordinances of the Interregnum, 1642–1660*, Firth CH and Rait RS (edd) London (1911), p 559–562. *British History Online* http://www.british-history.ac.uk/no-series/acts-ordinances-interregnum/

1663, *An Act for the Encouragement of Trade*, Charles II, 1663 in *Statutes of the Realm: Volume 5, 1628–80*, Raithby J (ed) (1819), 449, available from *British History Online* http://www.british-history.ac.uk/statutes-realm/vol5/pp449-452

1669 & 1668, *An Additional Act against the Importation of Foreign Cattle*, Charles II, 1667 and 1668, in *Statutes of the Realm: Volume 5, 1628–80*, Raithby J (ed) (1819), 641, available from *British History Online* http://www.british-history.ac.uk/statutes-realm/vol5/pp641-642

1680, *An Act Prohibiting the Importation of Cattle from Ireland*, Charles II, 1680. in *Statutes of the Realm: Volume 5, 1628–80*, Raithby J (ed) (1819), 94, available from *British History Online* http://www.british-history.ac.uk/statutes-realm/vol5/pp941-942

1706, *An Act for the Union of the Two Kingdoms of England and Scotland*, 1706 Chapter 11 6 Ann, available from the National Archives, on behalf of HM Government, from, http://www.legislation.gov.uk/aep/Ann/6/11

1774, *Importation Act,* 1774, George III c 7, *Statutes at Large*, Volume 43, J Bentham (1801) 3. https://play.google.com/books/reader?id=w6M3AAAAMAAJ&hl=en_US&pg=GBS.PA3

1774, *Importation Act,* 1774, George III c 8, *Statutes at Large*, Volume 43, J Bentham (1801) 157. https://play.google.com/books/reader?id=w6M3AAAAMAAJ&hl=en_US&pg=GBS.PA158

1787, *An Act for repealing the several Duties of* Customs *and Excise, and granting other Duties in lieu thereof, and for applying the said Duties, together with the other Duties composing the Public Revenue; for permitting the Importation of certain Goods, Wares and Merchandize, the Produce or Manufacture of the European Dominions of the French King, into this Kingdom; and for applying certain unclaimed Monies, remaining in the Exchequer for the Payment of Annuities on Lives, to the Reduction of the National Debt*, 1787, 27 George III, c. 13, *Statutes at Large*, Volume 36, Danby Pickering, Printed by John Archdeacon (1787), 23. https://play.google.com/books/reader?id=XrIuAAAAIAAJ&hl=en_GB&pg=GBS.PA23

1801, *Importation Act,* 1774, George III c 8, *Statutes at Large*, Volume 43, J Bentham (1801) 157. https://play.google.com/books/reader?id=w6M3AAAAMAAJ&hl=en_US&pg=GBS.PA158

1815, *An Act to Amend the Law now in Force for Regulating the Importation of Corn* (1815) Public General Act, 55 George III, c 26, ss II, III and IV, *Statutes of the United Kingdom, Great Britain and Ireland*, 55 George III 1815, Butterworth J and Son, London (1815) 93. https://books.google.co.uk/books?id=fhY3AQAAMAAJ

1842, Customs Acts—The Tariff, House of Commons Debate, 10 May 1842, Sir Robert Peel, HC Deb 10 May, (1842) 63 *Hansard*, cc351–412. https://api.parliament.uk/historic-hansard/commons/1842/may/10/customs-acts-the-tariff

1848, *An Act to prohibit the Importation of Sheep, Cattle, or other Animals, for the Purpose of Preventing the Introduction of* Contagious *or Infectious Disorder*, 1848, Public General Act, 11 & 12 Victoria, c. 105, Collection of the Public General Statute Passed in the Eleventh and Twelfth Year of the Reign of Her Majesty Queen Victoria 1847–8 Printed by George Edward Eyre and William Spottiswoode and published by W Benning and Co, London (1848a), 882. https://play.google.com/books/reader?id=WntDAAAAcAAJ&hl=en_GB&pg=GBS.PA882

1848, Privy Council, Orders in Council, 4 September 1848, *The London Gazette*, 22 September 1848, Issue 20899, 3477, available from https://www.thegazette.co.uk/London/issue/20899/page/3477; *The Edinburgh Gazette,* 3 November 1848, Issue 5800, 549. https://www.thegazette.co.uk/Edinburgh/issue/5789/page/480

1849, *An Act to amend the Laws in force for the Encouragement of British Shipping and Navigation* 1849, s 1, *Public General Statutes*, Twelfth and Thirteenth Year, Queen Victoria, volume 37, 1849, Printed by George Edward Eyre and William Spottiswoode and published by W Benning and Co, London (1849), 179. https://books.google.co.uk/books?id=RKJKAAAAMAAJ

1878, *An Act for Making Better Provision Respecting Contagious and Infectious Diseases of Cattle and Other Animals and for Other Purposes*, (*Contagious Diseases Animals Act* 1878), 41 & 42 Victoria c 74, The Public General Statutes passed in the Forty First and Forty Second Years of the Reign of Her Majesty Queen Victoria 1878, George Edward Eyre and William Spottiswoode, London (1878a), 581. https://archive.org/details/statutesatlarge15unkngoog/page/n592

1878, Privy Council, The Foreign Animals Order, 10 December, 1878, *The London Gazette,* Issue 24655, page 7073. https://www.thegazette.co.uk/London/issue/24655/page/7073

The Enlightenment Casts a Shadow: Anti-cruelty in the Nineteenth Century

3

Abstract

The Enlightenment was a period of scientific and social progress where rationality and reason became the analytical lenses for scrutinising all levels of society. However, until the end of the eighteenth century, the focus on logic, reason and cognition underscored differences between humans and animals, reinforcing religious views that animals were not an appropriate subject for moral concern. Both the notion of animal use and the legal classification of animals as property were consistent with long-standing religious doctrine that was not seriously challenged by Enlightenment ideals. This cast a metaphorical shadow over attempts to include animals within humanity's social and moral spheres until the issue was reframed as one of sentience rather than reason. Nevertheless, while the debate on animal sentience led to the passage of a range of anti-cruelty legislation, in somewhat of a paradox, the focus on sentience stymied deeper evaluation of the legitimacy of using animals, especially in intensive agricultural systems. Against this backdrop, animal pain and suffering acquired nuanced meanings, as evinced by decisions subsequent to *Ford v Wiley*, which indicated that animal cruelty depended less on animal sentience and more on how humans used the animals in question.

Keywords

Animals as property · Animal sentience · Anti-cruelty · The Enlightenment · Ford v Wiley

S. Riley, *The Commodification of Farm Animals*, Animal Welfare 21,
https://doi.org/10.1007/978-3-030-85870-4_3

3.1 Introduction

Historians vary in their characterisation and timing of the Enlightenment.[1] Its commencement is dated anywhere from 1650 to the 1680s, with the movement itself largely taking place in the eighteenth century.[2] The period was one of scientific and social progress that paved the way for intellectual and political freedoms, searching for knowledge beyond the confines of theological doctrine.[3] This not only challenged accepted religious authority, but also questioned secular control.[4] Rationality and reason became the analytical lenses for increasing scrutiny of the relationship between individuals and civil society, as well as individuals and government.[5] For these reasons, common characterisations of the Enlightenment focus on science, human rationality and scepticism, giving the movement its alternative sobriquet, the Age of Reason.[6]

In accordance with prevailing Judaeo-Christian beliefs, humans were given dominion over animals, a concept that was not challenged by Enlightenment philosophies, which neither disputed humanity's role as steward of the Earth nor questioned tenets concerning humanity's entitlement to use animals.[7] Indeed, until the end of the eighteenth century, the focus on logic, reason and cognition underscored differences between humans and animals, reinforcing religious views

---

[1] Kieron O'Hara, *The Enlightenment, a Beginners Guide*, Oneworld Oxford, Oxford (2010), 23, 41–45; Jan Golinski, "Science in the Enlightenment, Revisited", (2011) 49 (2) *History of Science*, 217, 217.

[2] Tony Tinker, "The Enlightenment and Its Discontents Antinomies of Christianity, Islam and the Calculative Sciences", (2004) 17 (3) *Accounting, Auditing & Accountability*, 442, 443; Kieron O'Hara, *The Enlightenment, a Beginners Guide*, above 1, 41–45.

[3] Andrea T Baumeister, "Kant: The Arch-Enlightener", in Norman Geras and Robert Wokler (eds) *The Enlightenment and Modernity*, 50, 52, Macmillan Press Ltd., London (2000); William Bristow, "Enlightenment", in *The Stanford Encyclopedia of Philosophy,* Edward N. Zalta (ed.), (Fall 2017 Edition), available from, https://plato.stanford.edu/archives/fall2017/entries/enlightenment/;F Bailey Norwood and Jayson L Lusk, *Compassion, by the Pound: The Economics of Farm Animal Welfare*, Oxford University Press, New York (2011) 23.

[4] Norman Hampson, "The Enlightenment", in *Early Modern Europe, an Oxford History,* Euan Cameron (ed) 265, 266 (OUP) 1999; Kieron O'Hara, *The Enlightenment, a Beginners Guide*, above 1, 4.

[5] Ian Holliday, "English Conservatism and Enlightenment Rationalism", in Norman Geras and Robert Wokler (eds) *The Enlightenment and Modernity*, 117, 118–119, Macmillan Press Ltd., London (2000).

[6] Tony Tinker, "The Enlightenment and Its Discontents Antinomies of Christianity, Islam and the Calculative Sciences", above 2, 443.

[7] Clarence J Glacken, "Count Buffon on Cultural Changes of the Physical Environment", (1960) 50 (1) *Annals of the Association of American Geographers*, 1, 3; William Bristow, "Enlightenment", above 3, parag 2.3; Vittorio Bufacchi, "The Enlightenment, Contractualism and Moral Polity" in Norman Geras and Robert Wokler (eds) *The Enlightenment and Modernity*, 204, 205, 206, Macmillan Press Ltd., London (2000).

that animals were not an appropriate subject for moral concern.[8] This cast a metaphorical shadow over attempts to include animals within humanity's social and moral spheres, until the issue was re-framed as one of sentience rather than reason.[9]

From the nineteenth century, the moral source for human-animal relationships was utilitarianism, a theory based on stakeholders achieving maximum happiness.[10] In a legal context, this led to anti-cruelty legislation, which initially targeted farm animals. However, by the twentieth century, the protection afforded to these animals was eroded to a level that was lower than that applying to other domestic animals, such as companion animals.[11]

This chapter and the next two, discuss the historical rationales for these shifts, commencing with an examination of the animals-as-property paradigm before moving to a discussion of how law and policy facilitated an interpretation of anti-cruelty that balanced animal suffering against commercial expediency. It is argued that animal suffering was largely subordinated to the dictates of the market, not only reinforcing the status of farm animals as property but also their position as commodities.

3.2 Animals, Property and the Social Contract

The Enlightenment unleashed a range of intellectual freedoms, expanding the search for knowledge towards science, or natural philosophy, as it was then called, which also led to scientific reasoning gaining intellectual ground.[12] At the same time, the Enlightenment involved broader transformations, extending to the political, social and religious domains.[13] Nevertheless, rationality and cognition were predominantly used to justify the subjugation of animal interests for human advancement.[14] The

[8]John Passmore, "The Treatment of Animals", (1975) 36 (2) *Journal of the History of Ideas*, 195, 202.

[9]Gary Steiner, *Anthropocentrism and its Discontents*, University of Pittsburgh (2005), 153.

[10]Gary L Francione, "Animal Welfare and the Moral Value of Nonhuman Animals", (2010) 6 (1) *Law, Culture and the Humanities*, 24, 26.

[11]For example, *European Convention for the Protection of Pet Animals 1987*; the Convention was adopted on 13 November 1987, *ETS* No 125 (entered into force 1 May 1992) and had 24 ratifications as of July 2019, available from https://rm.coe.int/168007a67d . Preamble recognises the special relationship pets have with humans; Elizabeth Ann Overcash, "Unwarranted Discrepancies in the Advancement of Animal Law: the Growing Disparity in Protection Between Companion Animals and Agricultural Animals", (2012) 90 *North Carolina Review*, 837, 864–872.

[12]Kieron O'Hara, *The Enlightenment, a Beginners Guide*, above 1, 48.

[13]Jan Golinski, "Science in the Enlightenment, Revisited", above 1, 218.

[14]Ibid.

classification of animals as property supported this standpoint, dovetailing with more general discussions on society's natural rights to property.[15]

3.2.1 Animals and Property Rights

As already indicated, the notion of animal use was consistent with long-standing religious doctrine that was not seriously challenged by Enlightenment ideals. John Gregory (1724–1773), a Scottish philosopher and physician, had views that were typical for the period. He considered that medical practitioners had ethical duties to enhance their understanding of human physiology by undertaking studies in comparative anatomy using animals.[16] It is tacit in this approach that Gregory placed human interests above animal interests, a viewpoint also replicated in philosophies of natural history.[17]

One of the foremost natural historians, Georges-Louis Leclerc, Comte de Buffon (1707–1788), published 36 volumes of natural history, under the titles *Histoire Naturelle, Générale et Particulière*.[18] His work was extensive, in some cases challenging church doctrine and in other respects acknowledging God's will that humans are stewards of the Earth with obligations to conquer and improve nature.[19] These obligations directly linked to the use of domesticated plants and animals to advance civilisation, a theme that was influential with Scottish Enlightenment thinkers, such as Adam Smith (1723–1790).[20] Smith proposed that the road to civilisation matured through four stages, starting with hunter-gatherers moving to domestication of animals, then the establishment of agriculture and, finally engaging in commerce.[21]

Each stage entailed the progressive domination of nature, as well as the increasing use of animals.[22] Smith also highlighted the connections between agriculture,

---

[15] Christine Korsgaard, "Kantian Ethics, Animals and the Law, (2013) 33 (4), *Oxford Journal of Legal Studies*, 629, 637, https://doi.org/10.1093/ojls/gqt028

[16] John Gregory, *Observations on the Duties and Offices of a Physician and on the Method of Prosecuting Enquiries in Philosophy*, London Printed for W Strahan and T Cadell, (Successor to Mr. Millar) in the Strand (1770), 67, 71, 72, available from https://play.google.com/books/reader?id=nJlbAAAAQAAJ&hl=en_GB&pg=GBS.PA1

[17] Nathaniel Wolloch, *The Enlightenment's Animals: Changing Conceptions of Animals in the Long Eighteenth Century*, Amsterdam University Press (2019), 90.

[18] Georges-Louis Leclerc, Comte de Buffon, *Histoire Natrelle, Générnral et Particulière avec la description du Cabinet du Roi*, Aux Deux-Ponts Chez Sanson & Compagne, first volume published in 1785, available from, https://www.biodiversitylibrary.org/bibliography/60802#/summary

[19] Clarence J Glacken, "Count Buffon on Cultural Changes of the Physical Environment", above 7, 3.

[20] Ibid, 13.

[21] Nathaniel Wolloch, *The Enlightenment's Animals: Changing Conceptions of Animals in the Long Eighteenth Century*, above 17, 115.

[22] Ibid, 115–116.

foreign trade and manufacturing, demonstrating that trade in surpluses improved economic productivity, leading to the development of industry standards which supported commercial activity.[23] He further emphasised that prosperity hinged on the sale and acquisition of agricultural produce, including animals, which in turn rested on legal and economic systems that facilitated trade by classifying animals as property.[24]

Enlightenment philosophy regarded the acquisition of property as an essential human prerogative. John Locke (1632–1704), the foremost exponent of property rights, argued that people were born free and consequently had rights to life, liberty and property.[25] These entitlements were seen as natural rights not dependent on legal rules. Accordingly, property could be acquired by one's labour, such as felling a tree and crafting furniture from the wood.[26] Locke's philosophies, which subordinated property to human control, applied as much to animals as to other forms of property. In a nod to religious doctrine, he concluded that humans had a right to dominate animals, because animals were made for the use of humans.[27] Analogous views were accepted by other Enlightenment thinkers, such as David Hume (1711–1776), whose standpoints equated property in animals with plants and slaves.[28]

The concept of animals as property was, of course, not unique to the Enlightenment. It derived from the earliest days of western civilisation, when animals were hunted, captured and domesticated.[29] The functional links between property and animals are deftly illustrated by the fact that terms, such as chattel, capital and money, stem from the same word origin as cattle.[30] Moreover, the notion of animals as property was well-accepted within the English legal system long before the Enlightenment. By the fourteenth century, English revenue laws classified animals as property for the purposes of taxation,[31] and, by the sixteenth century, legislation

---

[23] Kwangsu Kim, "Adam Smith's Theory of Economic History and Economic Development", (2009) 16 (1) *The European Journal of the History of Economic Thought*, 41, 58–59.

[24] Ibid.

[25] Alex Tuckness, "Locke's Political Philosophy", in Edward N Zalta (ed), *The Stanford Encyclopedia of Philosophy* (2018) Summer, available from https://plato.stanford.edu/archives/sum2018/entries/locke-political/

[26] Ben Mepham, "The Ethical Matrix as a Decision-Making Tool, With Specific Reference to Animal Sentience", in Jacky Turner and Joyce D'Silva (eds), *Animals, Ethics and Trade*, Earthscan, 134, 141 (2006).

[27] John Locke, *Second Treatise of Government*, Jonathan Bennett (2017), paragraph 26, available from https://www.earlymoderntexts.com/assets/pdfs/locke1689a.pdf; discussion Gary Steiner, *Anthropocentrism and Its Discontents*, above 9, 159.

[28] Gary Steiner, *Anthropocentrism and its Discontents*, above 9, 160–162.

[29] Joyce Salisbury, *The Beast Within: Animals in the Middle Ages*, Routledge, London (2010), 10.

[30] Gary L Francione, *An Introduction to Animal Rights: Your Child or the Dog?* Temple University Press, Philadelphia, reprinted 2007, 50–51.

[31] *The King's Grant, that the foresaid Subsidy of the Ninth Lamb,&c. shall be no Example, nor Prejudicial to his Subjects: All shall be spent in his Wars*, 1340 c 1, available from (14 Edw 3 St 2), available from Justis One, https://app-justis-com.ezproxy.lib.uts.edu.au/statute/taxation-etc-act-1340-c-1/overview/cYyZm0ednYWca (subscription needed).

supervised the sale and purchase of livestock as a form of merchantable goods.[32] The common law also treated animals as "personal goods", affirming they could be the object of a lease,[33] as well as property in bankruptcy proceedings.[34] In the eighteenth century, Sir William Blackstone (1723–1780) confirmed the common law position, including material on ownership of wild and domestic animals in his commentaries.[35]

The classification of animals as property meant that they were legally an object of the law. This made it difficult for society to confront animal abuse because up to the nineteenth century, an owner could not be charged with cruelty towards their own animals.[36] At best, a non-owner could be prosecuted for hurting or injuring an animal because this amounted to damaging the property of another. John Lawrence (1753–1839), an English writer, was highly critical of this situation, describing several acts of cruelty that would have escaped punishment had the animals in question been the property of the person charged.[37] In the first case, a William Parker was tried in 1794 for tearing out the tongue of a live mare,[38] and in 1793, two butchers in Manchester were fined for chopping off the feet of live sheep before driving them through the city. Lawrence pointedly notes that if the sheep had been the property of the butchers they could "with impunity, either have dissected them alive, or burned them alive".[39] However, the operation of property law was not the only reason for lack of attention to animal cruelty. Animals had long been regarded as a means to human ends, a standpoint validated by philosophies which excluded animals from humanity's social contract, thereby relegating them to the status of moral outsiders.[40]

---

[32] *An Act for Buying and Selling of other Beasts and Cattle*, 1549, (3 & 4 Edw 6) C A XIX, available from Justis One https://app-justis-com.ezproxy.lib.uts.edu.au/statute/buying-cattle-act-1549/overview/cYyJm5iJnZWca (subscription needed).

[33] *Spencer's Case*, [1583] EWHC KB J53 (01 January 1583), available from http://www.bailii.org/ew/cases/EWHC/KB/1583/J53.html

[34] *Joseph Milles v Mat Davies, Evan Watts and Selby Price, alias Rees* [1792] EngR 1293, available from http://www.commonlii.org/int/cases/EngR/1792/1293.pdf

[35] William Blackstone, *Blackstone's Commentaries on the Laws of England Book the Second—Chapter the First: Of Property in General* 1765–1769, 5, 14, available from http://avalon.law.yale.edu/18th_century/blackstone_bk2ch1.asp

[36] This changed with An Act to Prevent the Cruel and Improper Treatment of Cattle 1822, (3 Geo IV c 71), s 1; *The Statutes of the United Kingdom of Great Britain and Ireland 3 George IV. 1822*, his Majesty's statute and law printers London, sold by Butterworths and son, 403, available from https://archive.org/details/statutesunitedk10britgoog/page/n436

[37] John Lawrence, *Philosophical and Practical Treatise on Horses and on the Moral Duties of Man Towards the Brute* Creation, Vol 1, H D Symonds, Paternoster-Row, London, (1802), 124–125, available from https://ia800200.us.archive.org/27/items/philosophicalpra01lawr/philosophicalpra01lawr.pdf

[38] Ibid.

[39] Ibid.

[40] Christine M Korsgaard, "A Kantian Case for Animal Rights", in *Animal Law—Tier and Rect: Developments and Perspectives in the 21st Century*, Margot Michael, Daniela Kühne, and Julia

3.2.2 The Social Contract

A social contract describes humanity's willingness to accept communal obligations in return for "social benefits".[41] The notion can be traced to antiquity, but it was rekindled and advanced in various forms during the seventeenth and eighteenth centuries by philosophers that included John Locke and Immanuel Kant (1724–1804).[42] The theory declined in prominence during the nineteenth century but was revived in the twentieth century with the publication of John Rawls' (1921–2002) *Theory of Justice* in 1971.[43]

During the Enlightenment, social contract theory evolved as a counterbalance to religious doctrine, whose moral theories had been based on "superstition, enthusiasm, fanaticism and supernaturalism".[44] Instead, social contract theory tapped into Enlightenment motifs of rationality and personal responsibility. Kant defined Enlightenment as daring to be wise or having "the courage to use your own understanding".[45] He also formulated the Categorical Imperative, which determined how individuals should behave towards each other: "act only according to a maxim by which you can at the same time will that it shall become a general law".[46] Expressed in this manner, the Categorical Imperative was also a moral imperative, because people, as rational beings, were morally bound to adhere to obligations they had freely assumed.[47] As a corollary, people were also held accountable for their opinions and beliefs, which ought to be based on logic and observation.[48]

The moral underpinnings of the Categorical Imperative prompted Kant to conclude that humans are an end in themselves because they owe each other obligations voluntarily assumed.[49] Animals, on the other hand, could not rationally assume

Hänni (eds), 3, 5, Zurich: Dike Verlag (2012), available from http://nrs.harvard.edu/urn-3:HUL.InstRepos:34903186

[41] Ibid.

[42] Fred D'Agostino, Gerald Gaus, and John Thrasher, "Contemporary Approaches to the Social Contract", in *The Stanford Encyclopedia of Philosophy,* Edward N. Zalta (ed), (Spring 2019 Edition), available from https://plato.stanford.edu/archives/spr2019/entries/contractarianism-contemporary/

[43] Wayne Gabardi, *The Next Social Contract: Animals, the Anthropocene, and Biopolitics*, Temple University Press, Philadelphia (2017), 125.

[44] William Bristow, "Enlightenment", above 3, parag 2.3; Vittorio Bufacchi, "The Enlightenment, Contractualism and Moral Polity" above 7, 205, 206.

[45] Maurizio Passerin d'Entrèves, "Critique and Enlightenment: Michel Foucault on 'Was ist Aufklärung?'" in Norman Geras and Robert Wokler (eds) *The Enlightenment and Modernity*, 184, 187, Macmillan Press Ltd., London (2000).

[46] Kieron O'Hara, *The Enlightenment, a Beginners Guide*, above 1, 73–74.

[47] Wayne Gabardi, *The Next Social Contract: Animals, the Anthropocene, and Biopolitics*, above 43, 124.

[48] Fred D'Agostino, Gerald Gaus, and John Thrasher, "Contemporary Approaches to the Social Contract", above 42; Robert Garner, "Rawls, Animals and Justice: New Literature, Same Response" (2012) 18 *Res Publica*, 159, 160, DOI https://doi.org/10.1007/s11158-011-9173-z

[49] Christine Korsgaard, "Kantian Ethics, Animals and the Law, above 15, 631.

obligations, so were placed outside the social contract, which also meant that humans did not owe animals direct duties or moral consideration.[50] Other philosophers, such as Hume, indicated that humans and animals shared similarities, but animals' inferior cognitive abilities kept them "out of the realm of justice, which supposes equality and is founded on the utility of equals".[51]

Notwithstanding the moral separation of humans and animals, Kant's philosophies proscribed animal cruelty, because those who mistreated animals could also be cruel to humans.[52] Accordingly, humans owed direct obligations to each other not to be cruel to animals, leading to indirect duties towards animals.[53] This relegated animals to an object of morality, mirroring their legal classification which similarly rendered them an object of the law. Consequently, in neither the legal nor the moral sphere were animals regarded as subjects, so that they had no liberties, privileges, legal and/or moral status.[54] Yet for all its shortcomings, Kant's reasoning was a progression from earlier philosophies, such as those of René Descartes (1596–1650), who had argued that animals lacked rationality and therefore also lacked sentience.

3.3 Sentience

Descartes was a French philosopher and mathematician whose writings pre-date the Enlightenment but whose views were critical to the growth of utilitarianism, because utilitarianism challenged Descartes' dichotomy of mind and body.[55] Descartes reasoned that when he was thinking he also existed; therefore, *cogito, ergo sum*.[56] From this idea he extrapolated the duality of mind and body, arguing that because the two are separate, the body is similar to a machine that merely responds to the mind.[57] These conclusions meant that because animals lacked the capacity for thought and rationality, they also lacked a soul, precluding moral consideration.[58] In addition, Descartes concluded that the duality of mind and body signified that animals were not sentient because rationality and understanding were pre-requisites to feeling

[50] Wayne Gabardi, *The Next Social Contract: Animals, the Anthropocene, and Biopolitics*, above 43, 124.

[51] Michael J Seidler, "Hume and the Animals", (1977) 15 (3) *The Southern Journal of Philosophy*, 361, 369.

[52] Christine M Korsgaard, "A Kantian Case for Animal Rights" above 40, 14, 18.

[53] Ibid.

[54] Gary Steiner, *Anthropocentrism and its Discontents*, above 9, 161.

[55] Kieron O'Hara, *The Enlightenment, a Beginners Guide*, above 1, 41–42.

[56] William Bristow, "Enlightenment", above 3, parag 1.4, I think, therefore I am.

[57] Kieron O'Hara, *The Enlightenment, a Beginners Guide*, above 1, 41–42.

[58] F Bailey Norwood and Jayson L Lusk, *Compassion, by the Pound: The Economics of Farm Animal Welfare*, above 3, 24.

pain, as "pain exists only in the understanding".[59] Descartes thus regarded animals as little more than machines, so that cries and yelps were mechanical responses to stimuli, rather than indications of pain.[60]

While the question whether animals could feel pain continued to be disputed well into the twentieth century,[61] Descartes' views were challenged by philosophers such as Kant, Hume and Jeremy Bentham (1748–1832). These philosophers agreed that animals were sentient, yet they differed on the extent to which sentience could form the basis of moral concern.

As just discussed, Kant acknowledged that animals could feel pain and suffer, although his moral obligations derived from duties that humans owed to each other, which applied irrespective of animal sentience. Empiricists such as Hume (who was also a utilitarian) acknowledged that animals were sentient but did not consider sentience to be morally relevant.[62] Instead, empiricists regarded the type of cognitive abilities possessed by humans as an essential pre-condition for moral regard, leading to harsh views towards animals.[63] Accordingly, while empiricists acknowledged humans and animals shared characteristics, which included the capacity to reason and feel pain, animals' inferior intellectual capabilities led to a moral deficit that was not alleviated by their sentience.[64] In this regard, while Hume contemplated whether animal suffering ought to be taken into account in the weighing and balancing of pleasures and pains, he failed to advance this thought, leaving a moral lacuna that came to be filled by utilitarians such as Bentham.[65]

Bentham's philosophies built on the works of other Enlightenment thinkers, including Hume and Francis Hutcheson (1694–1746), both of whom developed moral principles based on happiness, or "utility", and the greater good of humankind.[66] Bentham's theories were founded on the "the greatest happiness" principle where moral integrity was equated with making the most of pleasure and minimising pain.[67] He was also one of the first to critique the British legal system for its lack of

[59] Erica Fudge, "The Animal Face of Early Modern England" (2013) 30 (7/8) *Theory, Culture and Society*, 177, 182, DOI: https://doi.org/10.1177/0263276413496122

[60] Ibid.

[61] Bernard Rollin, *The Unheeded Cry: Animal Consciousness, Animal Pain, and Science*. Oxford University Press, New York (1989), 150–155.

[62] Gary Steiner, *Anthropocentrism and its Discontents*, above 9, 160, 162.

[63] Gordon Graham, "Morality and Feeling in the Scottish Enlightenment" (2001) 76 (296) *Philosophy*, 271, 273.

[64] A T Nuyen, "Hume on Animals and Morality", (1998) 27 (2) *Philosophical Papers*, 93, 93; Gary Steiner, *Anthropocentrism and its Discontents*, University of Pittsburgh (2005), 153, 160, 162.

[65] Gary Steiner, *Anthropocentrism and its Discontents*, above 9, 162.

[66] Julia Driver, "The History of Utilitarianism", *The Stanford Encyclopedia of Philosophy* Edward N Zalta (ed), (2014) Winter Edition), available from https://plato.stanford.edu/archives/win2014/entries/utilitarianism-history/

[67] Kieron O'Hara, *The Enlightenment, a Beginners Guide*, above 1, Oneworld Oxford, Oxford (2010), 62.

moral concern for animals.[68] In his publication, *An Introduction to the Principles of Morals and Legislation*, he included a lengthy footnote, titled "Interests of the inferior animals improperly neglected in legislation", where he drew attention to the fact that legislation had no regard for the plight of animals.[69] Unlike other Enlightenment philosophers, Bentham did not consider that lack of reason or lower cognitive skills were sufficient to withhold "those rights [from animals] which never could have been withholden from them but by the hand of tyranny".[70] Bentham underscored the fact that animals can suffer, placing sentience at the core of the debate on society's interactions with animals, calling upon humans to maximise animals' happiness and reduce their suffering.[71] Bentham was not alone in calling for kinder treatment of animals; his views were part of an anti-cruelty movement that gained momentum from the late eighteenth century.

Participants in this movement included Humphrey Primatt (c 1735–1776/7) and several English women, who campaigned against animal cruelty as a means of censuring societal abuse in its many forms.[72] Primatt was an Anglican minister who blended anti-cruelty views with eighteenth-century religion. In his *Dissertation on the Duty of Mercy and the Sin of Cruelty to Brute Animals*,[73] Primatt acknowledged that animals are inferior to humans yet questioned why this should justify humans excluding animals from the type of mercy and compassion that humans apply to themselves.[74] The dissertation contains quotes from the Bible on almost every page, arguing that all animals have value and that cruelty to animals, particularly where it involves gratuitous pain, is an act of injustice.[75] Primatt also pointed out that animals feel pain as much as humans do and the fact that humans own animals does not justify mistreating them by inflicting unnecessary pain.[76]

[68] David S Favre and Vivien Tsang, "The Development of Anti-Cruelty Laws During the 1800s" (1993) Spring (1) *Detroit College of Law Review*, 1, 3.

[69] Jeremy Bentham, *An Introduction to the Principles of Morals and Legislation*, Clarendon Press, Oxford (1781), 143–144, Edited by Jonathon Bennett, available from https://www.earlymoderntexts.com/assets/pdfs/bentham1780.pdf

[70] Ibid, 144.

[71] Ibid.

[72] Discussion in Moira Ferguson, *Animal Advocacy and Englishwomen 1780–1900*, The University of Michigan Press (2001). The women included Elizabeth Heyrick, Anna Sewell and Frances Power Cobbe.

[73] Humphry Primatt (c 1735–1776/7), was an Anglican minister who wrote *Dissertation on the Duty of Mercy and the Sin of Cruelty to Brute Animals*, R Hett (printer) MDCCLXXVI (1776), (ii), available from, https://ia802305.us.archive.org/0/items/adissertationon00primgoog/adissertationon00primgoog.pdf

[74] Ibid, (ii).

[75] Ibid, 5–6, 8, 13, 73, 80, 86–104, 109–113, 125–128, 159–165, 206–208, 259–260, 274–275, 311–312, 314–315.

[76] Ibid, 13, 52, 61–67, 252–253.

Significantly, he also concluded that pain itself is an evil, thus making cruelty a sin, especially in view of the fact that animals cannot speak for themselves.[77]

In common with Bentham, Primatt envisaged that animals could be used, but these uses were constrained by animal sentience.[78] Consequently, the greatest happiness principle inherent in utilitarian philosophies did not extend to granting animals a life. Rather, utilitarianism promoted animals' greatest happiness, but this was balanced against human utility. As will be discussed later in this chapter, this balancing came to be sorely tested throughout the nineteenth century as the trade in animals and their products increased.

Among the group of English women who joined anti-cruelty causes, Elizabeth Heyrick (née Coltman, 1769–1831) supported "peace for animals and humans alike".[79] In 1823, she wrote and published a pamphlet titled "Immediate Not Gradual Abolition", referring to abolition of the slave trade.[80] In the minds of abolitionists, slaves were vulnerable and unable to protect themselves, a situation replicated with community violence towards animals.[81] The debates surrounding animal cruelty became especially pronounced following Lord Mansfield's seminal decision in *Somerset v Stewart* in 1772, where he held that slavery was neither "allowed [n]or approved by the laws of England".[82] While the decision did not abolish slavery, it drew attention to institutionalised cruelty and abuse at large, fuelling further discussion on humanity's relationship to animals.[83]

Societies in Western Europe had long accepted a high level of "cultural violence" towards animals, manifesting in animal baiting, cat burning, overworking of farm animals and inhumane slaughter practices.[84] Moreover, while church doctrine proscribed cruelty, a degree of animal suffering was considered unavoidable if society were to use animals in accordance with its God-given directive.[85] Yet, the dividing line between unavoidable suffering and cruelty was often a fine one, leading to abuse becoming assimilated into everyday life. In the case of animals used for labour or

---

[77] Ibid, 36–37, 52, 61–67, 252–253, 287–288.

[78] Ibid, 8–10.

[79] Moira Ferguson, *Animal Advocacy and Englishwomen 1780–1900*, above 72, 27, 28.

[80] Ibid.

[81] Ibid, 28, 45.

[82] *Somerset v Stewart* (1772) 98 ER 499, 510, available from http://www.commonlii.org/int/cases/EngR/1772/57.pdf

[83] Slavery was abolished in 1833, by An Act for the Abolition of Slavery throughout the British Colonies; for promoting the Industry of the manumitted Slaves; and for compensating the Persons hitherto entitled to the Services of such Slaves, 3 and 4 Will. IV C 73, available from http://www.pdavis.nl/Legis_07.htm; Moira Ferguson, *Animal Advocacy and Englishwomen 1780–1900*, above 72, 45.

[84] Linda Kalof, *Looking at Animals in Human History*, Reaktion Books (2007), 51–52, 89; Esther Cohen, "Animals in Medieval Perceptions: The Image of the Ubiquitous Other", in Aubrey Manning and James Serpell, James, (eds) *Animals and Human Society Changing Perspectives*, Routledge (1994), 66.

[85] Gary Steiner, *Anthropocentrism and its Discontents*, above 9, 153.

food, cruelty grew with market expansion.[86] Farm animals served an economic purpose and abuse directed towards them invariably aimed at forcing animals to fulfil that purpose. As Bargheer indicates:

> Coaches had to arrive in time, carts had to be packed to the limit of their capacity no matter how heavy, and cattle and sheep had to be directed to walk to the market. Finally, each of these animals was slaughtered in the city for the production of meat, leather, and other products made from their bodies.[87]

It was, therefore, common in London to hear the bleats and bellows of frightened animals as they were taken to market or for slaughter.[88] This led to increasing debate and discussion on animal cruelty, which by the beginning of the nineteenth century had reached the UK Parliament in the form of proposals to legislate for anti-cruelty. It was a move that proved exceedingly controversial, not only because the concept of animal cruelty was contentious but also because of concern that regulation would be used as a form of social control.

3.4 Anti-cruelty Legislation

Laws regulating society's use of animals have existed for centuries in England, yet few were designed to deal with animal cruelty. As just indicated, it was common for a level of cruelty to be normalised and not all communities related to animals in identical ways. Working class people regarded animals as essential for their livelihoods, while wealthier people, especially those in urban areas, increasingly regarded animals as pampered pets.[89] Elsewhere, the wealthy and aristocratic might have viewed animals as part of the lower orders to be used at will, while at all levels of society, sport and recreation, including hunting and animal baiting, incorporated substantial elements of abuse and cruelty.[90] Consequently, when Parliament turned its mind to proscribing animal cruelty, it faced considerable policy challenges in navigating this wide range of perspectives.

3.4.1 Early Laws

Early legislation primarily sought to manage human-animal interactions for the betterment of society. Accordingly, statutes from the fifteenth and seventeenth

[86] Stefan Bargheer, "The Fools of the Leisure Class: Honor, Ridicule, and the Emergence of Animal Protection Legislation in England, (2006) 47 (1) *European Journal of Sociology*, 3, 5, 25.

[87] Ibid, 28.

[88] Piers Beirne, "Hogarth's Animals" (2013) 3 (2) *Journal of Animal* Ethics, 133, 137–138.

[89] Stefan Bargheer, "The Fools of the Leisure Class: Honor, Ridicule, and the Emergence of Animal Protection Legislation in England, above 86, 5.

[90] Moira Ferguson, *Animal Advocacy and Englishwomen 1780–1900*, above 72, 28, 45.

centuries regulated the slaughter of animals for public health reasons[91] and outlawed bear baiting and bull-baiting on Sundays "for the better observation of the Lords Day".[92] In the sixteenth century, the Tudors resolved that agricultural land could be used more productively by destroying birds who ate grain and crops.[93] Therefore, the *Act for the Destruction of Crows and Rooks 1532* (also referred to as the *Preservation of Grain Act 1532*) proffered bounties for killing crows, rooks or choughs,[94] while elsewhere, "vert and venison laws" protected game animals and their habitat.[95] The objectives of these statutes thus had more to do with social control than preventing cruelty. In fact, anti-cruelty legislation was not a common feature of the regulatory landscape until the nineteenth century, although a handful of laws existed prior to that time.

In 1545, the English Parliament created an offence for cutting the tongue out of a living animal,[96] and, in 1635, the Irish Parliament prohibited attaching ploughs to animals' tails or tearing wool off living sheep.[97] In an analogous manner, across the Atlantic, the *Massachusetts Body of Liberties 1641* banned cruelty towards domestic

---

[91] An Act that No Butcher Flea any Manner of Beast within the Walls of London, 1488, 4 and 5 Henry VII C 3, Statutes of the Realm Printed by Command of his Majesty, George the Third, in pursuance of an address of the House of Commons of Great Britain Volume the Second, 1816, Reprinted 1963 Dawson of Pall Mall London, 527–528, available from https://babel.hathitrust.org/cgi/pt?id=pst.000017915526;view=1up;seq=553

[92] An Act For Punishing Divers Abuses Committed on The Lord's Day Called Sunday, 1625, First year of the reign of King Charles I, C1, The statutes of the Realm, Printed by command of his majesty King George the Third, in pursuance of an address of the House of Commons of Great Britain, Volume 5, 1819, 1, reprinted Dawsons of Pall Mall London (1963), available from https://babel.hathitrust.org/cgi/pt?id=pst.000017915564;view=1up;seq=29; An Act for the better observation of the Lords Day, 1657, *Acts and Ordinances of the Interregnum, 1642–1660*, C H Firth and R S Rait (eds), London, (1911), 1162, British History Online available from https://www.british-history.ac.uk/no-series/acts-ordinances-interregnum/pp1162-1170

[93] E Cudworth and S Hobden, "Civilisation and the Domination of the Animal", (2014) 42 (3) *Millennium—Journal of International Studies,* 746, 756.

[94] *An Act for the destruction of Crows and Rooks 1532*, 24 Henry VIII, C 10, s 7, Statutes at Large from the fifth year of K Richard III to the 31st year of K Henry VIII, inclusive, by Danby Pickering of Gray's Inn Esquire, Vol IV, Printed by Joseph Bentham, London (1736), 254, available from https://play.google.com/books/reader?id=UlMDAAAAQAAJ&hl=en_GB&pg=GBS.PA254; discussion in Roger Lovegrove, *Silent Fields: The Long Decline of a Nation's Wildlife*, OUP (2007), 80–81.

[95] *Order and Rules of the New Forest* 1537; discussion by John Langton, "Forest Law in the Landscape: Not the Clearing of the Woods, but the Running of the Deer?", in Karis Baker, Ruth Carden and Richard Madgwick (ed) *Deer and People,* Oxbow Books (2014), 222.

[96] Burning of Frames Act, 1545, 37 Henry VIII C 6, Recital and s III, Statutes of the Realm, Printed by Command of His Majesty, George the Third in pursuance of an address of the House of Commons of Great Britain, Volume the Third, 1817, Reprinted 1965, Dawsons of Pall Mall, London, 994, available from https://babel.hathitrust.org/cgi/pt?id=pst.000017915533;view=1up;seq=1060

[97] *Act Against Plowing by the Tayle, and pulling the Wooll off Living Sheep 1635* (Ireland), Statutes Passed in the Parliaments Held in Ireland, Vol 1 From the Third Year of Edward the Second, A.D. 1310, to the Fourteenth and Fifteenth Years of Charles the Second, A.D. 1662 inclusive,

Table 3.1 Anti-cruelty bills not passed

Bill	Objective	Margin of defeat
1800 Bill to Prevent the Practice of Bull-Baiting Bill	Prohibit bull-baiting	Defeated in the House of Commons 8 ayes to 59 nos.
1809 Cruelty to Animals Bill	Prohibit cruelty to all animals	Defeated in the House of Lords (introduced by Lord Erskine) 27 ayes to go to a committee to 37 nos. to go to a committee
1821 Ill-Treatment of Horses Bill in 1821	Prohibit cruelty to horses	Defeated 31ayes to 34 nos.
1823 A Bill to Prohibit Bull-Baiting and Dog-Fights	Prohibit bull-baiting and dog fights	Defeated 18 ayes to 47 nos.
1825 Cruelty to Animals Bill	Prohibit cruelty to animals generally	Defeated 32 ayes to 50 nos.
1826 Bear-Baiting and Dog-Fighting Bill	Prohibit bear baiting and dog fights	Defeated 37ayes to 76 nos.
1826 Cattle Ill-Treatment Bill	Extend 1822 Cattle Act to all domestic animals	Bill postponed

animals, with specific obligations to rest and refresh cattle if they became weary, hungry and lame or were otherwise driven long distances.[98] These laws operated in a piecemeal manner, targeting only a few cruel practices. Yet, by the end of the nineteenth century, the anti-cruelty movement had gained extraordinary prominence, counting the introduction of wide-ranging legislation among its successes.

3.4.2 The Nineteenth and Early Twentieth Centuries

The legislative journey towards anti-cruelty regulation was dotted with discussion, debate and numerous unsuccessful attempts, including the bills summarised in Table 3.1, which were not passed.[99] Nevertheless, those who sponsored anti-cruelty

Printed by George Grierson, Dublin (1794), 301, available from https://books.google.com.au/books?id=VYRRAAAAYAAJ&pg=PA301&redir_esc=y#v=onepage&q&f=false

[98]The Massachusetts Body of Liberties *1641*, ss 92–93. The document was a legal code adopted by colonists in New England, available from https://history.hanover.edu/texts/masslib.html

[99]These Bills are: Bill to Prevent the Practice of Bull Baiting, 2 April, 1800, Parliamentary history of England from the earliest period to the year 1803, Volume 35, T C Hansard (1819) column 202, available from https://babel.hathitrust.org/cgi/pt?id=mdp.39015087751502&view=1up&seq=125; For discussion, Marie Blosh, *The History of Animal Welfare and the Future of Animal Rights*, Electronic Thesis and Dissertation Repository. 803. https://ir.lib.uwo.ca/etd/803 (2012), 26; Cruelty to Animals Bill 1809, House of Lords, debate on Cruelty to Animals' Bill, *Historic Hansard*, HL Deb 15 May 1809, Series 1, Volume 14, cc553–71, available from https://api.parliament.uk/historic-hansard/lords/1809/may/15/cruelty-to-animals-bill; Ill-Treatment of Horses Bill 1821, *Historic Hansard*, HC Deb 1 June 1821, Volume 5, cc1098–9, available from https://api.

legislation were tenacious. They did not capitulate in the face of widespread ridicule and resistance, including against arguments that criminalisation of animal baiting would remove recreation from the poor.[100] Nor were they swayed by concerns that new laws could potentially open the floodgates and extend to other activities, such as hunting.[101] They were also not persuaded by cries that anti-cruelty laws would impinge against private property rights in animals. With respect to the last point, Lord Erskine detailed astonishingly cruel practices when he introduced the Cruelty to Animals Bill 1809, arguing that moral principles transcended private property rights:

> . . . the inhuman and wicked practice of houghing [disabling by cutting the hamstrings] and hamstringing cattle, of cutting the faces of sheep intended for future slaughter. . . the cutting and tearing out the tongue of so noble an animal as the horse. . . Hitherto, the matter had been perhaps too much considered only as it related to animals as the private property of individuals; but. . . the legislature should declare its opinion upon the morality of the principle, and on the duties that man owed not only to man, but to the lower world; duties so much connected with the general state of the moral feelings.[102]

However, as with the other bills listed in Table 3.1, not everyone agreed with Lord Erskine, and several members of the House of Lords expressed apprehension at the scope of the bill, seeking to limit its ambit to beasts of draught or burden, such as horses, asses, oxen and mules.[103] Others considered the bill unnecessary, eventually leading to its defeat by a substantial margin of ten votes, 27–37.[104] In reality, anti-cruelty legislation was so contentious that even suggestions to limit its scope proved

parliament.uk/historic-hansard/commons/1821/jun/01/ill-treatment-of-horses-bill; A Bill to Prohibit Bull-baiting and Dog-fights 1823, *Historic Hansard*, HC Deb 21 May 1823, Volume 9, cc432–5, 425 available from https://api.parliament.uk/historic-hansard/commons/1823/may/21/bull-baiting-and-dog-fights; Cruelty to Animals Bill, *Historic Hansard*, HC Deb 11 March 1825, Volume 12, cc1002–13, available from https://api.parliament.uk/historic-hansard/commons/1825/mar/11/cruelty-to-animals-bill; Bear-Baiting and Dog-Fighting Bill 1826, *Historic Hansard*, HC Deb 21 February 1826, Volume 14, cc647–52, available from https://api.parliament.uk/historic-hansard/commons/1826/feb/21/bear-baiting-and-dog-fighting-bill; House of Commons, Cattle Ill-Treatment Bill, *Historic Hansard* HC Deb 21 February 1826 Volume 14, cc653–7, Martin of Galway sought leave to amend "An Act to Prevent the Cruel and Improper Treatment of Cattle" 3rd George I, c 71, by extending its provisions to all domestic animals, available from https://api.parliament.uk/historic-hansard/commons/1826/feb/21/cattle-ill-treatment-bill

[100] Bill to Prevent the Practice of Bull Baiting, 2 April, 1800, debate on 18 April 1800, above 99, the speech extends through columns 203–209, especially at 204, 207–208, 214.

[101] House of Commons, Ill-Treatment of Horses Bill 1821, above 99, c1099, Mr. Scarlett.

[102] Bill introduced, House of Lords, debate on Cruelty to Animals' Bill, *Historic Hansard*, HL Deb 15 May 1809, above 99; Lord Erskine, House of Lords, debate on Cruelty to Animals' Bill, *Historic Hansard*, HL Deb 31 May 1809, Series 1, Volume 14, cc804–8, 805, available from https://api.parliament.uk/historic-hansard/lords/1809/may/31/cruelty-to-animals-bill

[103] The Earl of Liverpool, House of Lords, debate on Cruelty to Animals' Bill, *Historic Hansard*, HL Deb 31 May 1809, above 102, c807.

[104] Earl Darnley, House of Lords, debate on Cruelty to Animals' Bill, *Historic Hansard*, HL Deb 31 May 1809, above 102, c807; House of Commons Debate, Cruelty to Animals' Bill, *Historic*

unsuccessful, as illustrated by the defeat of narrowly drafted legislation, such as the Ill-Treatment of Horses Bill in 1821.[105] These early struggles not only demonstrated the challenges inherent in criminalising cruelty but also the danger in prohibiting recreational activities, such as animal baiting, which hitherto had been widespread, popular and legal.

Notwithstanding these hurdles, the UK Parliament passed its first anti-cruelty statute on 22 July 1822, the *Act to Prevent the Cruel and Improper Treatment of Cattle 1822* (1822 Cattle Act).[106] The legislation was limited to farm animals, including horses, cattle and sheep, and created offences for anyone who "wantonly and cruelly beat, abuse or ill-treat" those animals.[107] Although the purview of the act was restricted to a small range of domestic animals, it was still ground-breaking. In particular, it applied to anyone who committed an act of cruelty, so that offenders could no longer hide behind the shield of animal proprietorship.[108]

However, implementation of the act was inconsistent. An owner was still needed to file a complaint, and the position of animals not listed in the legislation was tenuous. Reports in *The Times* during the 1830s demonstrate that police were somewhat arbitrary in their approach, sometimes laying charges without an owner's complaint yet insisting on such a complaint in other cases.[109] Author and librarian Edward Nicholson (1849–1912) recounted the situation of a woman charged with skinning cats alive, on 17 Sept 1832. The woman was discharged by the judge, Sir Peter Laurie, because without a complaint filed by the owner, his honour held he had no jurisdiction.[110] This case was particularly concerning because according to the report, the defendant had been charged with the same offence in the past and indicated she would continue skinning cats, as this was her livelihood.[111]

Hansard, HC Deb 15 June 1809, Series 1, Volume 14, c1070, available from https://api.parliament.uk/historic-hansard/commons/1809/jun/15/cruelty-to-animals-bill

[105] House of Commons, Ill-Treatment of Horses Bill 1821, House of Commons, Ill-Treatment of Horses Bill, above 99. c1099.

[106] An Act to Prevent the Cruel and Improper Treatment of Cattle 1822, (3 Geo IV c 71) *The Statutes of the United Kingdom of Great Britain and Ireland 3 George IV. 1822*, his Majesty's statute and law printers London, sold by Butterworths and son, 403, available from https://archive.org/details/statutesunitedk10britgoog/page/n436

[107] 1822 Cattle Act, s I.\

[108] 1822 Cattle Act, s I.

[109] Police, Untitled, report of cats skinned alive, 27 June 1831, *The Times Digital Archive*, page 6, available http://tinyurl.galegroup.com/tinyurl/9oFYt4 (subscription required); Police, "William Mills Charged with Skinning cats Alive", 30 October, 1832, *The Times Digital Archive*, page 4, available http://tinyurl.galegroup.com/tinyurl/9oFR85 (subscription required).

[110] Edward Byron Nicholson, *The Rights of an Animal: A New Essay in Ethics*, C Kegan Paul and Co, London (1879), 11, available from https://ia902700.us.archive.org/29/items/rightsananimala01lawrgoog/rightsananimala01lawrgoog.pdf

[111] Ibid.

To capture this type of abuse, additional attempts were made to expand the ambit of the 1822 Cattle Act,[112] as well as to incorporate new prohibitions on animal baiting and dog fighting.[113] By the mid-1830s, these efforts led to new anti-cruelty statutes, including the *Cruelty to Animals Act 1835* (UK),[114] *Cruelty to Animals Act 1849* (UK)[115] and the *Protection of Animals Act 1911* (UK).[116] Each of these statutes broadened the scope of previous legislation, by extending prohibitions and the range of animals covered.[117] The *Cruelty to Animals Act 1849* (UK) did not apply to Scotland,[118] which had its own legislation, the *Cruelty to Animals (Scotland) Act 1850* (UK).[119] This act differed from the 1849 UK legislation in a number of respects, more particularly because the Scottish legislation targeted acts of "wanton" cruelty,[120] a requirement found in the 1822 Cattle Act and the *Cruelty to Animals Act 1835* (UK), but omitted from the *Cruelty to Animals Act 1849* (UK) and reinstated in the *Protection of Animals Act 1911* (UK). Table 3.2 provides a summary of the main anti-cruelty provisions applying to domestic animals, up to 1911.[121]

---

[112] For example, House of Commons Debate, Cruelty to Animals Bill, *Historic Hansard*, HC Deb 11 March 1825, above 99, c1013 defeated 32–50; House of Commons Debate, Ill-treatment of Animals Bill, *Historic Hansard*, HC Deb 24 March 1825, Volume 12, cc1160–2, 1162 defeated 23–33, available from https://api.parliament.uk/historic-hansard/commons/1825/mar/24/ill-treatment-of-animals-bill

[113] For example, House of Commons Debate, A Bill to Prohibit Bull-baiting and Dog-fights, above 99, c425, defeated 18–47. House of Commons Debate, Bear-Baiting and Dog-Fighting Bill, *Historic Hansard*, HC Deb 21 February 182, above 99, c652 defeated 37–76.

[114] *An Act to Consolidate and Amend the Several Laws Relating to the Cruel and Improper Treatment of Animals and the Mischiefs Arising from the Driving of Cattle, and to Make Other Provisions in Regard Thereto, 1835*, 5 & 6 William 4 c59, Cruelty to Animals Act, available from The Statutes Project, http://statutes.org.uk/site/the-statutes/nineteenth-century/1835-5-6-william-4-c-59-cruelty-to-animals-act/

[115] *An Act for the more Effectual Prevention of Cruelty to Animals, 1849*, 12 &13 Vic c 92, *Public General Statutes*, Twelfth and Thirteenth Year, Queen Victoria, 1849, Printed by George Edward Eyre and William Spottiswoode and published by W Benning and Co, London (1849), 592, available from https://play.google.com/books/reader?id=RKJKAAAAMAAJ&hl=en_GB&pg=GBS.PA592

[116] *Protection of Animals Act 1911* (UK), 1 & 2 Geo 5 c 271, available from http://www.bailii.org/uk/legis/num_act/1911/1069356.html

[117] *Cruelty to Animals Act 1835* (UK) ss 1–4; *Cruelty to Animals Act 1849* (UK), ss 3, 29; Protection of Animals Act 1911 (UK), s 1.

[118] *Cruelty to Animals Act 1849* (UK), s 30.

[119] *An Act for the More Effectual Prevention of Cruelty to Animals in Scotland 1850* (UK) 13 & 14 Vic c 59, *Public General Statutes*, Thirteenth and Fourteenth Year, Queen Victoria, 1850, Printed by George Edward Eyre and William Spottiswoode and published by W Benning and Co, London (1850), 801. available from https://play.google.com/books/reader?id=Z6pKAAAAMAAJ&hl=en_GB&pg=GBS.PA801

[120] *Cruelty to Animals (Scotland) Act 1850* (UK), s 1.

[121] Another important piece of legislation was the *Cruelty to Animals Act 1876* (UK), which applied to vivisection and as such is not relevant to the production of farm animals for food, *Cruelty to Animals Act 1876* (UK), 39 & 40 Vic c 77, available from, The Statutes Project, http://statutes.org.uk/site/the-statutes/nineteenth-century/1876-39-40-victoria-c77-cruelty-to-animals-act/

Table 3.2 Anti-cruelty legislation

Legislation	Animals	Offences
1822 Cattle Act	Section 1 Horses, mares, geldings, mules, asses, cows, heifers, steers, oxen, sheep and other cattle	Section 1 Wantonly and cruelly beat, abuse
1835 Cruelty to Animals Act	Section 1 Bears, cocks, badgers, fighting animals Section 2 Horses, mares, geldings, bulls, oxen, cows, heifers, steers, calves, mules, asses, sheep, lambs, dogs or any other cattle or domestic animal	Sections 1 and 3 Baiting or fighting of bears, cocks, badgers or other animals Sections 1 and 2 Cruelty in driving and conveying animals, keeping them at slaughterhouses without food and water Section 2 Wantonly and cruelly beat, ill-treat, abuse or torture Sections 1 and 4 Need to supply confined animals with sufficient food and nourishment
1849 Cruelty to Animals Act (s 30, did not apply to Scotland)	Section 29 Horses, mares, geldings, bulls, oxen, cows, heifers, calves, mules, asses, sheep, lamb, hogs, pigs, sows, goats, dogs, cats or other domestic animals	Section 2 Cruelly beat, ill-treat, over-drive, abuse or torture or cause or procure to be cruelly beaten, ill-treated, over-drive, abused or tortured Section 3 Prohibitions on keeping a place for animal baiting Section 8 Need to feed and water animals in slaughterhouses. Section 12 Conveying animals in a manner to cause unnecessary pain or suffering
1850 Cruelty to Animals (Scotland) Act	Section 1 Horses, mares, geldings, bulls, oxen, cows, heifers, steers, calves, mules, asses, sheep, lamb, hogs, pigs, sows, goats, dogs, cats or other domestic animals Section 1 Bulls, bears, badgers, dogs, cocks or other animals (baiting)	Section 1 Preamble refers to objective of preventing wanton cruelty Section 1 Cruelly beat, ill-treat, over-drive, abuse or torture or cause or procure to be cruelly beaten, ill-treated, over-driven, abused or tortured Section 2 Prohibitions on keeping a place for animal baiting
1911 Protection of Animals Act	Section 15 Animal means any domestic or captive animal Domestic animal "means any horse, ass, mule, bull, sheep, pig, goat, dog, cat or fowl or any other animal of whatsoever kind or species and whether a quadruped or not which is tame or which has been or is being sufficiently tamed to serve some purpose for the use of man" Captive animal means any animal (not being a domestic animal) of	Section 1(1)(a) "cruelly beat, kick, ill-treat, over-ride, over-drive, over-load, torture, infuriate, or terrify any animal, or shall cause or procure, or, being the owner, permit any animal to be so used, or shall, by wantonly or unreasonably doing or omitting to do any act, or causing or procuring the commission or omission of any act, cause any unnecessary suffering, or, being the owner, permit any

(continued)

Table 3.2 (continued)

Legislation	Animals	Offences
	whatsoever kind or species and whether a quadruped or not, including any bird, fish or reptile, which is in captivity	unnecessary suffering to be so caused to any animal" Section 1(1)(b)–(e) Offences for animal baiting, conveying an animal so as to cause unnecessary suffering, administering substances to animals, without reason and subjecting any animal to an operation which is performed without due care and humanity

A number of general comments can be made about these statutes. First, failings in enforcing the 1822 Cattle Act led to the formation in 1824, of the Society for the Prevention of Cruelty to Animals, which later became the Royal Society for the Prevention of Cruelty to Animals (RSPCA).[122] Second, the *Cruelty to Animals Act 1835* (UK) and *Cruelty to Animals Act 1849* (UK) stipulated that failing to feed and water animals confined for sale or slaughter amounted to a breach of the legislation, as did transporting animals in a cruel manner.[123] In addition, the legislation established offences for the practice of re-selling diseased or worn-out horses, who would otherwise have been destined for slaughter, as well as creating a licensing system for slaughterhouses, prohibiting those who slaughtered horses from being horse dealers.[124]

These developments indicate that the animal product sector was becoming more tightly regulated by legislation targeting cruelty along the supply chain. Transporting cattle, for example, often involved beating, pushing and prodding animals, either to get them to and from market or to load them onto railway cars.[125] This resulted in many injuries, which was not only cruel but also led to financial losses.[126] Frightened and stressed animals shed weight, with some estimates indicating that cows lost as much as seven kilograms of body weight from mishandling during transportation.[127] Legislation which pinpointed and dealt with wide-ranging cruelty demonstrated the potential for high-level interventions. This was significant in

[122] Christopher Otter, "Cleansing and Clarifying: Technology and Perception in Nineteenth-Century London", (2004) 43 (1) *Journal of British Studies*, 40, 45.

[123] *Cruelty to Animals Act 1835* (UK), ss 1, 2, 4; *Cruelty to Animals Act 1849* (UK) ss 8, 12.

[124] *Cruelty to Animals Act 1835* (UK), ss 7, 8; *Cruelty to Animals Act 1849* (UK) ss 7, 11.

[125] Richard Perren, *The Meat Trade in Britain 1840–1914*, Routledge & Kegan Paul, (1978), 33–34.

[126] Ibid, 35.

[127] Christopher Otter, "Cleansing and Clarifying: Technology and Perception in Nineteenth-Century London", above 122, 47.

view of the increasing trade in animals and their products that was discussed in Chap. 2 of this book. As trade increased, accepted practices were replicated many times over, a development that would be problematic if these practices involved cruelty.

Yet, two factors militated against the legislation achieving its full potential and both derived from the status of animals as property. To start with, anti-cruelty laws operated as a restriction on property rights by limiting the harm that owners could inflict on their animals.[128] As such, regulation infrequently led to optimum care. This is apparent from the material set out in Table 3.2, which details offences for abusing animals or treating them cruelly. The legislation responds to an act of cruelty by creating an offence; yet, this post-dates acts of cruelty and the pain suffered by animals. It is rare that parliament imposed positive duties, although it did in one or two instances, such as providing food and water for animals confined or sent to slaughter.[129] This approach illustrates the reluctance of government to interfere to any extent with property rights, especially in a commercial setting. It was a standpoint that was consistent with the second factor limiting the operation of anti-cruelty legislation, namely, judicial interpretation, which laid the foundation for determining an acceptable level of animal suffering in an economic context.

3.5 Judicial Interpretation of Anti-cruelty

The judiciary faced many challenges interpreting anti-cruelty legislation in the nineteenth century, the majority of which derived from balancing animal suffering against commercial uses of animals. These challenges are adroitly illustrated by a series of Scottish, Irish and English cases decided between 1884 and 1891 that focussed on whether "dishorning" cattle amounted to cruelty within the meaning of the legislation. A summary of the cases is set out in Table 3.3. The English and Irish cases were decided pursuant to the *Cruelty to Animals Act 1849* (UK), while the Scottish cases were decided pursuant to the *Cruelty to Animals Act 1850* (UK).

Dishorning, or dehorning as it is known today, involved immobilising the animal, placing its head in a vice and then sawing off its horns. Historically, no pain relief was administered, the animals lost a considerable amount of blood and according to accounts the animals' cries could be heard "a mile off".[130] As the cases signify, the practice was not widespread in England, but was common in Ireland and some counties in Scotland. Dishorning made animals easier to handle and allowed them to be raised on concentrated feed in winter, either indoors or in confined spaces, a

[128] Mike Radford, *Animal Welfare Law in Britain, Regulation and Responsibility*, Oxford University Press (2001), 102.

[129] *Cruelty to Animals Act 1835* (UK) s 8; *Cruelty to Animals Act 1849* (UK) s 8.

[130] *Ford v Wiley* 23 QBD 203, 212.

Table 3.3 Historical cases on dishorning (dehorning) of cattle

Date	Case	Decision	Reason
1884 (April)	Brady v McArgle 14 LR Ir 174 (Ireland)	Dishorning cattle was cruel within the meaning of the statute – 12 & 13 Vict c 92 (s 2 – cruelly ill-treat, torture and abuse)	Dishorning cattle needs to be necessary or reasonable for the purposes of general convenience. "Lower animals are not to be entirely subordinated to man. If men have their rights in respect of these animals, they also have their duties". Dishorning was done for the convenience of feeding them in a confined space, which is not necessary, pursuant to the legislation
1885 (June)	Callaghan v Mc'Evoy 16 LR Ir 325 (Ireland)	Dishorning did not amount to cruelty within the meaning of the statute – 12 & 13 Vict c 92 (s 2 -cruelly ill-treat, torture and abuse)	Dishorning was reasonable and necessary for straw-yard feeding in winter as was commonly practised in Ireland, as long as the operation was properly performed
1888 (July)	Renton v Wilson 15 RJ 84 (Scotland)	Dishorning did not amount to cruelty within the meaning of the statute – 13 and 14 Vic c 92 (s 1 cruelly beat, ill-treat, over-drive, abuse or torture or cause or procure to be cruelly beaten, ill-treated, over-driven, abused or tortured)	The famers are the ones best placed to determine how their interests are served. Dishorning does not amount to cruelty within the meaning of the legislation. The legislation applies to wanton cruelty where the purpose of inflicting pain is to cause suffering to the animal
1889 (May)	Ford v Wiley 23 QBD 203 (England)	Dishorning cattle was cruel within the meaning of the statute – 12 & 13 Vict c 92 (s 2 – cruelly ill-treat, torture and abuse)	Mere infliction of pain, even if extreme pain, is not by itself sufficient to bring the act within the meaning of the legislation. The pain must be unnecessary, namely, the pain is inflicted without adequate and reasonable objects. Economic objectives such as making cattle more valuable in the marketplace do not satisfy the threshold test of necessity: • Followed in Waters v Braithwaite • Not followed in Todrick v Wilson and The Queen v McDonagh • Distinguished in Tucker v Hazelhurst • Mentioned in Hart v Police SC Auckland

(continued)

Table 3.3 (continued)

Date	Case	Decision	Reason
1891 (March)	Todrick v Wilson 18 RJ 41 (Scotland)	Dishorning did not amount to cruelty within the meaning of the statute – 13 & 14 Vic c 92 (s 1 cruelly beat, ill-treat, over-drive, abuse or torture or cause or procure to be cruelly beaten, ill-treated, over-driven, abused or tortured)	Intention of the legislature is to prevent the infliction of "wanton cruelty", that is, purposeless cruelty. Dishorning cattle does not amount to cruelty when performed with skill and for the legitimate purpose of preventing the cattle from injuring one another
1891 (May)	The Queen v McDonagh 28 LR Ir 204 (Ireland)	Dishorning did not amount to cruelty within the meaning of the statute – 12 & 13 Vict c 92 (s 2 – cruelly ill-treat, torture and abuse)	To amount to cruelty something more is needed than mere infliction of pain. The question of what is necessary cannot be determined by what is necessary for the animal itself. Humans have dominion over cattle to fatten, sell and eat them. An act is not cruel if is "performed under an honest and reasonable belief of its usefulness for these purposes and with reasonable care and skill and is not attended with such suffering as men of ordinary humanity would consider disproportionate to the object sought to be attained". Ford v Wiley and Brady v McArgle not followed. Callaghan v Mc'Evoy, *Renton v Wilson and* Todrick v Wilson followed

practice described as "straw-yard feeding".[131] There is no question that animals suffered an appreciable amount of pain, the issue being whether dishorning amounted to statutory cruelty.[132]

In the first case, *Brady v McArgle*, the court interpreted cruelty by introducing a test of necessity, holding that "the cruelty intended by the statute is the unnecessary abuse of the animal".[133] The Court relied on earlier decisions, which had primarily

[131] *Callaghan v Mc'Evoy* 16 LR Ir 325, 332–333.

[132] Generally, Kevin J Stafford and David J Mellor, "Addressing the Pain Associated with Disbudding and Dehorning in Cattle", (2011) 35 *Applied Animal Behaviour Science*, 226.

[133] *Brady v McArgle* 14 LR Ir 174, 183.

dealt with animal baiting, to interpret cruelty widely, and necessity narrowly. The latter was to be determined by asking whether the act was done for the benefit of the animal.[134] On the facts, the Court found that dishorning provided no benefit to the animal, it being carried out purely for the convenience of the owner, to enable feeding in a confined space.[135] Accordingly, dishorning was deemed unnecessary, as well as unreasonable, cruel and illegal.[136] At first glance, *Brady v McArgle* appears to set a precedent for moral monism, where one moral theory, based on sentience, is capable of applying, irrespective of the relationship of the animals to humans. However, a closer reading of the case indicates that the Court may have been prepared to find dishorning legal, but there was insufficient evidence that the practice "makes the animal more serviceable for the use of man".[137] This comment may have been made to explain why the Court dismissed the appellant's claim that there were good commercial reasons for dishorning. However, the statement also appears to be at odds with the finding that necessity equates to a benefit to the animal.

Subsequent to *Brady v McArgle*, the concept of necessity became a lightning rod for dissension and argument, as the Courts' interpretations varied from case to case and across the jurisdictions. *Callaghan v Mc'Evoy* held that to establish whether a procedure was cruel involved undertaking a two-staged test: first, evaluating whether the pain inflicted is substantial and then determining whether the suffering of the animal is necessary.[138] The Court elaborated that the notion of necessity does not solely hinge on the animal's pain and suffering but also depends on the animal's use to humans.[139] In contrast to this approach, *Ford v Wiley* approved a test of proportionality based on animal sentience.[140] Although the mere infliction of pain would not have been sufficient to classify a procedure as cruel, pain and suffering had to be necessary, in the sense that the means by which the objective was achieved should be weighed against the animal's pain and suffering.[141]

While the test of "necessity" featured in each of these cases, the focus differed, leading to differing results. In *Brady v McArgle*, the emphasis lay on animal suffering and lack of benefit to the animal, so dishorning was held to be cruel. In *Callaghan v Mc'Evoy*, the focus lay on the animals' utility to humans, so the commercial advantages of dishorning outweighed the animals' pain and suffering.

---

[134] Ibid, 181, 182, 183; The cases the Court referred to included *Budge v Parsons 122 ER 144 that dealt with cock-fighting and Murphy v Manning 2 Ex Div 307, where the comb of a cock was cut off to make the bird more suitable for fighting.*

[135] *Brady v McArgle* 14 LR Ir 174, 181, 182, 183.

[136] Ibid, 182, 183.

[137] Ibid, 183.

[138] *Callaghan v Mc'Evoy* 16 LR Ir 325.

[139] Ibid, 333.

[140] *Ford v Wiley* 23 QBD 203, 209–210, 215; discussion Mike Radford, *Animal Welfare Law in Britain, Regulation and Responsibility*, above 128, 247–248.

[141] *Ford v Wiley* 23 QBD 203, 209.

It was also central to the Court's decision that straw-yard feeding was a common practice in Ireland. It was held to be a cost-effective way of maintaining animals in good condition during winter, allowing them to be sent to market earlier than had they been fed in the field.[142] The Court also accepted that dishorning stopped cattle from injuring themselves during confined feeding and permitted animals to be transported and handled safely.[143] Consequently, the Court found that dishorning was reasonable and necessary, provided that the operation was properly performed.[144] In effect, the animals' pain and suffering were deemed necessary because objectives flowing from the practice of dishorning were commercially reasonable. This contrasts with *Ford v Wiley*, which was decided after Callaghan's case and held that economic objectives such as making cattle more valuable in the marketplace did not satisfy the threshold test of proportionality.[145]

Judicial analysis on the necessity of dishorning continued in *Renton v Wilson*,[146] *Todrick v Wilson*[147] and *The Queen v McDonagh*.[148] *Renton v Wilson* (a decision of the Scottish courts) followed analogous reasoning to Callaghan's case. The judges noted that dishorning was customary in the districts of Fife and Forfar, which were "... considerable district[s] of the country, and [the procedure was] performed with a view to a rational purpose, and under the belief that it is necessary for the wellbeing and control of the animals".[149] This reasoning indicates that the Court was equating economic convenience with a form of animal wellbeing, which emphasised production and merchantability. In a similar manner, *Todrick v Wilson*, which was decided almost 2 years after *Renton v Wilson*, upheld the practice of dishorning, finding that it not only makes cattle more valuable to the owner but also makes the animals easier to handle so that they do not injure each other.[150]

The Queen v McDonagh,[151] decided in the same year as *Todrick v Wilson*, held that to amount to statutory cruelty, something more than the infliction of pain is needed.[152] The Court noted that other painful operations, such as castration, are routinely carried out on animals to make them easier to handle and these procedures have not attracted charges of cruelty.[153] For these reasons, Chief Justice O'Brien indicated that necessity cannot be determined by what is necessary for the animal

[142] *Callaghan v Mc'Evoy* 16 LR Ir 325, 332–333, 334.

[143] Ibid, 329, 323–333, 334.

[144] Ibid, 329, 331, 333.

[145] *Ford v Wiley* 23 QBD 203, 209.

[146] *Renton v Wilson 15 RJ 84.*

[147] *Todrick v Wilson* 18 RJ 41.

[148] *The Queen v McDonagh* 28 LR Ir 204.

[149] *Renton v Wilson 15 RJ 84, 88.*

[150] *Todrick v Wilson* 18 RJ 41, 44, 46.

[151] *The Queen v McDonagh* 28 LR Ir 204.

[152] Ibid, 210, 212.

[153] Ibid, 210, 214.

itself, a conclusion that clearly contradicts the findings in *Brady v McArgle*.[154] His Honour emphasised that humans have dominion over cattle, to fatten, sell and eat them.[155] Consequently, any operation which furthered these purposes did not amount to statutory cruelty, provided three criteria were satisfied: first that the operation was "performed under an honest and reasonable belief of its usefulness" for the purpose of rearing cattle; second, that the procedure was conducted with reasonable care and skill; and, third, that the suffering was not disproportionate to the objective to be achieved.[156]

Ford v Wiley also favoured a test of proportionality; however, that test and the one espoused in *The Queen v McDonagh* differed. Whereas *Ford v Wiley* referred to proportionality with respect to the means by which farmers achieved their objectives, *The Queen v McDonagh* referred to proportionality with respect to the objective itself. This is clear from the finding that the test in McDonagh centred on whether "men of ordinary humanity would consider [the pain and suffering] disproportionate to the object sought to be attained".[157] The practical difference between the two cases lies in the fact that human interests, particularly commercial interests, can justify pain and suffering under McDonagh test, whereas animal interests such as pain and suffering can supersede human interests under *Ford v Wiley*.[158]

The status of *Ford v Wiley* is unsettled. Notwithstanding the fact it was not followed in a number of Irish and Scottish decisions, the 1905 edition of *Stone's Justices Manual* stipulated that *Ford v Wiley* was still a good law in England, a view shared by Mike Radford.[159] Yet, the influence of the case is mixed. Its reasoning was not followed in *Todrick v Wilson* and *The Queen v McDonagh*; the decision was followed in *Waters v Braithwaite;*[160] and, outside of the UK, the New Zealand courts have distinguished the decision in *Tucker v Hazelhurst SC Palmerston North*[161] and mentioned it in *Hart v Police SC Auckland*.[162] Radford reconciles these differences

[154] Ibid, 211.

[155] Ibid, 212.

[156] Ibid.

[157] Ibid.

[158] Mike Radford, *Animal Welfare Law in Britain, Regulation and Responsibility*, above 128, 249.

[159] J R Roberts Esq (ed) *Stone's Justices Manual* 37th Edition, Shaw and Sons, Butterworth and Co, London (1905), 236; also, Mike Radford, *Animal Welfare Law in Britain, Regulation and Responsibility*, above 128, 257.

[160] *Waters v Braithwaite*, (1914) 78 JP 124; the case involved taking cows to market in an unmilked state so that potential buyers could assess the cows' productivity, which was held to amount to cruelty.

[161] *Tucker v Hazelhurst SC Palmerston North* [1906] NZGazLawRp 200; the case determined that shooting at pigeons to kill them did not amount to cruelty.

[162] *Hart v Police SC Auckland* [1965] New Zealand Law Reports, 666; the case held that shooting and wounding a dog without killing it amounted to cruelty, even though the reason for shooting was to protect livestock.

by concluding that *Ford v Wiley* stands for the proposition that there must be "a good reason to cause an animal to suffer" and this will vary according to circumstances.[163] The necessity of animal suffering also varies according to the wording of the legislation, particularly if the word "wanton" is used. For example, the second limb of Section 1(1)(a) of the *Protection of Animals Act 1911* (UK) refers to "wanton' cruelty, which centres on whether the suffering is inevitable, rather than unnecessary.[164] Accordingly, in an agricultural setting, suffering may be inevitable and thus not amount to wanton cruelty, where farmers follow accepted husbandry practices, even if these practices cause pain and suffering.[165]

Yet ultimately, the importance of *Ford v Wiley*, as well as the other cases listed in Table 3.3, extends beyond their precedent value. The Courts' determinations signalled what they, as well as society, considered to be legally and socially appropriate practices. Each decision acknowledged the inherent tension between commercial reality and animal suffering, yet as Francione cautions, many cases are "decided not by reference to some moral ideal but by reference to norms of exploitation already deemed legitimate".[166] Admittedly, this did not occur in every case, and decisions such as *Waters v Braithwaite* proscribed commonly accepted practices.[167] However, the cases taken as a whole create a pattern of advantage given to procedures commonly adopted for commercial convenience. This pattern becomes clearer upon examining the reasoning in the cases.

Ford v Wiley, which held that dishorning was cruel, was an English case, a jurisdiction where dishorning had not been used for almost two decades.[168] This contrasts with decisions upholding dishorning which came from Ireland and Scotland where the practice was relatively common. In *The Queen v McDonagh*, the Court noted that dishorning was necessary for the profitability of the cattle industry, which provided economic benefits to Irish communities, adding that there was no reasonable alternative to the practice.[169] The Court also placed a great deal of reliance on industry custom, observing that dishorning was increasing, rather than decreasing in popularity. Had demand been declining, this would have provided "evidence that it was not useful for any purpose, or that it was opposed to the moral sense of the community as involving the infliction of great pain without an adequate object".[170] The importance of existing practice was also underscored in

[163] Mike Radford, *Animal Welfare Law in Britain, Regulation and Responsibility*, above 128, 257.

[164] Ibid.

[165] Ibid, 251, discussing *Hall v RSPCA* [1993] QBD 11.

[166] Gary L Francione, *Rain without Thunder, the Ideology of the Animal Rights Movement*. Temple University, reprint (2007), 130; discussion Mike Radford, *Animal Welfare Law in Britain, Regulation and Responsibility*, above 128, 250.

[167] *Waters v Braithwaite*, 78 JP 124.

[168] *Ford v Wiley* 23 QBD 203, 210.

[169] *The Queen v McDonagh* 28 LR Ir 204, 215, 216.

[170] Ibid, 213.

Renton v Wilson, where the Court documented that dishorning was common in the Scottish districts of Fife and Forfar.[171]

The justifications provided in these decisions tend to support Francione's contention that decisions frequently lack meaningful moral and ethical engagement. Indeed, this point is reinforced by Justice Murphy's passing reference to Bentham's utilitarianism in *Callaghan v Mc'Evoy*.[172] His Honour interpreted Bentham as proscribing gratuitous cruelty, but held that dishorning was not gratuitous because it conveys benefits to the owner.[173] His Honour concluded by saying that "... self-interest would prevent any farmer from resorting to a practice... if the result were merely to cause useless pain or torture to the animals... [the farmer] would greatly reduce the condition of the animals... [and] lose in the sale".[174] However, husbandry practices are unlikely to be implemented solely for the purpose of cruelty, and the observation misses the point that cruelty and economic benefit often go hand in hand.[175]

By upholding dishorning, the cases illustrate the emergence of a form of moral pluralism where cruelty varies in accordance with the animal and the uses humans have for it. In those jurisdictions where dishorning was relatively widespread and commercially beneficial, the Courts maximised the valency of economic benefit, supporting farmers who aimed at producing as many animals as quickly as possible. This mirrors developments in the nineteenth century, where increased demand for meat led to a focus on producing surpluses for profit.[176] It is also consistent with the Courts' reluctance to adjudicate on industry standards or commercial practice, areas they saw as more appropriate to the realm of public policy, to be determined by government.[177] In *Callaghan v Mc'Evoy*, Justice Morris went as far as criticising organisations such as the RSPCA for bringing industry-related practices to court, accusing the society of using the Courts to pursue their "own views", rather than working towards legislative change.[178] More philosophically, *Lord Young in Renton v Wilson* questioned how far regulation should interfere in the daily lives of farmers, pointing out that farmers, rather than the legislature, are best placed to determine how to run their businesses.[179]

The call for legislative rather than curial intervention not only came from the Bench but also from reformers who felt that the current regime did not go far

[171] *Renton v Wilson 15 RJ 84, 88.*

[172] *Callaghan v Mc'Evoy* 16 LR Ir 325, 335.

[173] Ibid.

[174] Ibid.

[175] *Ruth Harrison, Animal Machines, CABI UK (2013), 175.*

[176] Section 2.7, Chap. 2 of this book.

[177] *Renton v Wilson 15 RJ 84, 87;* The Queen v McDonagh 28 LR Ir 204, 213, 218–219.

[178] *Callaghan v Mc'Evoy* 16 LR Ir 325, 330–331.

[179] *Renton v Wilson 15 RJ 84, 87, 88.*

enough.[180] During the nineteenth century, utilitarianism had become the underlying paradigm for conceding moral recognition to animals, yet utilitarianism did not prohibit the use of animals, it merely proscribed unnecessary suffering.[181] This permitted qualified use of animals, which in the context of food production, became arrogated to commercial objectives, obscuring a deeper analysis of humanity's relationship to farm animals. This then became the shadow cast by the Enlightenment.

3.6 The Enlightenment's Shadow

As discussed, philosophers such as Descartes and Kant emphasised the differences between humans and animals, resulting in a separation of animals from humanity's social and moral spheres. Although other philosophers attempted to attenuate this disjunction, they met with mixed success.[182] Jean-Jacques Rousseau (1712–1778) argued that humans were a category of animal, concluding that philosophies that underscored the differences between humans and animals were flawed.[183] Rousseau also indicated that as humans and animals share similar characteristics, it was logical to apply concepts of natural law to both, meaning that humans owed direct duties to animals.[184] Yet, these alternative approaches never totally succeeded in breaking down human-animal barriers, because humans were invariably placed at the top of a hierarchy.[185] Accordingly, humans could still use animals and utilitarianism, the predominant paradigm, considered it acceptable for humans to eat animals, because it gave humans pleasure and animals "are never the worse" for it.[186] Bentham, for example, justified this position by arguing that animals are spared from:

> long-protracted anticipations of future misery that we have; and the death they suffer at our hands usually is and always could be speedier and thus less painful than what would await them in the inevitable course of nature. There is also very good reason why we should be

---

[180]Henry S Salt, *Animals' Rights: Considered in Relation to Social Progress,* (1894), revised edition G Bell and Sons Ltd. (1922), 7, available from, https://ia600901.us.archive.org/32/items/cu31924030305332/cu31924030305332.pdf

[181]Ian MacLachlan, "Humanitarian Reform Slaughter Technology, and Butcher Resistance in Nineteenth-Century Britain", in *Meat, Modernity and the Rise of the Slaughterhouse* (Paula Young Lee ed), University of New Hampshire Press (2008) 107, 109.

[182]Primarily French philosophers, Denis Diderot (1713–1784), Étienne Bonnot de Condillac (1714–1780) and Jean-Jacques Rousseau (1712–1778), discussion Jean-Luc Guichet and Donald Nicholson-Smith, "Animal of the Enlightenment to the Animal of Postmodernism" (2015) 127 *French Yale Studies*, 69, 70, 73, 76.

[183]Keith Tester, *Animals and Society, The Humanity of Animal Rights*, Routledge, London (1991), 123, 125, 126, 133.

[184]Ibid, 123, 131.

[185]Jean-Luc Guichet and Donald Nicholson-Smith, "Animal of the Enlightenment to the Animal of Postmodernism", above 188, 73.

[186]Jeremy Bentham, *An Introduction to the Principles of Morals and Legislation*, above 69, 143.

> allowed to kill ones that attack us: we would be the worse for their living, and they are not the worse of being dead.[187]

Indeed, much of Enlightenment philosophy that was involved with ameliorating the position of animals was more concerned with rejecting gratuitous cruelty than questioning underlying assumptions about the appropriateness of capitalising on animals.[188] Even philosophers who employed rights language interwove their ideologies with healthy doses of utilitarianism, permitting the use of animals for the greatest happiness of humans.

Hutcheson's work in the 1720s was one of the earliest to argue in favour of animal rights.[189] His theories applied to domesticated animals and were broadly based on utilitarian foundations of the common good.[190] He argued that domesticated animals were members of a system created by God and, contra Kant, were connected to human beings through a web of happiness and justice, creating a "moral community" that could support rights and obligations.[191] At the same time, Hutcheson also classified animals as property and believed in a hierarchy of existence, so that humans, who were higher up the hierarchy, could eat animals, because they were lower down the hierarchy.[192] The rights that Hutcheson granted animals were rights to happiness, which in practical terms meant they should not be subjected to "useless pain or misery".[193] These rights were complemented by corresponding duties, so that humans should not abuse animals.[194] Consequently, while Hutcheson used rights language, he had much in common with utilitarians. He did not question the assumption that humans could use and eat animals, nor did his rights-based approach depend on individual happiness; instead, rights depended on the happiness of the entire system.[195]

---

[187] Ibid.

[188] Aaron Garrett, "Francis Hutcheson and the Origin of Animal Rights", (2007) 45 (2) *Journal of the History of Philosophy*, 243, 262.

[189] Francis Hutcheson, *Inquiry into the Original of Our Ideas of Beauty and Virtue*, Printed for D Midwinter, A Bettesworth and C Hitch, J and J Pemberton, R Ware, C Rivington, F Clay and A Ward, J and P Knapton, T Longman, R Hett and J Wood, London, Fourth Edition, 1738, (originally published 1725), available from https://archive.org/details/inquiryintoorigi00hutc/page/n6; Francis Hutcheson, *An Essay on the Conduct of the Passions and Affections, with Illustrations on the Moral Sense* Printed for W Innys and J Richardson S Birt, C Hitch and L Hawes, T Astley and R Baldwin, J Hinton, J and J Rivington, J Ward, T Field and R Withy, London, Fourth Edition 1756 (1726 original edition), available from https://archive.org/details/essayonnaturecon1756hutc/page/n4

[190] Aaron Garrett, "Francis Hutcheson and the Origin of Animal Rights", above 194, 249–250, 252.

[191] Ibid, 258.

[192] Ibid, 257, 260.

[193] Ibid, 260.

[194] Ibid, 260.

[195] Ibid, 262.

It was not until the nineteenth century that these tenets were challenged, as vegetarianism and a different genre of animal rights entered the discourse.[196] One of the most important publications was Henry Salt's (1851–1939) treatise, *Animals' Rights: Considered in Relation to Social Progress*, published in 1894; Salt defended this publication in an article he wrote in 1900 for the *International Journal of Ethics*.[197] Three core themes differentiate Salt's views from traditional utilitarian philosophies:

- Arguments in favour of breaking down the barriers between humans and animals
- Critiquing the status of animals as property
- Formulating rights for animals, which leads to a paradox resolved by humans being duty-bound to leave animals alone

With regard to the first theme, Salt argued that humans and animals differ in degree but not in kind. Therefore, to talk of "man and the animals" is illogical because "there is no such antithesis, and we ought rather to speak of 'man and the other animals'".[198] In this way, Salt concluded that the circle of humanity extends to "lower animals", so that human-animal relationships clearly involve questions of morality.[199] Although this part of Salt's argument was similar to Hutcheson's, unlike Hutcheson, Salt was highly critical of the classification of animals as property, the second theme that differentiates Salt from utilitarians. Salt pointed to the fact that categorising animals as property promotes their treatment as objects, which normalises cruelty towards them.[200] Thus, sheep, oxen and goats are seen "as mere live-stock", while smaller animals such as pigs, poultry and rabbits are

[196] Examples of publications on vegetarianism include, Joseph Ritson, *An Essay on Abstinence from Animal Food, as a Moral Duty*, printed for Richard Phillips no 71 St Pauls, Church-Yard, (1802), available from https://books.googleusercontent.com/books/content?req=AKW5Qae1cmFyHzLza6Gdm8g6JRBV9S_LJU3ASSge7B-2djGbdXYtnVw1XsVkFDk82VtYHd9Zj8iKtI-LTcz87Kdt8KXVwJjeDUz9DbgMTTuaysgzFIi5NfvE475EtG7Whz_1mQ7pKnj4fco0VZnuUTAQyc_EEEfy5GkAGXYLV7nk9dSo8WGJ4uksgfUVBKKhC9eAdKbqaNH_K2PfXzlRtEri85TUPLKsjcybMJBxnbaImomNVsm9SG39SFrMYfnuIyumpeW_N680SgsLrINg1J-UJwoBJdgf-pqcWqsV4JqAfCeK5KPvHVk; Percy Bysshe Shelley, *On the Vegetable System of Diet*, (1813) reprint, Folcroft Library Editions (1975);

H S Salt, *A Plea for Vegetarianism and Other Essays*, The Vegetarian Society, Manchester, 75 Princess Street, Manchester, John Heywood, Deansgate and Ridgefield, (1886), available from https://ia802504.us.archive.org/30/items/pleaforvegetaria1886salt/pleaforvegetaria1886salt.pdf.

[197] Henry S Salt, *Animals' Rights: Considered in Relation to Social Progress*, above 186, 29; Henry Salt, "The Rights of Animals" (1900) 10 (2) *International Journal of Ethics*, 206.

[198] Henry Salt, "The Rights of Animals", above 203, 207–208. Keith Tester discusses these views, as initially set out in *Animals' Rights: Considered in Relation to Social Progress*, Keith Tester, *Animals and Society, The Humanity of Animal Rights*, above 189, 123, 125, 126, 133.

[199] Henry S Salt, *Animals' Rights: Considered in Relation to Social Progress*, above 186, 82–83.

[200] Ibid, 29.

regarded as "farm-produce".[201] In effect, what Salt was critiquing was the characterisation of farm animals as commodities.[202]

Salt also agreed with Hutcheson and Bentham that animals had a right to be treated with "gentleness and consideration"[203] but drew on the work of Henry Spencer (1820–1903) who wrote:

> Every man is free to do that which he wills, provided he infringes not the equal liberty of any other man . . . Whoever admits that each man must have a certain restricted freedom, asserts that it is right he should have this restricted freedom. . . . And hence the several particular freedoms deducible may fitly be called, as they commonly are called, his rights.[204]

From this philosophy Salt developed a notion of animal rights, which sat outside strict legal rights but also extended beyond a moral direction of gentle use.[205] This led to the third differentiation from utilitarians, where Salt used the concept of "restricted freedom", established by Spencer, to argue in favour of individual animal rights that would allow animals to live to their potential, although this still depended on the wellbeing of the community as a whole.[206] On this point, in similarity with the utilitarians, Salt does not prohibit killing animals, but specifies it needs to be done quickly and only when necessary.[207] However, Salt also considers that necessity does not encompass abusing animals or eating them.[208]

The proscription against eating animals crystallised the connection between rights and vegetarianism, because being cruel to, or eating an animal, encroaches on the animal's restricted freedom.[209] Tester has pointed out that restricted freedom leads to a paradox because if humans and animals are individuals with the same right to restricted freedom, the only way each can exercise this right is to remain separated.[210] The upshot being that humans should not do anything to animals, other than leave them alone.[211] The practicalities of this approach were the antithesis of occurrences throughout the nineteenth century as trade in animal products continued to expand. It is therefore a matter of some irony that even though the earliest anti-cruelty laws targeted farm animals, by the late nineteenth century when market expansion was in full swing, anti-cruelty regulation had become constrained by industry customs designed to gain commercial advantages in the marketplace.

---

[201] Ibid, 2.

[202] Ibid, 29.

[203] Ibid, 6.

[204] Quoted in Henry S Salt, Ibid.

[205] Henry Salt, "The Rights of Animals", above 203, 211–212.

[206] Ibid, 22.

[207] Ibid, 6, 22.

[208] Ibid.

[209] Keith Tester, *Animals and Society, The Humanity of Animal Rights*, above 189, 157.

[210] Ibid, 155.

[211] Ibid, 155–156.

3.7 Conclusion

The Enlightenment ushered in an era of widespread intellectual freedom and debate, which also incorporated discussion of society's relationship to animals. Although humans' lack of knowledge of animals' cognitive abilities initially excluded animals from humanity's social and moral spheres, by the end of the eighteenth century, philosophies of animal sentience were gaining traction. In response, anti-cruelty regulation limited how society used animals. However, this legislation also intersected with increasing urbanisation, industrialisation and the growing appetite for meat and other animal products.

Farmers turned to production methods that would make animals ready for market as quickly as possible. Where this involved methods, such as dishorning, which resulted in considerable pain and suffering, it did not automatically lead to a determination of statutory cruelty. This depended on whether the practice was deemed necessary, so that industry mechanisms assumed a high degree of significance, although they were not necessarily conclusive. Anti-cruelty thus became tailored to market conditions, obscuring a deeper analysis of humanity's relationship to animals.[212] As a result, the debate side-stepped meaningful discussion of key principles, such as animals' classification as property and the morality of humans using animals. By 1894, when Salt proposed animal rights, he faced a society that had come to expect easy access to animal products, supported by a regulatory system which interpreted cruelty in a way that facilitated enhanced production. This made it difficult and impracticable for society to engage with Salt's views.

Yet, there was more to come. Anti-cruelty laws created offences for animal abuse that had already occurred, but largely did not stipulate obligations of care. It was an omission that was to become critical as increasing trade led to the introduction and spread of epizootics, a matter briefly discussed in Chap. 2 of this book. By the end of the nineteenth century, the veterinary profession had developed a good understanding of how animal diseases were introduced and spread, knowledge that was important to supporting the animal trade. However, regulators had difficulty separating veterinary health measures from the philosophical and moral foundations of anti-cruelty.[213] The successes and failings of this approach are elaborated in the next three chapters, commencing with Chap. 4, which analyses the links between disease control and commerce in farm animals.

[212] Cassandra D Williams, "Liberating the Enlightenment: how a Transformed Relationship with Animals can help us Transcend Modernity", (2003) 29 1 *Religious Education*, 95, 98–99.

[213] John McEldowney, Wyn Grant, Graham Medley et al, *The Regulation of animal Health and Welfare*, Routledge, Taylor & Francis Group, Abingdon, Oxfordshire (2013), 20.

Bibliography

Agostino FD', Gaus G Thrasher J (2019) Contemporary approaches to the social contract. In: Zalta EN (ed) The Stanford encyclopedia of philosophy, Spring edn. https://plato.stanford.edu/archives/spr2019/entries/contractarianism-contemporary/

Bargheer S (2006) The fools of the leisure class: honor, ridicule, and the emergence of animal protection legislation in England. Eur J Sociol 47(1):3

Baumeister AT (2000) Kant: the arch-enlightener. In: Geras N, Wokler R (eds) The enlightenment and modernity. Macmillan Press Ltd, London, p 50

Beirne P (2013) Hogarth's animals. J Anim Ethics 3(2):133

Bentham B (1781) In: Bennett J (ed) An introduction to the principles of morals and legislation. Clarendon Press, Oxford. https://www.earlymoderntexts.com/assets/pdfs/bentham1780.pdf

Blackstone W (1765–1769) Blackstone's commentaries on the Laws of England Book the second—chapter the first: of property in general. http://avalon.law.yale.edu/18th_century/blackstone_bk2ch1.asp

Blosh M (2012) The history of animal welfare and the future of animal rights. Electronic Thesis and Dissertation Repository. https://ir.lib.uwo.ca/etd/803

Bristow W (2017) Enlightenment. In: Zalta EN (ed) The Stanford encyclopedia of philosophy, Fall edn. https://plato.stanford.edu/archives/fall2017/entries/enlightenment/

Bufacchi V (2000) The enlightenment, contractualism and moral polity. In: Geras N, Wokler R (eds) The enlightenment and modernity. Macmillan Press Ltd, London, p 204

Cohen E (1994) Animals in medieval perceptions: the image of the ubiquitous other. In: Manning A, Serpell J (eds) Animals and human society changing perspectives. Routledge, London, p 66

de Comte B, Leclerc G-L Histoire Natrelle, Génénral et Particulière avec la description du Cabinet du Roi, Aux Deux-Ponts Chez Sanson & Compagne, first volume published in 1785. https://www.biodiversitylibrary.org/bibliography/60802#/summary

Cudworth E, Hobden S (2014) Civilisation and the domination of the animal. Millennium J Int Stud 42(3):746

Driver J (2014) The history of utilitarianism. In: Zalta EN (ed) The Stanford encyclopedia of philosophy, Winter edn. https://plato.stanford.edu/archives/win2014/entries/utilitarianism-history/

Entrèves MPD' (2000) Critique and enlightenment: Michel Foucault on 'was ist Aufklärung?'. In: Geras N, Wokler R (eds) The enlightenment and modernity. Macmillan Press Ltd, London, p 184

Favre DS, Tsang V (1993) The development of anti-cruelty Laws during the 1800s. Detroit Coll Law Rev 1:1

Ferguson M (2001) Animal advocacy and Englishwomen 1780–1900. The University of Michigan Press

Francione GL (2007a) Rain without thunder, the ideology of the animal rights movement. Temple University, Philadelphia, reprint

Francione GL (2007b) An introduction to animal rights: your child or the dog? Temple University, Philadelphia, reprint

Francione GL (2010) Animal welfare and the moral value of nonhuman animals. Law Cult Human 6(1):24

Fudge E (2013) The animal face of early modern England. Theory Cult Soc 30(7/8):177–182. https://doi.org/10.1177/0263276413496122

Gabardi W (2017) The next social contract: animals, the Anthropocene, and biopolitics. Temple University Press, Philadelphia, p 125

Garner R (2012) Rawls, animals and justice: new literature, same response. Res Publica 18:159–160. https://doi.org/10.1007/s11158-011-9173-z

Garrett A (2007) Francis Hutcheson and the origin of animal rights. J Hist Philos 45(2):243

Glacken CJ (1960) Count Buffon on cultural changes of the physical environment. Ann Assoc Am Geogr 50(1):1

Golinski J (2011) Science in the enlightenment, revisited. Hist Sci 49(2):217

Graham G (2001) Morality and feeling in the Scottish enlightenment. Philosophy 76(296):271

Gregory J (1770) Observations on the duties and offices of a physician and on the method of prosecuting enquiries in philosophy, London Printed for Strahan W and Cadell T (Successor to Mr Millar) in the Strand, pp 67, 71, 72. https://play.google.com/books/reader?id=nJlbAAAAQAAJ&hl=en_GB&pg=GBS.PA1

Guichet J-L, Nicholson-Smith D (2015) Animal of the enlightenment to the animal of postmodernism. French Yale Stud 127:69

Hampson N (1999) The enlightenment. In: Cameron E (ed) Early modern Europe, an Oxford history. OUP, Oxford, p 265

Harrison R (2013) Animal machines. CABI, Wallingford

Holliday I (2000) English conservatism and enlightenment rationalism. In: Geras N, Wokler R (eds) The enlightenment and modernity. Macmillan Press Ltd, London, p 117

Hutcheson F (1756) An essay on the conduct of the passions and affections, with illustrations on the moral sense. Printed for W Innys and J Richardson S Birt, C Hitch and L Hawes, T Astley and R Baldwin, J Hinton, J and J Rivington, J Ward, T Field and R Withy, London, 4th edn (1726 original edition). https://archive.org/details/essayonnaturecon1756hutc/page/n4

Hutcheson F (1738) Inquiry into the original of our ideas of beauty and virtue. Printed for D Midwinter, A Bettesworth and C Hitch, J and J Pemberton, R Ware, C Rivington, F Clay and A Ward, J and P Knapton, T Longman, R Hett and J Wood, London, 4th edn (originally published 1725). https://archive.org/details/inquiryintoorigi00hutc/page/n6

Kalof L (2007) Looking at animals in human history. Reaktion Books, London

Korsgaard CM (2012) A Kantian case for animal rights. In: Michael M, Kühne D, Hänni J (eds) Animal law—tier and rect: developments and perspectives in the 21st century. Dike Verlag, Zurich, p 3. http://nrs.harvard.edu/urn-3:HUL.InstRepos:34903186

Korsgaard C (2013) Kantian ethics, animals and the law. Oxf J Leg Stud 33(4):629. https://doi.org/10.1093/ojls/gqt028

Kim K (2009) Adam Smith's theory of economic history and economic development. Eur J Hist Econ Thought 16(1):41

Langton J (2014) Forest law in the landscape: not the clearing of the woods, but the running of the deer? In: Baker K, Carden R, Madgwick R (eds) Deer and people. Oxbow Books, Oxford, p 222

Lawrence J (1802) Philosophical and practical treatise on horses and on the moral duties of man towards the brute creation, Vol 1, H D Symonds, Paternoster-Row, London. https://ia800200.us.archive.org/27/items/philosophicalpra01lawr/philosophicalpra01lawr.pdf

Locke J (2017) Second treatise of government. Jonathan Bennett. https://www.earlymoderntexts.com/assets/pdfs/locke1689a.pdf

Lovegrove R (2007) Silent fields: the long decline of a nation's wildlife. OUP, Oxford

MacLachlan I (2008) Humanitarian reform slaughter technology, and butcher resistance in nineteenth-century Britain. In: Lee PY (ed) Meat, modernity and the rise of the slaughterhouse. University of New Hampshire Press, Durham, p 107

McEldowney J, Grant W, Medley G et al (2013) The regulation of animal health and welfare. Routledge/Taylor & Francis Group, Abingdon, p 20

Mepham B (2006) The ethical matrix as a decision-making tool, with specific reference to animal sentience. In: Turner J, D'Silva J (eds) Animals, ethics and trade. Earthscan, UK, p 134

Nicholson EB (1879) The rights of an animal: a new essay in ethics. C Kegan Paul and Co, London, p. 11. https://ia902700.us.archive.org/29/items/rightsananimala01lawrgoog/rightsananimala01lawrgoog.pdf

Norwood FB, Lusk JL (2011) Compassion, by the pound: the economics of farm animal welfare. Oxford University Press, New York

Nuyen AT (1998) Hume on animals and morality. Philos Pap 27(2):93

O'Hara K (2010) The enlightenment, a beginners guide. Oneworld Oxford, Oxford

Otter C (2004) Cleansing and clarifying: technology and perception in nineteenth-century London. J Br Stud 43(1):40

Overcash EA (2012) Unwarranted discrepancies in the advancement of animal law: the growing disparity in protection between companion animals and agricultural animals. North Carolina Rev 90:837

Passmore J (1975) The treatment of animals. J Hist Ideas 36(2):195

Perren R (1978) The meat trade in Britain 1840–1914. Routledge & Kegan Paul

Radford M (2001) Animal welfare law in Britain, regulation and responsibility. Oxford University Press, Oxford

Ritson J (1802) An essay on abstinence from animal food, as a moral duty, printed for Richard Phillips no 71 St Pauls, Church-Yard. https://books.googleusercontent.com/books/content?req=AKW5Qae1cmFyHzLza6Gdm8g6JRBV9S_LJU3ASSge7B-2djGbdXYtnVw1XsVkFDk82VtYHd9Zj8iKtI-LTcz87Kdt8KXVwJjeDUz9DbgMTTuaysgzFIi5NfvE475EtG7Whz_1mQ7pKnj4fco0VZnuUTAQyc_EEEfy5GkAGXYLV7nk9dSo8WGJ4uksgfUVBKKhC9eAdKbqaNH_K2PfXzlRtEri85TUPLKsjcybMJBxnbaImomNVsm9SG39SFrMYfnuIyumpeW_N680SgsLrINg1J-UJwoBJdgf-pqcWqsV4JqAfCeK5KPvHVk

Roberts JR (ed) (1905) Stone's justices manual, 37th edn. Shaw and Sons, Butterworth and Co, London

Rollin B (1989) The unheeded cry: animal consciousness, animal pain, and science. Oxford University Press, New York

Salisbury J (2010) The beast within: animals in the middle ages. Routledge, London

Salt HS (1886) A plea for vegetarianism and other essays. The Vegetarian Society, Manchester, 75 Princess Street, Manchester, John Heywood, Deansgate and Ridgefield. https://ia802504.us.archive.org/30/items/pleaforvegetaria1886salt/pleaforvegetaria1886salt.pdf

Salt HS (1894) Animals' rights: considered in relation to social progress. revised edition G Bell and Sons Ltd (1922). https://ia600901.us.archive.org/32/items/cu31924030305332/cu31924030305332.pdf

Seidler MJ (1977) Hume and the animals. South J Philos 15(3):361

Shelley PB (1975) On the vegetable system of diet (1813) reprint. Folcroft Library Editions

Stafford KJ, Mellor DJ (2011) Addressing the pain associated with disbudding and dehorning in cattle. Appl Anim Behav Sci 35:226

Steiner G (2005) Anthropocentrism and its discontents. University of Pittsburgh, Pitsburgh

Tester K (1991) Animals and society, the humanity of animal rights. Routledge, London

The Times Digital Archive, 1831, Police, untitled, report of cats skinned alive, 6, 27 June. http://tinyurl.galegroup.com/tinyurl/9oFYt4 (subscription required)

The Times Digital Archive, 1832, Police, William Mills charged with skinning cats alive, 4, 30 October. http://tinyurl.galegroup.com/tinyurl/9oFR85 (subscription required)

Tinker T (2004) The enlightenment and its discontents antinomies of Christianity, Islam and the calculative sciences. Account Audit Account 17(3):442

Tuckness A (2018) Locke's political philosophy. In: Zalta EN (ed) The Stanford encyclopedia of philosophy. https://plato.stanford.edu/archives/sum2018/entries/locke-political/

Williams CD (2003) Liberating the enlightenment: how a transformed relationship with animals can help us transcend modernity. Relig Educ 29(1):95

Wolloch N (2019) The enlightenment's animals: changing conceptions of animals in the long eighteenth century. Amsterdam University Press, Amsterdam

Legislation, Regulations, Gazettes and Hansard

1340, *The King's Grant, that the Foresaid Subsidy of the Ninth Lamb,*&c. *shall be no Example, nor Prejudicial to his Subjects: All shall be spent in his Wars*, 1340 c 1, available from (14 Edw 3 St 2), available from Justis One, https://app-justis-com.ezproxy.lib.uts.edu.au/statute/taxation-etc-act-1340-c-1/overview/cYyZm0ednYWca (subscription needed)

1488, *An Act that No Butcher Flea any Manner of Beast within the Walls of London, 1488*, 4 and 5 Henry VII C 3, Statutes of the Realm Printed by Command of his Majesty, George the Third, in pursuance of an address of the House of Commons of Great Britain Volume the Second, 1816, Reprinted 1963 Dawson of Pall Mall London, 527–528. https://babel.hathitrust.org/cgi/pt?id=pst.000017915526;view=1up;seq=553

1532, *An Act for the Destruction of Crows and Rooks 1532*, 24 Henry VIII, C 10, s 7, Statutes at Large from the fifth year of K Richard III to the 31st year of K Henry VIII, inclusive, by Danby Pickering of Gray's Inn Esquire, Vol IV, Printed by Joseph Bentham, London (1736), 254. https://play.google.com/books/reader?id=UlMDAAAAQAAJ&hl=en_GB&pg=GBS.PA254

1549, *An Act for Buying and Selling of Other Beasts and Cattle*, 1549, (3 & 4 Edw 6) C A XIX, available from Justis One https://app-justis-com.ezproxy.lib.uts.edu.au/statute/buying-cattle-act-1549/overview/cYyJm5iJnZWca (subscription needed)

1545, *Burning of Frames Act*, 1545, 37 Henry VIII C 6, Recital and s III, Statutes of the Realm, Printed by Command of His Majesty, George the Third in pursuance of an address of the House of Commons of Great Britain, Volume the Third, 1817, Reprinted 1965, Dawsons of Pall Mall, London, 994. https://babel.hathitrust.org/cgi/pt?id=pst.000017915533;view=1up;seq=1060

1625, *An Acte For Punishing Divers Abuses Committed on The Lord's Day Called Sunday, 1625*, First year of the reign of King Charles I, C1, The statutes of the Realm, Printed by command of his majesty King George the Third, in pursuance of an address of the House of Commons of Great Britain, Volume 5, 1819, 1, reprinted Dawsons of Pall Mall London (1963). https://babel.hathitrust.org/cgi/pt?id=pst.000017915564;view=1up;seq=29

1635, *Act Against Plowing by the Tayle, and pulling the Wooll off Living Sheep 1635* (Ireland), Statutes Passed in the Parliaments Held in Ireland, Vol 1 From the Third Year of Edward the Second, A.D. 1310, to the Fourteenth and Fifteenth Years of Charles the Second, A.D. 1662 inclusive, Printed by George Grierson, Dublin (1794), 301. https://books.google.com.au/books?id=VYRRAAAAYAAJ&pg=PA301&redir_esc=y#v=onepage&q&f=false

1641, *The Massachusetts Body of Liberties* 1641. https://history.hanover.edu/texts/masslib.html

1657, *An Act for the Better Observation of the Lords Day, 1657, Acts and Ordinances of the Interregnum, 1642–1660*, C H Firth and R S Rait (eds), London (1911), 1162, British History Online. https://www.british-history.ac.uk/no-series/acts-ordinances-interregnum/pp1162-1170

1800 Bill to Prevent the Practice of Bull Baiting, 2 April, 1800, Parliamentary history of England from the earliest period to the year 1803, Volume 35, *T C Hansard* (1819) column 202. https://babel.hathitrust.org/cgi/pt?id=mdp.39015087751502&view=1up&seq=125

1809, Cruelty to Animals Bill -Cruelty to Animals Bill 1809, House of Lords, debate on Cruelty to Animals' Bill, *Historic Hansard*, HL Deb 15 May 1809, Series 1, Volume 14, cc553–71. https://api.parliament.uk/historic-hansard/lords/1809/may/15/cruelty-to-animals-bill

1809, Cruelty to Animals Bill-Lord Erskine, House of Lords, debate on Cruelty to Animals' Bill, *Historic Hansard*, HL Deb 31 May 1809, Series 1, Volume 14, cc804–8. https://api.parliament.uk/historic-hansard/lords/1809/may/31/cruelty-to-animals-bill

1809, Cruelty to Animals Bill-House of Commons Debate, Cruelty to Animals' Bill, *Historic Hansard*, HC Deb 15 June 1809, Series 1, Volume 14, c1070. https://api.parliament.uk/historic-hansard/commons/1809/jun/15/cruelty-to-animals-bill

1821, Ill-Treatment of Horses Bill, House of Commons, *Historic Hansard*, HC Deb 1 June 1821, Volume 5, cc1098–9. https://api.parliament.uk/historic-hansard/commons/1821/jun/01/ill-treatment-of-horses-bill

1822, *An Act to Prevent the Cruel and Improper Treatment of Cattle 1822*, (3 Geo IV c 71), *The Statutes of the United Kingdom of Great Britain and Ireland 3 George IV. 1822*, his Majesty's statute and law printers London, sold by Butterworths and son, 403. https://archive.org/details/statutesunitedk10britgoog/page/n436

1823, A Bill to Prohibit Bull-baiting and Dog-fights, *Historic Hansard*, HC Deb 21 May 1823, Volume 9, cc432–5, 425. https://api.parliament.uk/historic-hansard/commons/1823/may/21/bull-baiting-and-dog-fights

1825, Cruelty to Animals Bill, *Historic Hansard*, HC Deb 11 March 1825, Volume 12, cc1002–13. https://api.parliament.uk/historic-hansard/commons/1825/mar/11/cruelty-to-animals-bill

1825, Ill-Treatment of Animals Bill, House of Commons Debate, *Historic Hansard*, HC Deb 24 March 1825, Volume 12, cc1160–2. https://api.parliament.uk/historic-hansard/commons/1825/mar/24/ill-treatment-of-animals-bill

1826, Bear-Baiting and Dog-Fighting Bill, *Historic Hansard*, HC Deb 21 February 1826, Volume 14, cc647–52. https://api.parliament.uk/historic-hansard/commons/1826/feb/21/bear-baiting-and-dog-fighting-bill

1826, Cattle Ill-Treatment Bill House of Commons, *Historic Hansard* HC Deb 21 February 1826 Volume 14, cc653–7. https://api.parliament.uk/historic-hansard/commons/1826/feb/21/cattle-ill-treatment-bill

1833, *An Act for the Abolition of Slavery throughout the British Colonies*; 1833 for promoting the Industry of the manumitted Slaves; and for compensating the Persons hitherto entitled to the Services of such Slaves, 3 and 4 Will. IV C 73, available from http://www.pdavis.nl/Legis_07.htm

1835, *An Act to Consolidate and Amend the Several Laws Relating to the Cruel and Improper Treatment of Animals and the Mischiefs Arising from the Driving of Cattle, and to Make Other Provisions in Regard Thereto,* 1835, 5 & 6 William 4 c59, Cruelty to Animals Act, available from The Statutes Project, http://statutes.org.uk/site/the-statutes/nineteenth-century/1835-5-6-william-4-c-59-cruelty-to-animals-act/

1849, *An Act for the more Effectual Prevention of Cruelty to Animals*, 1849, 12 &13 Vic c 92, *Public General Statutes*, Twelfth and Thirteenth Year, Queen Victoria, 1849, Printed by George Edward Eyre and William Spottiswoode and published by W Benning and Co, London (1849), 592. https://play.google.com/books/reader?id=RKJKAAAAMAAJ&hl=en_GB&pg=GBS.PA592

1850, *An Act for the More Effectual Prevention of Cruelty to Animals in Scotland 1850* (UK) 13 & 14 Vic c 59, *Public General Statutes*, Thirteenth and Fourteenth Year, Queen Victoria, 1850, Printed by George Edward Eyre and William Spottiswoode and published by W Benning and Co, London (1850), 801. https://play.google.com/books/reader?id=Z6pKAAAAMAAJ&hl=en_GB&pg=GBS.PA801

1876, *Cruelty to Animals Act 1876* (UK), 39 & 40 Vic c 77, available from, The Statutes Project, http://statutes.org.uk/site/the-statutes/nineteenth-century/1876-39-40-victoria-c77-cruelty-to-animals-act/

1911, *Protection of Animals Act 1911* (UK), 1 & 2 Geo 5 c 271. http://www.bailii.org/uk/legis/num_act/1911/1069356.html

Cases

Brady v McArgle 14 LR Ir 174, 183

Budge v Parsons 122 ER 144

Callaghan v Mc'Evoy 16 LR Ir 325

Ford v Wiley 23 QBD 203

Hall v RSPCA [1993] QBD 11

Hart v Police SC Auckland [1965] New Zealand Law Reports, 666

Joseph Milles v Mat Davies, Evan Watts and Selby Price, alias Rees [1792] EngR 1293. http://www.commonlii.org/int/cases/EngR/1792/1293.pdf

Murphy v Manning 2 Ex Div 307

Renton v Wilson 15 RJ 84

Somerset v Stewart (1772) 98 ER 499, 510, available from http://www.commonlii.org/int/cases/EngR/1772/57.pdf

Spencer's Case, [1583] EWHC KB J53 (01 January 1583). http://www.bailii.org/ew/cases/EWHC/KB/1583/J53.html

The Queen v McDonagh 28 LR Ir 204
Todrick v Wilson 18 RJ 41
Tucker v Hazelhurst SC Palmerston North [1906] NZGazLawRp 200
Waters v Braithwaite, (1914) 78 JP 124

International Treaties

European Convention for the Protection of Pet Animals 1987, adopted on 13 November 1987, *ETS* No 125 (entered into force 1 May 1992) and had 24 ratifications as of July 2019. https://rm.coe.int/168007a67d

4 Animal Disease as a Trade Issue: Cattle Plagues and the Veterinary Profession

Abstract

Animal disease, particularly the cattle plagues of the 1860s, had significant consequences for farm animal commodification. During the latter part of the nineteenth century, the veterinary profession came under increasing scrutiny, especially for its early failures to control cattle plague, which led to calls for improved training and stricter licensing conditions. By the time the plagues were brought under control, veterinarians had positioned themselves as gatekeepers, acquiring formal roles as government inspectors and certifiers of animal health. This represented a rise in the status of the profession, which also saw the initiation of professional conferences and gatherings, designed to share knowledge and ideas. The first veterinary conference, held in Hamburg, Germany, in 1863, was hugely influential, vindicating the contagion theory of animal disease and becoming the forerunner of the World Veterinary Association. At the same time, these developments occurred against the backdrop of growing trade and commerce in the sector, leading to animal disease being regarded as a trade issue. Consequently, animal health became subsumed into the practicalities of market transactions and the merchantability of animals as goods, providing a further steppingstone towards their commodification.

Keywords

Animal commodification · Animal disease · Cattle plagues · John Gamgee · Veterinary profession

4.1 Introduction

This chapter discusses the regulatory consequences of animal disease, predominantly during the latter part of the nineteenth century, concentrating on the United Kingdom (UK). The discussion evaluates the transformation of law and policy,

S. Riley, *The Commodification of Farm Animals*, Animal Welfare 21,
https://doi.org/10.1007/978-3-030-85870-4_4

which resulted from government softening its laissez-faire stance in response to the cattle plagues of the 1860s. High mortality rates led regulators to take more active roles in disease management, shifting administration and oversight from the private to the public sphere.[1] It was a transformation that initially faltered, partly because of uncertainties regarding the cause of disease and partly because of the unpopularity of strict measures. However, once authorities were forced to accept that contagion was responsible for illness, there was little room for argument.[2]

In the course of this transformation, the veterinary profession came under increasing scrutiny, particularly for its early failures. Calls for improved training at veterinary colleges were later accompanied by stricter licensing conditions.[3] Although these reforms highlighted deficiencies within the profession, they also tacitly acknowledged that veterinarians, with their science-based training, were best positioned to act as gatekeepers against the introduction and spread of animal diseases. This standpoint was reinforced by regulation which gave veterinarians formal roles as government inspectors and certifiers of animal health.

At the time of the cattle plagues, veterinarians initiated professional gatherings, where they came together to share knowledge and ideas. The first of these was arranged by John Gamgee (1830–1894), an English veterinarian, and was held in July 1863 in Hamburg.[4] The conference was hugely influential, vindicating the contagion theory of animal disease and then becoming the forerunner of the World Veterinary Association.[5] Thereafter, veterinary and industry conferences were held regularly, providing stakeholders with expertise and cutting-edge knowledge to inform their undertakings. The importance of these developments lay in the fact that veterinary medicine, as well as official disease regulation, evolved against the backdrop of growing trade and commerce in animals. It is argued that this led to animal disease being regarded as a trade issue, subsuming animal health into the practicalities of market transactions. In consequence, the regulation of animal disease formed an important steppingstone in the commodification of farm animals.

---

[1] Paul Brassley, "Animal Health and Veterinary Medicine", in Joan Thirsk (general editor) *The Agrarian History of England and Wales* Volume VII 1850–1914, Part 1, 587, 588–589, 591–592, Cambridge University Press, Cambridge (2000).

[2] Michael Worboys, "Germ Theories of Disease and British Veterinary Medicine, 1860–1890", (1991) 35 *Medical History*, 308, 318.

[3] Veterinary Surgeons Act 1881, 44 & 45 Victoria c 62, Preamble, The Public General Statutes Passed in the Forty Fourth and Forty Fifth Years of the Reign of Her Majesty Queen Victoria, Printed by G.E. Eyre and W. Spottiswoode, printers to the Queen and W. Clowes and Sons, printers to the Council of Law Reporting, London, (1881), 371, available from https://archive.org/details/publicgeneralst05walegoog/page/n385;discussion, Paul Brassley, "Animal Health and Veterinary Medicine", above 1, 591.

[4] *The Cattle Plague, Official Reports, of the International Veterinary Congress, held in Hamburg, 1863, and in Vienna 1865*, by John Gamgee, Robert Hardwicke, London (1866).

[5] World Veterinary Association, "History", available from http://www.worldvet.org/about.php

4.2 Animal Disease and the Early Veterinary Profession

Coping with animal diseases has long been an important feature of the human-animal relationship.[6] An early reference from 3000 BC praises Urlugaledinna of Mesopotamia for his success in treating sick animals.[7] The ancient Greeks, Romans and Egyptians were also accomplished animal healers, and the term veterinarian is thought to have derived from the Roman word "veterinarius", describing an animal physician.[8] The expertise of the ancients lapsed after the fall of the Roman Empire, although their knowledge was partially preserved by Persian, Arabian and Islamic scribes who recognised the importance of treating valuable species, such as horses.[9]

Up to the late eighteenth century and the emergence of veterinary professionals, animal care was largely the purview of animals' owners, or species-specific attendants, such as cow leeches and shepherds.[10] These personnel were drawn from the farm animal sector and gained practical experience on the job.[11] Nevertheless, the causes of animal disease were poorly understood, and in Western Europe, diseases were blamed on witchcraft and "noxious exhalates from the earth".[12] In some cases, these fallacies persisted into the nineteenth century, such as occurred in Hungary in 1894, where a woman was crucified for witchcraft following the death of several cattle.[13]

Infectious diseases of farm animals, particularly cattle and horses, had harsh consequences for agricultural communities, galvanising searches for cures. However, most efforts were unsuccessful, reflecting the limited knowledge of the times.[14] Animals were subjected to bleeding, administration of purgatives, applying tar to their noses or burning both sides of their necks with a hot iron and placing roots of the hellebore plant in the blisters.[15] Plagues were attributed to divine punishment, resulting in calls for drastic and disconcertingly cruel measures, such as burying a calf alive or driving cattle through fire in the hope it would stop the spread of

[6] George Fleming, *Animal Plagues: Their History, Nature, and Prevention*, Chapman and Hall (1871) (reprint), (xv–xvi).

[7] James Yeates, *Veterinary Science: A Very Short Introduction*, OUP, Oxford (2018), 1.

[8] Frederick Smith, "The Work of the British Army Veterinary Corps at the Fronts", (1919) 75 (1) *The Veterinary Journal*, 8, 9–10.

[9] James Yeates, *Veterinary Science: A Very Short Introduction*, above, 7, 1–6.

[10] R M Hartwell, *the Industrial Revolution and Economic Growth*, Methuen and Co Ltd., London (1971), 218.

[11] Ibid, 218.

[12] George Fleming, *Animal Plagues: Their History, Nature, and Prevention, from BC 1490 to AD 1800,* above, 6, 106.

[13] C A Spinage, *Cattle Plague, A History*, Kluwer Academic, New York (2003) 341.

[14] Joanna Swabe, *Animals, Disease and Human Society, Human-animal Relations and the Rise of Veterinary Medicine*, Routledge, London (1999), 81.

[15] C A Spinage, *Cattle Plague, A History*, above, 13, 342, 348, 355.

disease.[16] In some instances, authorities conferred with surgeons and doctors, as occurred in 1711 when Pope Clement XI called in his personal doctor, Giovanni Lancisi (1654–1720), to report on the best way of containing cattle plague.[17] Lancisi avoided attempts to effect a cure, instead focussing on slaughtering animals who were diseased or had been exposed to disease.[18]

The Enlightenment, which persistently challenged superstitions, led to calls for more systematic and scientific approaches to treating animals, including the training and education of veterinary surgeons.[19] The first veterinary school in Western Europe opened in 1762, in the French town of Lyon.[20] Formal training in England commenced in 1792 with the opening of a school in London, followed some 30 years later, in 1832, by the second institution in Edinburgh.[21] The London school received a royal charter in 1844, which established it as the Royal College of Veterinary Surgeons, a status it enjoys to the present day.[22] However, the 1844 Charter was not underpinned by legislation, so that unqualified people could still call themselves veterinarians. This changed with the enactment of the *Veterinary Surgeons Act* 1881, which distinguished between qualified and unqualified practitioners, prohibiting the unqualified from asserting they were veterinarians.[23]

The initial focus of animal care, both before and after the establishment of veterinary colleges, centred on horses.[24] In England, they were the most valuable domestic animal, providing draught power and transportation, not only in farm production but also in times of war.[25] A range of specialised occupations derived from the care of horses, including farriers, "blacksmiths, grooms and horse

---

[16] Peter A Koolmees, "Epizootic Diseases in the Netherlands, 1713–2002", in, Karen Brown and Daniel Gilfoyle (eds)¸ *Healing the Herds, Disease, Livestock Economics, and the Globalization of Veterinary Medicine*, 19, 25, Ohio University Press, Athens, Ohio (2010).

[17] George Fleming, *Animal Plagues: Their History, Nature, and Prevention, from BC 1490 to AD 1800,* above, 6, 200–201.

[18] Joanna Swabe, *Animals, Disease and Human Society, Human-animal Relations and the Rise of Veterinary Medicine*, above, n 14, 84.

[19] Ibid, 88.

[20] Some commentators place this date a little earlier, in 1761 or even 1760, Vanessa Carli Bones and James W Yeates, "The Emergence of Veterinary Oaths: Social, Historical and Ethical Considerations", (2012) 2 (1) *Journal of Animal Ethics*, 20, 22.

[21] Ibid, 22, 23.

[22] Ibid; The Royal Charter of the Royal College of Veterinary Surgeons 1844, available from the Royal College of Veterinary Surgeons, https://www.rcvs.org.uk/document-library/royal-charter-1844/

[23] Veterinary Surgeons Act 1881, 44 & 45 Victoria c 62, Preamble; discussion, Paul Brassley, "Animal Health and Veterinary Medicine", above 1, 591.

[24] Lise Wilkinson, *Animals and Disease, An introduction to the History of Comparative Medicine*, Cambridge University Press, Cambridge (1992), 70, 101; George Fleming, *Animal Plagues: Their History, Nature, and Prevention, from BC 1490 to AD 1800,* above, 6, 76.

[25] Lise Wilkinson, *Animals and Disease, An introduction to the History of Comparative Medicine*, above 24, 69, 70–71.

doctors".[26] Farriers were accomplished in medical matters, gaining proficiency in castrating horses, docking their tails and treating them with poultices and ointments, which later came to be used on sheep, cattle and pigs.[27] In addition, knowledge and literature on horse care was augmented by the expertise of army surgeons who frequently treated horses.[28]

At the same time, the status of the veterinary profession, apart perhaps from army veterinarians, was not as high as for other professions.[29] Outside of the military, veterinary surgeons were an unknown quantity, struggling to earn the trust and respect of farmers and regulators alike.[30] The situation was not helped by disagreement among scientists and medical personnel, including doctors and veterinarians as to the causes of disease.

4.3 Animal Diseases and Contagion Theory

Since ancient times, an abundance of theories has attempted to explain the origin of disease. The existence of cattle plague, for example, was noted in antiquity, with ancient Greeks attributing outbreaks to germs already existing in the air which grew upon finding a host.[31] Beyond antiquity, human and animal diseases were explained by miasmas (poisonous particles in the air that caused disease), "witchcraft, demons, gods, comets, earthquakes", as well as spontaneous generation, whereby living organisms could develop from non-living material.[32] The Enlightenment largely discounted supernatural causes, so that by the early eighteenth century the range of possibilities had narrowed to chemical agents, "animalcules" (microscopic organisms),[33] spontaneous generation as well as "infectious effluvia" which could spread from host to host.[34]

In the case of livestock, the likelihood of disease being transmitted from animal to animal had been debated from the thirteenth century, and by the 1600s, the theory of

[26] Vanessa Carli Bones and James W Yeates, "The Emergence of Veterinary Oaths: Social, Historical and Ethical Considerations", above, 20, 21.

[27] Ibid.

[28] Lise Wilkinson, *Animals and Disease, An introduction to the History of Comparative Medicine*, above 24, 52.

[29] J R Fisher, "Not quite a Profession: the Aspirations of Veterinary Surgeons in England in the Mid Nineteenth Century", (1993) *Institute of Historical Research*, 284, 287.

[30] Paul Brassley, "Animal Health and Veterinary Medicine", above 1, 588–589, 591–592.

[31] C A Spinage, *Cattle Plague, A History*, above, 13, 59.

[32] Marianna Karamanou, George Panayiotakopoulos, Gregory Tsoucalas, Antonis A Kousoulis and George Androutsos, "From Miasmas to Germs: A Historical Approach to Theories of Infectious Disease Transmission", (2012) 1 *Le Infenzioni in Medicina*, 52, 59, 60.

[33] Melvin Santer, "Richard Bradley: A Unified, Living Agent Theory of the Cause of Infectious Diseases of Plants, Animals and Humans in the First Decades of the 18th Century", (2009) 52 (40 *Perspectives in Biology and Medicine*, 566, 567, 658, 575.

[34] Ibid, 573.

contagium vivum (contagion) had gained a strong foothold.[35] Proponents argued that living organisms or some other "common cause" were responsible for disease, even though they could neither identify the cause nor explain how it was transmitted.[36] This led to confusion and error, illustrated by Lancisi's conclusions during the 1711 cattle plague, that although disease spread from animal to animal, this could not be attributed to contagion caused by living organisms or germs.[37]

Clearer explanations had to wait until the early 1860s when Louis Pasteur (1822–1895) demonstrated the links between germs and disease, discoveries that Robert Koch (1843–1910) expanded in detail a decade later when he identified how specific bacteria caused disease.[38] These discoveries all but settled what by the nineteenth century had become a battleground between rival theories of spontaneous generation and contagion. In the context of animal disease, this conflict mirrored the friction between James Beart Simonds (1810–1904) and John Gamgee, two leading veterinarians of the day.

Simonds had graduated from the College of Veterinary Surgeons in 1829[39] and in 1842 became the first lecturer of cattle diseases at the same college.[40] He was active in promoting veterinary practice, successfully lobbying for the College to be granted its royal charter in 1844; he also acted as editor for the journal, *The Veterinarian*.[41] Simonds, however, was an anti-contagionist, who came to this position after analysing historical data on bovine pleuropneumonia and foot-and-mouth (FDM) disease in England, concluding that disease was attributable to spontaneous

[35] Vivian Nutton, "The Seeds of Disease: an Explanation of Contagion and Infection from the Greeks to the Renaissance." (1983) 27 (1) *Medical History*, 1, 5–6; Melvin Santer, "Richard Bradley: A Unified, Living Agent Theory of the Cause of Infectious Diseases of Plants, Animals and Humans in the First Decades of the 18th Century", above, 33, 573.

[36] Vivian Nutton, "The Seeds of Disease: an Explanation of Contagion and Infection from the Greeks to the Renaissance", above, 35, 5–6; Melvin Santer, "Richard Bradley: A Unified, Living Agent Theory of the Cause of Infectious Diseases of Plants, Animals and Humans in the First Decades of the 18th Century", above, 33, 573.

[37] C A Spinage, *Cattle Plague, A History*, above, 13, 60.

[38] Marianna Karamanou, George Panayiotakopoulos, Gregory Tsoucalas, Antonis A Kousoulis and George Androutsos, "From Miasmas to Germs: A Historical Approach to Theories of Infectious Disease Transmission" above, 32, 52, 59, 60.

[39] Iain Patterson, *A Great British Veterinarian Forgotten: James Beart Simonds 1810–1904*, J A Allen & Co, London (1990), 5.

[40] Ibid, 21, Michael Worboys places the date at 1843, Michael Worboys "Germ Theories of Disease and British Veterinary Medicine, 1860–1890", (1991) 35 *Medical History*, 308, 310; The date 1842 is preferred for the appointment of Simonds, although the subject started being taught in 1843 and was the target of criticism, J F Bickford, "On the Exclusiveness of the Publication of the Veterinary Transaction—The False Ground on which it was Founded—Valuable Introduction of Cattle Pathology—the Strange Ignorance of Many Country Practitioners—The Present State of Cattle Pathology- On Paraplegia in Cattle, Mr. T Turner's Oration—And the Strange Duplicity Practised by Many uncertified Men", (1843) XVI (181) *The Veterinarian*, 13, 15.

[41] Iain Patterson, *A Great British Veterinarian Forgotten: James Beart Simonds 1810–1904*, above, 39.

generation.[42] As such, he also considered the threat of disease being introduced into the UK by the importation of animals to be largely overstated.[43] These views brought him into conflict with Gamgee, a contagionist who drew on European practices to advocate for stricter veterinary measures.[44]

Gamgee, who was some 20 years younger than Simonds, had completed his studies at the Royal Veterinary College London in 1852.[45] While still a student, he was appalled by unsanitary conditions in the London dairies as well as the fact that meat from sick cattle was used to make sausages or dressed in ways that obscured signs of disease.[46] After completing his studies, Gamgee travelled throughout Europe, observing that continental veterinary methods were more progressive than those of the UK.[47] On his return to the UK, he was determined to use his newfound knowledge and opened the New Veterinary College in 1857 in Edinburgh.[48] Although the college closed in 1868 and Gamgee travelled to the United States of America (USA), in the interim he continued to campaign for European-type measures.[49]

European regulation, however, proved to be somewhat of a double-edged sword because it also fuelled debate on the extent to which European animals posed a risk to agriculture in the UK. In 1859, following an inspection of the European cattle industry, Simonds concluded that cattle plague was limited to the steppes of Russia.[50] This was significant because the UK did not source cattle from this region, preferring to buy from Central and Western Europe where cattle plague had not been recorded for almost 42 years.[51] For these reasons Simonds concluded that cattle plague was unlikely to enter the UK.[52] This squarely contradicted Gamgee's standpoint, which argued that after the progressive liberalisation of trade in the middle nineteenth century, most animal diseases could be ascribed to diseased livestock

---

[42] Michael Worboys, "Germ Theories of Disease and British Veterinary Medicine, 1860–1890", above 1, 310.

[43] Ibid.

[44] Ibid.

[45] Richard Perren, "Filth and Profit, Disease and Health: Public and Private Impediments to Slaughterhouse Reform in Victorian Britain", in *Meat, Modernity and the Rise of the Slaughterhouse* (Paula Young Lee ed), 127, 133, University of New Hampshire Press (2008).

[46] Ibid, 133.

[47] Ibid.

[48] John Francis, "John Gamgee (1831–1894): Our Greatest Veterinarian", (1962) 118, *The British Veterinary Journal*, 430. This college should not be confused with the Royal (Dick) School of Veterinary Studies which still forms part of the University of Edinburgh.

[49] Ibid, 431.

[50] Lise Wilkinson, *Animals and Disease, An introduction to the History of Comparative Medicine*, above 24, 104.

[51] Ibid.

[52] Ibid.

imported from Europe.[53] That fact, coupled with the growth of railways, meant that it was possible to transport sick animals across great distances, not only spreading disease but also allowing sick animals to be slaughtered and sold to an unsuspecting public.[54]

Gamgee estimated that after the middle of the nineteenth century, approximately one-quarter of the national herd was disease-ridden.[55] He, therefore, argued for tighter controls, restricting the movement of animals, mandating veterinary inspections at slaughter and calling on the government to shut down private slaughterhouses because they lacked even minimum standards of hygiene.[56] Although his proposals were cogent, they were also problematic, because if diseased animals stopped being traded, farmers would suffer economically, and consumers would face food shortages.[57] Consequently, Gamgee's views were met with strong opposition and ridicule, targeting not only Gamgee personally but also contagion theory itself, which formed the basis of his proposals.[58] A pamphlet published in 1866 by Henry Strickland Constable (1821–1909) critiqued contagion, Gamgee and two international veterinary congresses Gamgee had organised:

> [the congresses] held at Hamburg and Vienna. . .[are] exceedingly entertaining. Such a charming farrago of antiquated ideas about contagion, infection, Government interference, official inspections, police regulations, veterinary supervision, restrictions of traffic, quarantine, inoculation, &c. &c., I could not have believed possible to find extant in any civilized country in this the nineteenth century.[59]

The antagonism to strict measures was also evident from the fact that although animal disease continued to appear in the UK, authorities were reluctant to use their powers pursuant to the *Contagious Disorders (Sheep) Act* 1848 (Sheep Pox Act) and order the destruction of diseased animals imported into the UK.[60] Gamgee's

53 Richard Perren, "Filth and Profit, Disease and Health: Public and Private Impediments to Slaughterhouse Reform in Victorian Britain", above, 45, 135.

54 Abigail Woods, "A Historical Synopsis of Farm Animal Disease and Public Policy in Twentieth Century Britain", (2011) 366 *Philosophical Transactions of the Royal Society B*, 1943, 1944; Richard Perren, "Filth and Profit, Disease and Health: Public and Private Impediments to Slaughterhouse Reform in Victorian Britain", above, 45, 135.

55 Richard Perren, *The Meat Trade in Britain 1840–1914*, Routledge & Kegan Paul (1978), 53.

56 Abigail Woods, "The Construction of and Animal Plague, Foot and Mouth Disease in Nineteenth-Century Britain", (2004) 17 (1) *Social History of Medicine*, 23, 29.

57 Richard Perren, "Filth and Profit, Disease and Health: Public and Private Impediments to Slaughterhouse Reform in Victorian Britain", above, 45, 135.

58 Joanna Swabe, *Animals, Disease and Human Society, Human-animal Relations and the Rise of Veterinary Medicine*, above, n 14, 99.

59 Henry Strickland Constable, *Observations Suggested by the Cattle Plague: About Witchcraft, Credulity. . .*, Dalton and Lucy, Charing Cross (1866), 11.

60 Abigail Woods, "A Historical Synopsis of Farm Animal Disease and Public Policy in Twentieth Century Britain", above 54, 1944.

frustration with this regulatory lethargy reached a crisis point in the summer of 1865, when infected cattle entered Britain at Hull, from the Baltic port of Reval (now Tallinn).[61]

4.4 Cattle Plagues of the 1860s

The term "cattle plague", or "distemper of horned cattle", was used in the eighteenth and nineteenth centuries to describe the disease known today as rinderpest.[62] Until its worldwide eradication in 2011, cattle plague was an extremely contagious and commonly fatal disease of ruminants, including cattle, deer, goats and sheep.[63] It was, however, not the only animal disease to reach the UK during the nineteenth century. In 1839, FDM was discovered at Smithfield markets in London and identified by Simonds, in Surrey, the same year.[64] In addition, bovine pleuropneumonia and sheep pox had devastated UK herds prior to 1850, resulting in high levels of infection and/or animal mortality.[65]

Notwithstanding this range of animal diseases, the catalyst for change stemmed from the heavy livestock losses of the 1860s.

4.4.1 Cattle Plague and Government Responses

Cattle plague had been present in England from at least 800 AD and was pervasive in Europe throughout the eighteenth century, largely concentrated in three epidemics, 1709–1720, 1742–1760 and 1768–1786.[66] Although the outbreaks varied in

[61] Richard Perren, "Filth and Profit, Disease and Health: Public and Private Impediments to Slaughterhouse Reform in Victorian Britain", above, 45, 141.

[62] John Broad, "Cattle Plague in Eighteenth-Century England", (1983) 31 (2) *The Agricultural History Review*, 104, 104.

[63] C A Spinage, *Cattle Plague, A History*, above, 13, 3; "Rinderpest has now been eradicated", unanimous resolution, OIE, "Declaration of Global Eradication of Rinderpest and Implementation of Follow-up Measures to Maintain World Freedom from Rinderpest", 79th General Session, Resolution 18, May 2011, Paris, 151, available from http://www.oie.int/fileadmin/Home/eng/Animal_Health_in_the_World/docs/pdf/Disease_cards/SHEEP_GOAT_POX.pdf

[64] J Penberthy, "Foot-and-Mouth-Disease", (1901) 14 *Journal of Comparative Pathology and Therapeutics* 16, 16–18.

[65] J R Fisher, "Not quite a Profession: the Aspirations of Veterinary Surgeons in England in the Mid Nineteenth Century", above 29, 293.

[66] C A Spinage, *Cattle Plague, A History*, above, 13, 3, 59 George Fleming, *Animal Plagues: Their History, Nature, and Prevention, from BC 1490 to AD 1800,* above, 6, 264, 267; John Broad, "Cattle Plague in Eighteenth-Century England" above, 62, 104. Others place the outbreaks between 1711–1717, 1745–1757 and 1769–1786; Karen Brown and Daniel Gilfoyle¸ *Healing the Herds*, Ohio University Press, Athens, Ohio (2010), 76.

intensity, they resulted in an overall mortality rate of between 70 and 90%.[67] In the UK cattle plague appeared in 1714 and then again in 1745, with the latter outbreak not abating until 1759.[68] The disease appeared once more in 1781 and then spared the UK for more than half of the nineteenth century, until it was introduced by way of an infected shipment from the continent.[69]

After landing at Hull, importers conveyed their animals inland, including to markets at Wakefield-Manchester and Smithfield in London.[70] This allowed the disease to spread quickly. Simonds, who was the Chief Veterinary Inspector at the time, became aware of early cases of cattle plague in June 1865, but in miscalculating the danger, he declined to implement measures.[71] The effect was dramatic, killing an estimated 3 million animals prior to the introduction of stringent measures in February 1866 and resulting in the loss of approximately "7% of the national herd".[72]

The severity of the summer epidemic of 1865 led to the establishment, on 29 September 1865, of the first of three Royal Commissions, appointed to inquire into the origin and nature of the cattle plague.[73] The initial Commission handed down its report on 31 October 1865 noting that in Europe cattle plague was known as rinderpest, a very contagious disease, strictly controlled by the immediate slaughter of infected animals.[74] However, the report also observed that wholesale slaughter involved great economic sacrifices, which the UK was not prepared to accept.[75] Instead, the Commission noted that pursuant to the 1848 Sheep Pox Act, the Privy Council had made a succession of Orders in Council, which avoided imposing wholesale slaughter by appointing inspectors to stop the movement of infected

---

[67] George Fleming, *Animal Plagues: Their History, Nature, and Prevention, from BC 1490 to AD 1800,* above, 6, 264, 267; John Broad, "Cattle Plague in Eighteenth-Century England" above, 62, 104.

[68] C A Spinage, *Cattle Plague, A History*, above, 13, 111, 121, 127–131.

[69] Ibid, 146, 162; C Knick, "Steers Afloat: The North Atlantic Meat Trade, Liner Predominance, and Freight Rates, 1870–1913", (2008) 68 (4) *Economic History Association*, 1028, 1034.

[70] Richard Perren, "Filth and Profit, Disease and Health: Public and Private Impediments to Slaughterhouse Reform in Victorian Britain", above, 45, 142.

[71] Lise Wilkinson, *Animals and Disease, An introduction to the History of Comparative Medicine*, above 24, 106.

[72] Richard Perren, "Filth and Profit, Disease and Health: Public and Private Impediments to Slaughterhouse Reform in Victorian Britain", above, 45, 142; C A Spinage, *Cattle Plague, A History*, above, 13, 3.

[73] Commissioners Appointed to Inquire Into the Origin and Nature, &c of the Cattle Plague, First-Third Reports of the Commissioners Appointed to Inquire Into the Origin and Nature, &c. of the Cattle Plague: with the minutes of evidence and an appendix, printed by George Edward Eyre and William Spottiswoode, Great Britain, H M Stationery Office, London (1866), available from https://babel.hathitrust.org/cgi/pt?id=coo.31924110082843&view=1up&seq=11

[74] Ibid, First Report of the Commissioners Appointed to Inquire Into the Origin and Nature, &c. of the Cattle Plague, vii–viii.

[75] Ibid.

animals.[76] These orders were consolidated on 25 September 1865.[77] Overall, the Commission concurred with the approach of the Privy Council, although the Commission also proposed forbidding the movement of all cattle and stipulating that foreign cattle be slaughtered upon landing.[78] The Commission's recommendations, however, were not followed, and the Orders in Council failed to halt the progress of the plague.[79]

A further wave of disease was recorded from October 1865, leading to the second Royal Commission, which handed down its report on 5 February 1866. The report indicated that between 7 October 1865 and 7 January 1866, notifiable cases of plague had doubled every 4 weeks although actual numbers were likely to be higher than those officially recorded.[80] This Commission again recommended stopping all movement of cattle, appointing inspectors who had veterinary training, giving inspectors powers to seize and destroy animals, as well as requiring imported cattle to be slaughtered at specified ports, unless they came from non-infected areas.[81] Once again, because of the potential damage to trade, these recommendations garnered much opposition, particularly the powers to destroy animals and restrict their transportation.[82]

However, continued livestock losses were so heavy that in February 1866 industry organisations agreed that the slaughter of infected animals was necessary.[83] By the end of that month, the government had passed legislation to control the transportation of cattle, giving special emphasis to slaughtering infected animals, requiring foreign animals to be killed on arrival and providing that imported animals could only be landed at designated foreign animal wharves.[84] Relevant regulation included: the *Cattle Diseases Prevention Act* of 20 February 1866, which did not apply to Ireland, but otherwise gave local authorities the power to slaughter diseased animals and pay compensation to owners;[85] the *Cattle Diseases Prevention Act* of

---

[76] First Report of the Commissioners Appointed to Inquire Into the Origin and Nature, &c. of the Cattle Plague, above, 73, xvi–xviii.

[77] Consolidated order of 25 September 1865, published in London Gazette Tuesday 26 September, 1865 Issue 23017 page 4585, available from https://www.thegazette.co.uk/London/issue/23017/page/4585

[78] First Report of the Commissioners Appointed to Inquire Into the Origin and Nature, &c. of the Cattle Plague, above, 73, xvi–xviii.

[79] C A Spinage, *Cattle Plague, A History*, above, 13, 288.

[80] Second Report of the Commissioners Appointed to Inquire Into the Origin and Nature, &c. of the Cattle Plague with the Minutes of Evidence and an Appendix, above, 73, iii.

[81] C A Spinage, *Cattle Plague, A History*, above, 13, 289.

[82] Ibid, 289–290.

[83] Paul Brassley, "Animal Health and Veterinary Medicine", above 1, 590.

[84] Richard Perren, "Filth and Profit, Disease and Health: Public and Private Impediments to Slaughterhouse Reform in Victorian Britain", above, 45, 142.

[85] *An Act to Amend the Law Relating to Contagious or Infectious Diseases in Cattle and Other Animals* (The Cattle Diseases Prevention Act 1866) 29 & 30 Victoria c 2, dated 20 February 1866, ss 2 and 12, A Collection of the Public General Statutes Passed in the Twenty Ninth and Thirtieth

6 March 1866 which applied to Ireland and gave the governor of Ireland the power to make necessary orders;[86] and, the *Cattle Diseases Prevention Act* of 23 April 1866, which confirmed the powers of the Privy Council to make whatever orders it deemed necessary.[87] As a result of these initiatives, cattle plague was finally brought under control in 1867.

The third and final report was handed down on 1 May 1866.[88] It concluded that the reduction of cattle plague up to that time was due to the implementation of stringent measures, as proposed in the second report.[89] The third Commission additionally observed that animal disease "is spread by multiplication in the bodies of sick animals", thus rejecting spontaneous generation as a viable theory of disease.[90] By focussing on contagion, this report had vindicated Gamgee's approach and brought British science in line with disease control on the continent.[91] The success of strict measures transformed government policy and stakeholder perceptions, demonstrating the need for government intervention as well as equipping veterinarians with formal roles.[92]

4.4.2 Effects of Cattle Plague: Law and Policy

The acceptance of contagion theory and the recognition that cattle plague had been introduced by trade also saw animal disease categorised as something that was "foreign [and] imported", requiring government involvement to prevent it from

Years of the Reign of Her Majesty Queen Victoria 1866, printed by George Edward Eyre and William Spottiswoode, London (1866), 2, https://play.google.com/books/reader?id=jJp0lRyIP4oC&printsec=frontcover&pg=GBS.PA2

[86] *An Act to amend the Law relating to Contagious Diseases amongst Cattle and other Animals in Ireland,* 1866, dated 6 March 1866, 29 & 30 Victoria c 4, ss 1–5, A Collection of the Public General Statutes Passed in the Twenty Ninth and Thirtieth Years of the Reign of Her Majesty Queen Victoria 1866. Printed by George Edward Eyre and William Spottiswoode, London (1866), 13, available from https://play.google.com/books/reader?id=jJp0lRyIP4oC&printsec=frontcover&pg=GBS.PA13

[87] *An Act to amend the Act of the Eleventh and Twelfth Years of Her present Majesty, Chapter One hundred and seven to prevent the spreading of contagious or infectious Disorders among Sheep, Cattle, and Other Animals,* dated 23 April 1866, 29 & 30 Victoria c 15, ss 1–8, A Collection of the Public General Statutes Passed in the Twenty Ninth and Thirtieth Years of the Reign of Her Majesty Queen Victoria 1866. Printed by George Edward Eyre and William Spottiswoode, London (1866), 149, available from https://play.google.com/books/reader?id=jJp0lRyIP4oC&printsec=frontcover&pg=GBS.PA149

[88] Third Report of the Commissioners Appointed to Inquire Into the Origin and Nature, &c. of the Cattle Plague with an Appendix, above, 73.

[89] Ibid, x–xiii, xiv.

[90] Ibid, vii.

[91] C A Spinage, *Cattle Plague, A History*, above, 13, 291.

[92] Paul Brassley, "Animal Health and Veterinary Medicine", above 1, 590.

gaining a foothold.[93] The losses of the 1860s were exacerbated a decade later by the introduction and/or spread of other animal diseases, such as FMD and bovine pleuropneumonia.[94] Importantly, processes implemented to manage cattle plague came to apply with equal force to other diseases, reinforced by the *Contagious Diseases (Animals) Act 1894* (UK) (Diseases of Animals Act 1894) that consolidated three decades of disease regulation.[95]

Among other things, the act mandated public control of FMD, bovine pleuropneumonia and swine fever, giving authorities the power to stop the movement of livestock as well as to order the slaughter of animals suspected of being infected.[96] The act also strengthened the concept of foreign animal wharves, by stipulating that unless an exception had been granted by the Board of Agriculture pursuant to Section 26, imported animals were at all times to be landed at special wharves and slaughtered soon after arrival.[97] In 1896, the Diseases of Animals Act 1894 was amended by the repeal of Section 26, so that henceforth all foreign animals had to be slaughtered on arrival unless they were intended for exhibition or some other exceptional purpose.[98] The Board of Agriculture also had the power to prohibit the landing of any foreign animal, whatsoever.[99]

The pattern of regulation, which prevented live animals from being brought into the UK, but still allowed them to be transported for their products, underscored the fact that animal disease was seen as a trade issue. Moreover, it was an issue, if not properly managed, that had the potential to devastate the profitability of the farm animal sector. Accordingly, veterinarians were important to preventative goals, because the regime was based on keeping disease at bay by inspections, certifications and slaughter, all of which required veterinary expertise.

---

[93] Abigail Woods, "The Construction of and Animal Plague, Foot and Mouth Disease in Nineteenth-Century Britain" above, 56, 30–31.

[94] C A Spinage, *Cattle Plague, A History*, above, 13, 3.

[95] *An Act to Consolidate the Contagious Diseases (Animals) Acts 1878 to 1893,* dated 25 August 1894, 57 & 58 Victoria c 57, provided by The Royal College of Surgeons of England, available from https://archive.org/details/b2241695x

[96] Diseases of Animals Act 1894, ss 9–11 (stopping movement of livestock), ss 14, 15, 19 (slaughter and compensation for foot-and-mouth-disease, pleuropneumonia and swine fever).

[97] Diseases of Animals Act 1894, ss 24–30; This requirement was consistent with the earlier, Foreign Animals Order, gazetted on 10 December 1878, Privy Council, The Foreign Animals Order, 10 December, 1878, *The London Gazette,* Issue 24655, page 7073, available from https://www.thegazette.co.uk/London/issue/24655/page/7073. The use of the word "slaughter" meant that animals were killed, their carcasses could be used for food, following veterinary inspection.

[98] *Diseases of Animals Act 1896*, dated 20 July 1896, 59 & 60 Victoria c 15, s 1, provided by The Royal College of Surgeons of England, available from https://archive.org/details/b22416961

[99] *Diseases of Animals Act 1896*, s 1.

4.4.3 Effects of Cattle Plague: Veterinarians

The reputation of the veterinary profession initially emerged from the cattle plagues in tatters due to its failure to deal effectively with the disease. However, veterinarians came to enjoy an elevated status, endowed with legitimacy, thanks to changes in government policy.[100]

Prior to the 1860s, traditional professions had centred on law, divinity and medicine, supplemented by engineering, teaching and journalism, deriving from society's needs for specialised knowledge and expertise following the industrial revolution.[101] Yet, veterinarians tended to lag behind, perhaps due to the limited nature of their services, which by the mid-nineteenth century had crystallised into two areas: treating animals with high individual worth, such as horses and breeding stock of cattle and sheep, or treating animals in collectives, such as "packs of hounds, sheep in flocks and...cattle in herds and dairies".[102] Since horses were the most valuable animal, horse care provided veterinarians with the bulk of their livelihood. Yet even before the cattle plagues of the 1860s, diseases such as sheep pox, bovine pleuropneumonia, FDM as well as the cattle plagues of the eighteenth century had already demonstrated the need for veterinary skills that extended beyond the treatment of horses.[103] In 1834, William Youatt (1776–1847), a veterinary surgeon from the UK, criticised the reluctance of veterinary colleges to include such studies in their curriculum:

> Veterinary schools, that owed their origin to the ravages of epidemics among cattle, and that were established for the express purpose of teaching 'a more systematic knowledge of the management of sheep and cows' have shamefully neglected their trust.[104]

Although Simonds had been appointed professor of cattle diseases at the Royal College of Veterinary surgeons in 1842, by the time of the cattle plagues, veterinarians remained incapable of treating cattle diseases or, indeed, other diseases introduced by freer trade.[105]

Accordingly, when individual cows or sheep became ill with FMD or bovine pleuropneumonia, professional veterinary treatment often fared no better than failed home remedies.[106] Largely, this occurred because livestock diseases could not be

[100]Michael Worboys, "Germ Theories of Disease and British Veterinary Medicine, 1860–1890", above 1, 319, 321, 327.

[101]R M Hartwell, *The Industrial Revolution and Economic Growth*, above, 10, 217–218.

[102]J R Fisher, "Not quite a Profession: the Aspirations of Veterinary Surgeons in England in the Mid Nineteenth Century", above 29, 292.

[103]Ibid, 293.

[104]W Youatt, *Cattle; Their Breeds, Management and Diseases; with an index,* Baldwin and Cradock, Paternoster-Row, 1834, 2, available from https://archive.org/details/Youatt1834ip29P

[105]Michael Worboys, "Germ Theories of Disease and British Veterinary Medicine, 1860–1890", above 1, 311.

[106]Richard Perren, *The Meat Trade in Britain 1840–1914*, above, 55, 66.

cured or treated beyond remedies already proffered by lay practitioners, so colleges saw little value in extending their programs.[107] The corollary was that farmers and producers also saw little value in consulting veterinarians, instead preferring non-professionals or taking the practical option of selling animals before illnesses became visible.[108] In some instance, such as bovine pleuropneumonia epidemics, stakeholders preferred to carry insurance, which provided compensation for stock losses, rather than rely on veterinary medicine.[109]

This created a disconnect between veterinarians and other stakeholders, circumscribing the number of applicants enrolled in veterinary studies.[110] In the 1850s, Gamgee commented that Scotland only had 120 veterinarians to look after almost 8 million animals.[111] Consequently, when cattle plague reached the UK, there were insufficient professionals to limit the contagion, hastening the spread of disease as producers and vendors relied on ineffective home remedies.[112] Although veterinarians may not have had better treatments at their disposal, they had scientific training that potentially could have limited the spread of disease had they been involved early on. Their involvement, however, depended on government policy.

Throughout the mid-nineteenth century, government clung to its political beliefs, based on laissez-faire philosophies, which repudiated government intervention in private matters.[113] The cattle plagues not only reshaped these policies by transferring disease control from private owners to the public sphere but also altered the way government perceived veterinary training and practice.[114] The cattle plagues forced regulators to improve educational standards for veterinarians so that by 1905 several universities in the UK offered veterinary degrees.[115] These advances made it easier to legislate against the unqualified calling themselves veterinarians, a significant step given that veterinarians were expected to play a crucial role in disease control.[116] Following stricter regulation that commenced in 1866, government employed hundreds of veterinarians as "part-time inspectors... enjoying new powers and

[107] J R Fisher, "Not Quite a Profession: the Aspirations of Veterinary Surgeons in England in the Mid Nineteenth Century", above 29, 293–294.

[108] Richard Perren, *The Meat Trade in Britain 1840–1914*, above, 55, 66.

[109] J R Fisher, "Not quite a Profession: the Aspirations of Veterinary Surgeons in England in the Mid Nineteenth Century", above 29, 293.

[110] Ibid, 293–294.

[111] John Francis, "John Gamgee (1831–1894): Our Greatest Veterinarian" above, 48, 436.

[112] J R Fisher, "Not quite a Profession: the Aspirations of Veterinary Surgeons in England in the Mid Nineteenth Century", above 29, 293.

[113] Ibid, 284.

[114] The Royal Economic Society, "The Economic Effects of Cattle Disease Legislation", (1905) 15 (58) *The Economic Journal*, 156, 157.

[115] Paul Brassley, "Animal Health and Veterinary Medicine", above 1, 592.

[116] J R Fisher, "Not quite a Profession: the Aspirations of Veterinary Surgeons in England in the Mid Nineteenth Century", above 29, 299; Veterinary Surgeons Act 1881, ss 4, 5, 11.

status".[117] In addition, local authorities, who carried out much of the groundwork in enforcing law and policy, also employed a large number of veterinary surgeons.[118]

The links between the profession and government were further strengthened by the establishment of formal veterinary departments. As an emergency measure, authorities created the Cattle Plague Department in 1865,[119] which later became the Veterinary Department of the Privy Council and then subsequently the State Veterinary Service.[120] In 1919, government established the Diseases of Animals Division within the Ministry of Agriculture, again underscoring the significance of disease control and the role of veterinarians.[121] Trained veterinary surgeons thus enjoyed enhanced opportunities to participate in the livestock sector as inspectors, as well as treating animals and contributing to the regulation of production, commerce and trade.[122] This led to veterinary knowledge and expertise being given substantial weight so that findings and outcomes from veterinary conferences became increasingly influential to the industry as well as to law and policy.

4.5 Veterinary and Industry Conferences

This next sub-section analyses a selection of resolutions, proceedings and conferences taken from veterinary and industry gatherings, convened in the nineteenth to middle-twentieth centuries. The discussion is not intended to provide a detailed analysis of every event or proceeding, but is designed to identify matters considered important to stakeholders and thus highlights what was prioritised in the treatment of animals. The terms "conference(s)" and "congress(es)" are used interchangeably in the discussion that follows, including all meetings and gatherings.

[117] Michael Worboys, "Germ Theories of Disease and British Veterinary Medicine, 1860–1890", above 1, 311.

[118] Abigail Woods, "The Construction of and Animal Plague, Foot and Mouth Disease in Nineteenth-Century Britain" above, 56, 30–32; Joanna Swabe, *Animals, Disease and Human Society, Human-animal Relations and the Rise of Veterinary Medicine*, above, n 14, 108.

[119] Peter Atkins and Paul Brassley, "Mad Cows and Englishmen", (1996) 46 (9) *History Today*, 14, 15.

[120] Alison Wilson, "One Hundred and Fifty Years of the State Veterinary Service, 1865–2015", (2015) 72 (3) *Science in Parliament*, 8, 8–9, available from http://www.scienceinparliament.org.uk/wp-content/uploads/2015/09/150-YEARS-OF-THE-STATE-VETERINARY-SERVICE-by-Dr-Alison-Wilson.pdf

[121] Ibid.

[122] J R Fisher, "Not quite a Profession: the Aspirations of Veterinary Surgeons in England in the Mid Nineteenth Century", above 29, 300.

4.5.1 Stakeholder Perspectives, Government Regulation and Veterinary Surgeons

Table 4.1 sets out a list of materials used in this study, commencing with the First Session of the Veterinary Medical Association held in 1836,[123] before advancing to 1863 and the First International Veterinary Congress held in Hamburg and then concluding with the Conference of the New Zealand Society of Animal Production held in 1950.[124] The materials were sourced from collections held at the Royal College of Veterinary Surgeons in London and were selected to provide a cross section of veterinary and industry meetings, which occurred close to, and after the regulatory upheavals in the UK of the 1860s, ending with the aftermath of World War II. The analysis addresses the following research questions: what was the focus of the conferences, how did stakeholders perceive their roles and what was the impact of these findings on the way animals were regulated.

Following the establishment of formal veterinary colleges in the UK, one of the earliest assemblies of veterinarians occurred in 1836, when the UK Veterinary Medical Association met to settle its constitution and rules. The association's objectives emphasised the "advancement and diffusion of veterinary knowledge", to be achieved by discussion, the establishment of a reference library and circulation of relevant materials.[125] The emphasis on veterinary knowledge was to be expected and not, surprisingly, also formed the focal point of future veterinary conferences.

John Gamgee played a crucial role in initiating the First International Veterinary Congress, held in Hamburg in July 1863, some 2 years before the cattle plague struck the UK. As already discussed, Gamgee was a contagionist who firmly believed germs were responsible for the disease and the conference brought together some 103 veterinarians from around the world who presented and discussed findings and research on cattle plague. The conference was immensely successful, laying a pathway for other similar conferences, including those listed in Table 4.1. An important finding from the first congress was that contagion was responsible for the introduction and spread of cattle plague, placing veterinarians at the forefront of disease control.[126] Thereafter, veterinary conferences not only underscored the

[123] The Veterinary Medical Association was established by the Royal College of Veterinary Surgeons, and succeeded the London Veterinary Medical Society; An Abstract of the proceedings of the Veterinary Medical Association During its First Session 1836–37, by Order of the Council, Veterinary Medical Association, Longman, Orme, Brown, Green and Longmans, London (1837).

[124] The Cattle Plague, Official Reports, of the International Veterinary Congress, held in Hamburg, 1863, and in Vienna 1865, by John Gamgee, above, 4; New Zealand Society of Animal Production, Proceedings of the Tenth Annual Conference held in the Lecture Room, Dominion Museum, Wellington on May 3rd, 4th and 5th 1950, Volume 10 of the Proceedings of the New Zealand Society of Animal Production (1950).

[125] An Abstract of the proceedings of the Veterinary Medical Association During its First Session 1836–37, above, 123, 2, Rules 1 and 2.

[126] The Cattle Plague, Official Reports, of the International Veterinary Congress, held in Hamburg, 1863, and in Vienna 1865, by John Gamgee, above, 4, 468–469, 496–497.

Table 4.1 Source documents

Date	Type of proceeding	Name of proceeding
1836	Veterinary Association (National UK)	An Abstract of the Proceedings of the Veterinary Medical Association During Its First Session, 1836–1837
1863	Veterinary Association (International)	International Veterinary Congress, Held in Hamburg, 1863
1865	Veterinary Association (International)	International Veterinary Congress, Held in Vienna, 1865
1883	Veterinary Association (National UK)	Proceedings of the First General Meeting of the National Veterinary Association
1891–1892	Veterinary Association (National USA)	Proceedings of the American Veterinary Medical Association Sessions of 1891–1892
1893	Veterinary Association (National USA)	Proceedings of the 30th Annual Convention of the US Veterinary Medical Association and First Veterinary Congress of America
1896	Veterinary Association (National USA)	Proceedings of the US Veterinary Medical Association, Session of 1896
1897	Veterinary Association (National USA)	Proceedings of the First General Meeting of the National Veterinary Association Session of 1897
1900	Veterinary Association (National USA)	Proceedings of the American Veterinary Medical Association Session of 1900
1901	Veterinary Association (National USA)	Proceedings of the American Veterinary Medical Association Session of 1901
1905	Veterinary Association (International)	Report of the 8th International Veterinary Congress Budapest, 1905
1905	Veterinary Association (National USA)	Proceedings of the American Veterinary Medical Association 42nd Annual Convention
1907	Industry (International Dairy Federation) (International)	Official Programme of the 3rd International Dairy Congress
1909	Veterinary Association (Regional)	Report of the Proceedings of the Pan-African Veterinary Conference Convened at Pretoria
1914	Veterinary Association (International)	Tenth International Veterinary Congress, London
1923	Veterinary Association (Regional)	Colony and Protectorate of Kenya, Report of the Proceedings of the Fifth Pan-African veterinary conference held at Nairobi
1923	Industry (World's Dairy Congress) (International)	Proceedings of the World's Dairy Congress, Washington DC, Syracuse
1930	Veterinary Association (International)	Eleventh International Veterinary Congress, London
1923	Legislative Report (National UK)	Report of Proceedings Under the Diseases Of Animals Acts for the Year 1932
1934	Veterinary Association (International)	Twelfth International Veterinary Congress, New York
1938	Veterinary Association (International)	Thirteenth International Veterinary Congress, Zurich

(continued)

Table 4.1 (continued)

Date	Type of proceeding	Name of proceeding
1938	Veterinary Association (International)	Report of the FImperial Veterinary Conference Weybridge
1938–1947	Legislative Report (National UK)	Ministry of Agriculture and Fisheries, Report of Proceedings Under the Diseases of Animals Acts for the Years 1938–1947
1940	Industry (the US Livestock Sanitary Association) (National USA)	Proceedings of the 44th Annual Meeting of the US Livestock Sanitary Association, Chicago
1944	Industry (British Society of Animal Production) (National UK)	British Society of Animal Production, Report of Proceedings Second Meeting
1945	Industry (British Society of Animal Production) (National UK)	British Society of Animal Production, Report of Proceedings of Third Meeting
1950	Industry (New Zealand Society of Animal Production) (National New Zealand)	New Zealand Society of Animal Production, Proceedings of the Tenth Annual Conference, Wellington

importance of managing disease and/or the role of veterinarians but also broadened the debate to consider a range of related impacts, including the effects of animal disease on trade, human health and government regulation.[127]

By way of illustration, the Second International Veterinary Congress, held in 1865, stressed the importance of veterinary knowledge for its own sake but also resolved that such knowledge was critical to the regulation of trade because trade was implicated in the introduction and spread of disease.[128] The Eighth International Veterinary Congress, held in 1905, extended discussion to the links between animal disease and human health.[129] Industry conferences similarly underscored the significance of veterinary knowledge to commerce and human health, although they additionally incorporated dialogue of how law and policy affected their product

[127]For example, International Veterinary Congress, The Tenth International Veterinary Congress vol 1 London, August 3–8, 1914, John Bale, Sons and Danielsson Ltd., London (1915), 39, 103, section 1 "Veterinary Science in Relation to Public Health"; African Transvaal Department of Agriculture, Report of the proceedings of the Pan-African Veterinary Conference Convened at Pretoria on the 12th, 13th, and 14th January, 1909 by His Excellency the Governor, at the request of General Botha, Prime Minister, Transvaal, for the Purpose of Considering Matters of Inter-colonial Importance in Connection with Diseases of Stock, The Government Printing and Stationery Office, Pretoria (1909), 4; Thirteenth international Veterinary Congress Zurich- Interlaken 1938, Vol 1 Zurich, August 21–27, 1938, Berne, Suisse, Impr. dr. G. Grunau (1939), xi-xiii.

[128]The Cattle Plague, Official Reports, of the International Veterinary Congress, held in Hamburg, 1863, and in Vienna 1865, by John Gamgee, above, 4, 530–531.

[129]International Veterinary Congress, Report of the 8th International Veterinary Congress vol 2 Budapest, 1905, Pátria" irodalmi vállat és nyomdai részvénytársaság nyomása, Budapest (1905), Vol III, v–viii; V Babes, "Recent Experiences of the Infection of Man by Animal Diseases with Especial Regard to Different Occupations", in International Veterinary Congress, Report of the 8th international veterinary congress vol 2 Budapest, 1905, 444, Pátria" irodalmi vállat és nyomdai részvénytársaság nyomása, Budapest (1905).

sectors.[130] At the 44th Annual Meeting of the US Livestock Sanitary Association, delegates discussed how livestock diseases overlapped with sanitary legislation which authorised inspections and the impacts on industry practices and profitability.[131]

In general, the consequences of disease were seen in terms of networks that linked the role of veterinarians with government policies that promoted commerce and, later, safeguarded public health.[132] These connections were particularly strong in the UK because the British Empire was largely agricultural in nature.[133] Consequently, meetings with British colonial representation, such as those occurring in Africa, specifically discussed the profitability of livestock production and the necessity of establishing colonial administrative systems for dealing with colonial animal diseases.[134] The nexus that linked animal disease to veterinary practice, animal production and wealth creation strengthened throughout the first half of the twentieth century. At the Thirteenth International Veterinary Congress, held in 1938, delegates concluded that the "enormous economic importance of infectious diseases in animals" indisputably justifies the study of those diseases.[135]

Conferences held in the USA likewise devoted attention to the relationship between disease control and commerce, finding that markets, livelihoods, consumers and "national prosperity" depended on healthy farm animals and economic growth in the farm animal product sector.[136] Significantly, wealth creation could not occur in the absence of a robust and informed veterinary profession, suggesting a strong dependency between agricultural markets and the profession.[137] In this respect, American conferences pointedly surmised it was no coincidence that European veterinarians had become successful in tandem with the rise in the value of farm

[130]International Dairy Federation, Official Programme of the 3rd International Dairy Congress The Hague, Scheveningen, September 15–20, 1907, General Secretary's Office (Brussels) Van Langenhuysen Brothers (1907), 42; United States Livestock Sanitary Association, Proceedings of the 44th Annual Meeting of the United States Livestock Sanitary Association, Hotel Morrison, Chicago, IL, December 4–6, 1940, United States Livestock Sanitary Association, Reese Press Baltimore (1941), iii–v.

[131]United States Livestock Sanitary Association, Proceedings of the 44th Annual Meeting of the United States Livestock Sanitary Association, above, 130, contents, iii–v.

[132]National Veterinary Association, Proceedings of the First General Meeting of the National Veterinary Association 8 May 1883, Bailliere, Tindall and Cox, London (1883), 12.

[133]Ibid.

[134]Colony and Protectorate of Kenya, Report of the Proceedings of the Fifth Pan-African Veterinary Conference Held at Nairobi from 6th to 14th April, 1923, Government Press, Nairobi (1924), 102–103.

[135]P J Du Toit, "General Conditions Governing the Origin and Development of Infectious Diseases", in Thirteenth International Veterinary Congress Zurich- Interlaken 1938, above, 127, 174, 188–189.

[136]Leonard Pearson, "A Review of the Field of Veterinary Medicine", Proceedings of the First General Meeting of the National Veterinary Association of 1897, 126, 131, 136–137, NVA, Kansas City (1897).

[137]Ibid, 131, 136–137.

animals, which resulted in greater appreciation and reliance by stakeholders on the veterinarian's skill and expertise.[138] The profession itself regarded its members as "the expert on all live stock questions", calling for veterinary education to be updated, including studies in animal husbandry for commercially important farm animals.[139]

Indeed, the training and education of veterinarians was a recurring topic, frequently debated in terms of gaps in the curriculum and responsiveness to the needs of stakeholders. Curricula differed across the jurisdictions and in the USA, unlike the UK, horses were largely ignored by veterinary colleges, notwithstanding the fact that they were economically indispensable. As late as 1896, the US Veterinary Medical Association was forced to concede that in many cases farriers were more knowledgeable about horseshoeing and the care of horses' feet than veterinarians, leading to pleas for formal education to be more receptive to the needs of horse owners.[140] Largely as a result of these gaps, American farmers became notoriously mistrustful of the science that underpinned animal testing and treatment, creating an appreciable degree of tension between themselves and the profession.[141]

Discussions on the quality and relevance of veterinary studies continued into the early part of the twentieth century, which saw calls to expand syllabi by including studies in colonial animal diseases,[142] as well as appeals to create courses on animal husbandry that could enhance food production.[143] This was regarded as an important mechanism in satisfying increased demand for animal products and supporting growth in the commercial value of livestock, which the profession hoped would lead to more demand for veterinary services.[144]

---

138 A Liautard, "Veterinary Education", in USAVMA, Proceedings of the 30th Annual Convention of the United States Veterinary Medical Association [AVMA] and First Veterinary Congress of America, Chicago, W Horace Hoskins (ed), 186, 187, USAVMA, Philadelphia (1894).

139 Leonard Pearson, "A Review of the Field of Veterinary Medicine", above, 136, 136–137.

140 H D Gill, "The Relation of Horseshoers to Veterinarians and Veterinary Colleges", in USAVMA, Proceedings of the United States Veterinary Medical Association [USAVMA]Session of 1896, W L Williams (ed), 90, 93, USAVMA, Philadelphia (1896).

141 Car W Gay, "The Attitude of the Farmer Toward the Tuberculin Test", in American Veterinary Association, Proceedings of the American Veterinary Medical Association [AVMA] session of 1901, M H Reynolds, Chairman (ed), 213, 214, 221, Webb Publishing Co St Paul, Minnesota (1902).

142 African Transvaal Department of Agriculture, Report of the proceedings of the Pan-African Veterinary Conference Convened at Pretoria on the 12th, 13th, and 14th January, above, 127, 40–41; J Share-Jones, "The Relationship of the Veterinary Surgeon to Animal Husbandry", in International Veterinary Congress, Eleventh International Veterinary Congress vol 2 London, Central Hall, Westminster, August 4–9, 1930, 234, 234, John Bale, Sons and Danielsson Ltd., London (1931).

143 Leonard Pearson, "A Review of the Field of Veterinary Medicine", above, 136, 136–137; J Share-Jones, "The Relationship of the Veterinary Surgeon to Animal Husbandry", above, 142, 234; Preliminary material, Thirteenth international Veterinary Congress Zurich- Interlaken 1938, above, 127, Vol 1, xi–xiii.

144 J Share-Jones, "The Relationship of the Veterinary Surgeon to Animal Husbandry", above, 142, 235–236.

Veterinarians became increasingly keen to secure their position in the sector, on the basis that they alone had sufficient skill and proficiency to stamp out disease, and to ensure that animal products were of sufficient quality and quantity.[145] However, a persistent thread from start to finish accentuated difficulties generated by a lack of uniform regulation.[146] In the UK, measures were largely promulgated by way of Orders in Council, which dealt with problems as they arose.[147] This resulted in a piecemeal regime without strategic direction, where government did not adequately manage overlapping issues deriving from animal production, animal health and human health, matters that were obvious to veterinarians.[148] Prior to the cattle plagues, a particular concern was the fact that a lack of bureaucratic foresight had left the profession in a precarious position, at the whim of industry that was already sceptical of veterinary involvement because of the potential to hamper trade and impinge on private property rights.[149]

Even where government promulgated measures, it could be challenging to implement them. Veterinarians, for example, might not be able to examine animals across the supply chain, as frequently occurred during transportation because conveyances such as railways were privately owned.[150] Moreover, in an age before biomedical testing, there were no guarantees that visual inspections could identify infected animals.[151] This led Gamgee to argue that the animal sector lacked adequate supervision because of inadequate government intervention.[152] He emphasised his arguments by observing that official policy left owners to deal with problems, invariably resulting in sick animals being sold as quickly as possible. This

---

[145] Generally, John W Gadsden, "Our Duty to the Cattle Interests of the Country", in AVMA, Proceedings of the American Veterinary Medical Association [AVMA] sessions of 1891–1892, W Horace Hoskins (ed), 342, AVMA, Philadelphia (1893); J Share-Jones, "The Relationship of the Veterinary Surgeon to Animal Husbandry", above, 142, 235–236.

[146] The Cattle Plague, Official Reports, of the International Veterinary Congress, held in Hamburg, 1863, and in Vienna 1865, by John Gamgee, above, 4, 496–497; Williamson Bryden, "The Transatlantic Cattle Trade and its Regulations from a Veterinary Point of View", in AVMA, Proceedings of the American Veterinary Medical Association [AVMA] sessions of 1891–1892, W Horace Hoskins (ed), 99, 106, AVMA, Philadelphia (1893); Colony and Protectorate of Kenya, Report of the Proceedings of the Fifth Pan-African Veterinary Conference, above, 134, 102–103, 124.

[147] The Cattle Plague, Official Reports, of the International Veterinary Congress, held in Hamburg, 1863, and in Vienna 1865, by John Gamgee, above, 4, 545–546.

[148] J Share-Jones, "The Relationship of the Veterinary Surgeon to Animal Husbandry", J Share-Jones, "The Relationship of the Veterinary Surgeon to Animal Husbandry", above, 142, 239.

[149] The Cattle Plague, Official Reports, of the International Veterinary Congress, held in Hamburg, 1863, and in Vienna 1865, by John Gamgee, above, 4, 544; This issue also manifested in the twentieth century, Colony and Protectorate of Kenya, Report of the Proceedings of the Fifth Pan-African Veterinary Conference, above, 134, 102–103, 124.

[150] The Cattle Plague, Official Reports, of the International Veterinary Congress, held in Hamburg, 1863, and in Vienna 1865, by John Gamgee, above, 4, 468–469, 471.

[151] Ibid, 468–469, 471.

[152] Ibid, 489–490.

contributed to the spread of disease, creating greater, long-term harm to commerce and human health when entire herds became infected.[153]

As discussed, the reluctance of government to intervene partly rested on fears that regulation would restrain trade and partly derived from the tenacity of laissez-faire policies that categorised animal disease as a private matter.[154] On one level, trade in animals and their products could be seen as something that was purely transactional, an economic arrangement between private individuals. This view was supported by discourses that centred on the financial benefits of animal production, where animals were described as "living articles of commerce".[155] These perspectives also meant that animals' economic value was seen as a private matter, solely the domain of parties to contractual arrangements.

Delegates at veterinary conferences, however, did not regard commerce in animals in such a legally limited way. Instead, they saw a role for veterinarians that was outside the scope of the primary contract between vendor and purchaser, a role that promoted veterinarians as defenders of animal health in the context of trade.[156] The difficulty was that such a role involved crossing the private property barrier, which required government authorisation. Moreover, it was not clear to what extent the profession ought to be involved. As veterinarians debated these issues, they confirmed that at the very least their role should be that of a gatekeeper, working alongside government and enforcing prohibitions that banned the sale of animals without a veterinary certificate that verified animals were sourced from disease-free areas.[157] This would provide safeguards to assist in stopping the spread of disease and also benefit the profession by advancing the authority of the veterinary surgeon, replacing the "interference of the lawyer".[158]

As discussed, the UK government did carve out such a role for veterinarians, by way of secondment as official inspectors. The involvement of government in this manner, via the veterinary profession, shifted disease control away from private owners towards the public arena. It was seen as a necessary step to the establishment of coordinated and uniform measures, which were important to defending the country's "economic welfare".[159] Yet, this regulatory transformation also carried several challenges. To start with, government needed to ensure that those charged

[153] Ibid, 484–485, 489–490.

[154] John McEldowney, Wyn Grant, Graham Medley et al. *The Regulation of animal Health and Welfare*, Routledge, Taylor & Francis Group, Abingdon, Oxfordshire (2013), 18.

[155] The Cattle Plague, Official Reports, of the International Veterinary Congress, held in Hamburg, 1863, and in Vienna 1865, by John Gamgee, above, 4, 718.

[156] Ibid, 718.

[157] Ibid, 496–497.

[158] Ibid, 726.

[159] Ibid. The Cattle Plague, Official Reports, of the International Veterinary Congress, held in Hamburg, 1863, and in Vienna 1865, by John Gamgee, above, 4, 496–497; J H Cox, "Human and Bovine Tuberculosis—its Communicability", in National Veterinary Association, Proceedings of the First General Meeting of the National Veterinary Association 8 May 1883, above 132, 23, 40–41; Resolutions of the Seventh International Congress of Veterinary Surgeons, Held

with carrying out procedures, such as the slaughter of entire herds, were not swayed by the impacts on individual producers, or the effects, in general, on commerce and trade.[160] Moreover, these concerns also recognised a practical problem, in that strict measures would prove difficult to enforce where the burden fell solely on cattle producers and traders because animals were still private property, whose destruction resulted in financial burdens.[161]

4.5.2 Animals, Disease and Commodification

As discussed in Chaps. 2 and 3 of this book, farm animals had long been treated as private property and commodities in the marketplace.[162] The government's response to the cattle plague in the UK reinforced these views by merging animal health with economic objectives, identifiable on two levels. The first was broadly based and aimed to protect the viability of the farm animal sector, and the second was a narrower objective that sought to recompense producers whose animals were destroyed.

With respect to the first objective, stricter policies from February 1866 targeted eradication of disease by tightening border controls and authorising a system of slaughter and prohibitions on the movement of livestock. There was very little in these policies related to animal wellbeing. Instead, farm animal health was equated with public control of disease, to safeguard the livestock industry, a goal that saw government policy:

> trapped within two interlocking vicious circles. If disease could not be brought under control, no livestock industry could prosper, but if pastoralism was not commercially productive it could not generate sufficient resources to fund veterinary schemes.[163]

Although these comments were made by Waller with respect to Kenya, his observations are equally relevant to regimes in other jurisdictions. In the UK, eradication of disease was achieved by slaughtering sick animals, who had contracted scheduled diseases listed in relevant legislation.[164] The goal was to attain disease-free herds, a concept equated with a "healthy livestock economy".[165]

---

at Baden-Baden in August 1899, (1899) 49 (10) *The Veterinary Journal and Annals of Comparative Pathology*, 236, 236–237.

[160] J H Cox, "Human and Bovine Tuberculosis—Its Communicability" above, 159, 36–37, 40–41.

[161] C A Spinage, *Cattle Plague, A History*, above, 13, 185–186.

[162] Discussion, in parts 2.4, 2.7, 3.2.1 and 3.4.2.

[163] Richard Waller, "'Clean' and 'Dirty': Cattle Disease and Control Policy in Colonial Kenya, 1900–40", (2004) 45 (1) *The Journal of African History*, 45, 56.

[164] Michael Worboys, "Germ Theories of Disease and British Veterinary Medicine, 1860–1890", above 1, 308

[165] Ibid, 319.

These policy goals were well-accepted among the veterinary profession, as demonstrated by these insights from George Fleming (1833–1901), an influential veterinarian:

> Agriculture must ever occupy a higher position than manufactures and the prevention of epizootic diseases should be regarded as a political question involving more or less the well-being of the whole community, not merely affecting those who own or who endeavour to derive profit from rearing animals, but also affecting the public at large, as regards health, the supply of food and other essentials.[166]

In considering the greater good of the sector, law and policy effectively relegated animals to the status of a food commodity. Regulation did not envisage safeguarding animals, either for their own sake or for the sake of alleviating their pain and suffering. However, while protection of the sector was of paramount importance, there were still limitations, such as the need to take into account the impact of policies on individual farmers and producers.

During the epidemics of the eighteenth century, the notion of compensation for livestock losses proved contentious.[167] The system was open to abuse and did not necessarily cover farmers and producers adequately for their losses.[168] Nevertheless, if eradication was the objective, it was important to engage with commercial solutions for commercial losses.[169] For this reason, by the turn of the twentieth century, recommendations to improve the regime incorporated economic yardsticks, such as extending existing insurance and compensation arrangements to cover horses, pigs and goats, in addition to cattle.[170] After initial hesitation, authorities allowed insurance payouts and provided state-sponsored compensation, as the quid pro quo for the destruction of herds.[171] Such payments replaced property in livestock, but they also took the place of veterinary treatment, such as it was. The upshot was a tendency, in some jurisdictions, to destroy animals as soon as they displayed signs of disease, without isolating them and waiting to see if they recovered.[172]

At the same time, delaying compensation claims to effect a cure could be a risky business.[173] European experience had taught that segregation and slaughter were the most effective ways to stop the cattle plague. The Third Royal Commission into

---

[166] Fleming, George, *Animal Plagues: Their History, Nature, and Prevention, from BC 1490 to AD 1800,* Chapman and Hall (1871), XXXIII. This copy is a reprint.

[167] C A Spinage, *Cattle Plague, A History*, above, 13, 125–127.

[168] Ibid.

[169] Ibid, 185–186.

[170] International Veterinary Congress, Report of the 8th international veterinary congress vol 1 Budapest, 1905, above, 129, 350–351.

[171] C A Spinage, *Cattle Plague, A History*, above, 13, 185–186, 301–304.

[172] African Transvaal Department of Agriculture, Report of the proceedings of the Pan-African Veterinary Conference Convened at Pretoria on the 12th, 13th, and 14th January, 1909 above, 127, 17.

[173] George Fleming, *Animal Plagues: Their History, Nature, and Prevention, from BC 1490 to AD 1800,* above, 6, XXIV.

Cattle Plague, for instance, had dismissed inoculation, observing that trials were not conclusive and the treatment, futile.[174] This approach, however, needed to be tempered by the fact that other developments indicated treatments could sometimes provide a cure.[175] A case in point derives from the management of bovine pleuro-pneumonia. The disease was endemic to Europe, and European veterinarians had developed a form of inoculation that was protective and which also came to be used in South Africa and Australia.[176] However, the process never caught on in the UK, even though authorities had investigated it in the 1850s.[177] Contrary to the UK approach, in 1881 US veterinarians lauded the development of anthrax vaccine, regarding it as the foreunner of an age of "preventative vaccination" that had already saved the cattle industry much money.[178] By the twentieth century, literature on the use of vaccines appeared regularly in veterinary journals, and veterinarians were advised to keep abreast of any innovations.[179]

Notwithstanding these initiatives, during the formative years of disease control in the UK, which roughly occurred from 1860 to 1890, government policy did not encourage veterinarians to research cures.[180] Instead, veterinarians were persuaded to maintain their role as gatekeepers and implement official policy.[181] In reality, veterinarians were trapped in the same sequence of events as regulators, because they felt they needed to justify their role, as well as earn a living.[182] Many veterinarians were thus content to play a part that gave them "professional legitimisation" but which also cemented their connection with government.[183] This led to, what Bernard Rollins has described as, the "historical subordination of

---

[174]Third Report of the Commissioners Appointed to Inquire into the Origin and Nature, &c. of the Cattle Plague with an Appendix, in First-Third Reports of the Commissioners Appointed to Inquire Into the Origin and Nature, &c. of the Cattle Plague, above, 73, x–xiii.

[175]John R Fisher, "Cattle Plagues Past and Present: The Mystery of Mad Cow Disease", (1998) 33 (2) *Journal of Contemporary History*, 215, 227.

[176]J R Fisher, "Not quite a Profession: the Aspirations of Veterinary Surgeons in England in the Mid Nineteenth Century", above 29, 295–296.

[177]Ibid, 296.

[178]W M Rushworth, "Live Stock Vaccines and Serum Therapy", in American Veterinary Association, Proceedings of the American Veterinary Medical Association [AVMA] session of 1900, M H Reynolds (ed), 243, 244, Pioneer Press Co, St Paul, Minnesota (1900).

[179]Richard Perren, "Veterinary Products", in Joan Thirsk (general editor) *The Agrarian History of England and Wales* Volume VII 1850–1914, Part II, 1045, 1050, Cambridge University Press (2000).

[180]Michael Worboys, "Germ Theories of Disease and British Veterinary Medicine, 1860–1890", above 1, 319.

[181]Ibid.

[182]Richard Waller, "'Clean' and 'Dirty': Cattle Disease and Control Policy in Colonial Kenya, 1900–40", above, 163, 56.

[183]Michael Worboys, "Germ Theories of Disease and British Veterinary Medicine, 1860–1890", above 1, 319, 321, 327.

veterinary medicine to agriculture".[184] The pathway that promoted this development also depended on government administration of farm animals and veterinary medicine within agricultural portfolios.

As discussed, in the 1860s the UK government had set up a Cattle Plague Department, which was renamed the Veterinary Department in 1870. In 1899, that department was established within the Board of Agriculture, and as part of its remit, the Department exercised a supervisory role, maintaining effective border controls against the importation of diseased animals and carrying out the destruction of animals in accordance with legislation.[185] Although farm animals generated the sector's wealth, their wellbeing was not holistically managed. The animal itself was regulated by the Board of Agriculture as a foodstuff, while animal health was regulated by the Veterinary Department as a trade issue, explicitly in an agricultural setting. In this way, animal wellbeing became subsumed into the niceties of production as well as the necessities of market transactions. This trajectory was squarely settled by the late nineteenth century and provided a ready-made pathway for the increasingly intensive production of animals that followed World War II.

4.6 Conclusion

The cattle plagues of the 1860s were a turning in the regulation of animal diseases, which were only brought under control in 1867, once regulators implemented strict orders that had been proposed by Gamgee. Measures were based on stamping out disease and involved stringent border controls, restrictions on the movement of animals and slaughter of infected animals. Although these measures succeeded in controlling the cattle plague, they were also reactive, responding to disease in a way that would benefit the sector, without considering broader aspects of animal wellbeing. Eradicating disease also had consequences for veterinary medicine, which similarly became shaped by initiatives that would benefit the sector.

In consequence, disease control and veterinary medicine came to be viewed through the lens of food security and trade protection. Moreover, because strict interventions had triumphed, this reinforced the desirability of government involvement, supported by associated law and policies.[186] In practical terms, this meant that owners exchanged their property rights in animals for benefits conferred by the state, including protection of the product sector from disease, and the right to receive

[184] Bernard E Rollin, "Putting the Horse before the Descartes: My Life's Work on Behalf of Animals", Temple University Press, Philadelphia (2011), 22.

[185] Paul Brassley, "Animal Health and Veterinary Medicine", above 1, 590–591.

[186] Abigail Woods, "The Construction of and Animal Plague, Foot and Mouth Disease in Nineteenth-Century Britain" above, 56, 30–32; Joanna Swabe, *Animals, Disease and Human Society, Human-animal Relations and the Rise of Veterinary Medicine*, above, n 14, 104, 108.

compensation for animals that were destroyed.[187] Farm animals thus became ever more fungible, a commodity interchangeable with money, either through market transactions or by way of compensation.

As trade and commerce in farm animals grew, so did the likelihood of disease being transmitted, a situation that was exacerbated where trade crossed international boundaries. This made it increasingly necessary to find an international solution to the problem of animal disease.[188] Yet, the challenges faced by domestic regulators would be magnified at the international level, making it more difficult to balance trade, animal health and animal wellbeing, challenges that are discussed in the next chapter.

Bibliography

Atkins P, Brassley P (1996) Mad cows and Englishmen. Hist Today 46(9):14

Bickford JF (1843) On the exclusiveness of the publication of the veterinary transaction—the false ground on which it was founded—valuable introduction of cattle pathology—the strange ignorance of many country practitioners—the present state of cattle pathology- on paraplegia in cattle, Mr T Turner's oration—and the strange duplicity practised by many uncertified men. Veterinarian XVI(181):13

Brassley P (2000) Animal health and veterinary medicine. In: Thirsk J (general ed) The Agrarian history of England and Wales Volume VII 1850–1914, Part 1, 587, 588–589, 591–592. Cambridge University Press, Cambridge

Broad J (1983) Cattle plague in eighteenth-century England. Agric Hist Rev 31(2):104

Brown K, Gilfoyle D (2010) Healing the herds. Ohio University Press, Athens

Carli Bones V, Yeates JW (2012) The emergence of veterinary oaths: social, historical and ethical considerations. J Anim Ethics 2(1):20

Commissioners Appointed to Inquire Into the Origin and Nature, &c of the Cattle Plague, First-Third Reports of the Commissioners Appointed to Inquire Into the Origin and Nature, &c. of the Cattle Plague: with the minutes of evidence and an appendix, printed by George Edward Eyre and William Spottiswoode, Great Britain, H M Stationery Office, London (1866). https://babel.hathitrust.org/cgi/pt?id=coo.31924110082843&view=1up&seq=11

Fisher JR (1993) Not quite a profession: the aspirations of veterinary surgeons in England in the mid nineteenth century. Hist Res 66(161):284. https://doi.org/10.1111/j.1468-2281.1993.tb01814.x

Fisher JR (1998) Cattle plagues past and present: the mystery of mad cow disease. J Contem Hist 33(2):215

Fleming G (1871) Animal plagues: their history, nature, and prevention. Chapman and Hall (reprint)

[187] Abigail Woods, "The Construction of and Animal Plague, Foot and Mouth Disease in Nineteenth-Century Britain" above, 56, 30–32; Joanna Swabe, *Animals, Disease and Human Society, Human-animal Relations and the Rise of Veterinary Medicine*, above, n 14, 108.

[188] Joanna Swabe, *Animals, Disease and Human Society, Human-animal Relations and the Rise of Veterinary Medicine*, above, n 14, 105.

Francis J (1962) John Gamgee (1831-1894): our greatest veterinarian. Br Vet J 118:430
Hartwell RM (1971) The industrial revolution and economic growth. Methuen and Co Ltd, London
Karamanou M, Panayiotakopoulos G, Tsoucalas G, Kousoulis AA, Androutsos G (2012) From miasmas to germs: a historical approach to theories of infectious disease transmission. Le Infen Med 1:52
Knick C (2008) Steers afloat: the North Atlantic meat trade, liner predominance, and freight rates, 1870-1913. Econ Hist Assoc 68(4):1028
Koolmess PA (2010) Epizootic diseases in the Netherlands, 1713–2002. In: Brown K, Gilfoyle D (eds) Healing the herds, disease, livestock economics, and the globalization of veterinary medicine. Ohio University Press, Athens, p 19
McEldowney J, Grant W, Medley G et al (2013) The regulation of animal health and welfare. Routledge/Taylor & Francis Group, Abingdon
Nutton V (1983) The seeds of disease: an explanation of contagion and infection from the Greeks to the renaissance. Med Hist 27(1):1
Patterson I (1990) A great British veterinarian forgotten: James Beart Simonds 1810–1904. J A Allen & Co, London
Penberthy J (1901) Foot-and-mouth-disease. J Comp Pathol Therap 14:16
Perren R (1978) The meat trade in Britain 1840–1914. Routledge & Kegan Paul, London
Perren R (2000) Veterinary products. In: Thirsk J (general ed) The Agrarian history of England and Wales Volume VII 1850–1914, Part II. Cambridge University Press, Cambridge, p 1045
Perren R (2008) Filth and profit, disease and health: public and private impediments to slaughter-house reform in Victorian Britain. In: Lee PY (ed) Meat, modernity and the rise of the slaughterhouse. University of New Hampshire Press, New Hampshire, p 127
Rollin BE (2011) Putting the horse before the Descartes: my life's work on behalf of animals. Temple University Press, Philadelphia
Santer M (2009) Richard Bradely: a unified, living agent theory of the cause of infectious diseases of plants, animals, and humans in the first decades of the 18th century. Perspect Biol Med 52 (4):566
Smith F (1919) The work of the British Army veterinary Corps at the Fronts. Vet J 75(1):8
Spinage CA (2003) Cattle plague, a history. Kluwer Academic, New York
Strickland Constable H (1866) Observations suggested by the cattle plague: about witchcraft, credulity.... Dalton and Lucy, Charing Cross
Swabe J (1999) Animals, disease and human society, human-animal relations and the rise of veterinary medicine. Routledge, London
The Royal Economic Society (1905) The economic effects of cattle disease legislation. Econ J 15(58):156
Waller R (2004) 'Clean' and 'dirty': cattle disease and control policy in colonial Kenya, 1900-40. J African Hist 45(1):45
Wilkinson L (1992) Animals and disease, an introduction to the history of comparative medicine. Cambridge University Press, Cambridge
Wilson A (2015) One hundred and fifty years of the state veterinary service, 1865-2015. Sci Parliam 72(3):8. http://www.scienceinparliament.org.uk/wp-content/uploads/2015/09/150-YEARS-OF-THE-STATE-VETERINARY-SERVICE-by-Dr-Alison-Wilson.pdf
Woods A (2004) The construction of and animal plague, foot and mouth disease in nineteenth-century Britain. Social Hist Med 17(1):23
Woods A (2011) A historical synopsis of farm animal disease and public policy in twentieth century Britain. Philos Trans R Soc B 366:1943
Worboys M (1991) Germ theories of disease and British veterinary medicine, 1860-1890. Med Hist 35(308):318
Yeates J (2018) Veterinary science: a very short introduction. OUP, Oxford
Youatt W (1834) Cattle; their breeds, management and diseases; with an index, Baldwin and Cradock, Paternoster-Row. https://archive.org/details/Youatt1834ip29P

Legislation, Regulations, Gazettes and Hansard

1881, *Veterinary Surgeons Act 1881* (UK), 44 & 45 Victoria c 62, *The Public General Statutes Passed in the Forty Fourth and Forty Fifth Years of the Reign of Her Majesty Queen Victoria*, Printed by G.E. Eyre and W. Spottiswoode, printers to the Queen and W. Clowes and Sons, printers to the Council of Law Reporting, London, (1881), 371. https://archive.org/details/publicgeneralst05walegoog/page/n385

1844, The Royal Charter of the Royal College of Veterinary Surgeons 1844, available from the Royal College of Veterinary Surgeons, https://www.rcvs.org.uk/document-library/royal-charter-1844/

1865, Consolidated Order of 25 September 1865, published in London Gazette Tuesday 26 September, 1865 Issue 23017 page 4585. https://www.thegazette.co.uk/London/issue/23017/page/4585

1866, *An Act to Amend the Law Relating to Contagious or Infectious Diseases in Cattle and Other Animals* 29 & 30 Victoria c 2, dated 20 February 1866, *A Collection of the Public General Statutes Passed in the Twenty Ninth and Thirtieth Years of the Reign of Her Majesty Queen Victoria 1866*, printed by George Edward Eyre and William Spottiswoode, London (1866). https://play.google.com/books/reader?id=jJp0lRyIP4oC&printsec=frontcover&pg=GBS.PA2

1866, A*n Act to amend the Law relating to Contagious Diseases amongst Cattle and other Animals in Ireland,* 1866, dated 6 March 1866, 29 & 30 Victoria c 4, *A Collection of the Public General Statutes Passed in the Twenty Ninth and Thirtieth Years of the Reign of Her Majesty Queen Victoria 1866*, Printed by George Edward Eyre and William Spottiswoode, London (1866). https://play.google.com/books/reader?id=jJp0lRyIP4oC&printsec=frontcover&pg=GBS.PA13

1866, *An Act to amend the Act of the Eleventh and Twelfth Years of Her present Majesty, Chapter One hundred and seven to prevent the spreading of contagious or infectious Disorders among Sheep, Cattle, and Other Animals,* dated 23 April 1866, 29 & 30 Victoria c 15, *A Collection of the Public General Statutes Passed in the Twenty Ninth and Thirtieth Years of the Reign of Her Majesty Queen Victoria 1866*, Printed by George Edward Eyre and William Spottiswoode, London (1866), available from https://play.google.com/books/reader?id=jJp0lRyIP4oC&printsec=frontcover&pg=GBS.PA149

1878, Privy Council, Foreign Animals Order, gazetted on 10 December 1878, *The London Gazette*, Issue 24655, page 7073. https://www.thegazette.co.uk/London/issue/24655/page/7073

1894, *An Act to Consolidate the Contagious Diseases (Animals) Acts 1878 to 1893,* dated 25 August 1894, 57 & 58 Victoria c 57, provided by The Royal College of Surgeons of England. https://archive.org/details/b2241695x

1896, *Diseases of Animals Act 1896*, dated 20 July 1896, 59 & 60 Victoria c 15, s 1, provided by The Royal College of Surgeons of England, available from https://archive.org/details/b22416961

Conferences, Congresses and Meetings

1837, An Abstract of the Proceedings of the Veterinary Medical Association During its First Session 1836–37, by Order of the Council, Veterinary Medical Association, Longman, Orme, Brown, Green and Longmans, London (1837)

1865, *The Cattle Plague, Official Reports, of the International Veterinary Congress, held in Hamburg, 1863, and in Vienna 1865*, by John Gamgee, Robert Hardwicke, London (1866)

1883, Cox JH 1883 Human and bovine tuberculosis—its communicability. In: National Veterinary Association, Proceedings of the First General Meeting of the National Veterinary Association 8 May 1883, Bailliere, Tindall and Cox, London, p 23

1891–1892, Bryden W (1893) The transatlantic cattle trade and its regulations from a veterinary point of view. In: Hoskins WH (ed) Proceedings of the American Veterinary Medical Association [AVMA] sessions of 1891–1892. AVMA, Philadelphia, p 99

1891–1892 Gadsden, JW (1893), Our Duty to the Cattle Interests of the Country. In: Hoskins WH (ed), Proceedings of the American Veterinary Medical Association [AVMA] sessions of 1891–1892. AVMA, Philadelphia, p 342

1894, Liautard A (1894) Veterinary education. In: Hoskins WH (ed) Proceedings of the 30th annual Convention of the United States Veterinary Medical Association [AVMA] and First Veterinary Congress of America, Chicago. USAVMA, Philadelphia, p 186

1896, Gill HD (1896) The relation of horseshoers to veterinarians and veterinary colleges. In: Williams WL (ed), USAVMA, Proceedings of the United States Veterinary Medical Association [USAVMA] Session of 1896. USAVMA, Philadelphia, p 90

1897, Pearson L (1897) A review of the field of veterinary medicine. In: Proceedings of the first general meeting of the National Veterinary Association of 1897. *NVA*, Kansas City, p 126

1899, (1899) Resolutions of the seventh international congress of veterinary surgeons, held at Baden-Baden in August 1899, *The Veterinary Journal and* Ann Compar Pathol 49(10):236

1900, Rushworth WM (1900) Live stock vaccines and serum therapy. In: Reynolds MH (ed) Proceedings of the American Veterinary Medical Association [AVMA] session of 1900, American Veterinary Association. Pioneer Press Co, St Paul, Minnesota, p 243

1901, Gay CW (1902) the attitude of the farmer toward the tuberculin test. In: Reynolds MH, Chairman (ed) American Veterinary Association, Proceedings of the American Veterinary Medical Association [AVMA] session of 1901. Webb Publishing Co St Paul, Minnesota, p 213

1905, International Veterinary Congress, Report of the 8th international veterinary congress vol 2 Budapest, 1905, Pátria" irodalmi vállat és nyomdai részvénytársaság nyomása, Budapest (1905)

1907, International Dairy Federation, Official Programme of the 3rd International Dairy Congress The Hague, Scheveningen, September 15–20, 1907, General Secretary's Office (Brussels) Van Langenhuysen Brothers (1907)

1909, African Transvaal Department of Agriculture, Report of the proceedings of the Pan-African Veterinary Conference Convened at Pretoria on the 12th, 13th, and 14th January, 1909 by His Excellency the Governor, at the request of General Botha, Prime Minister, Transvaal, for the Purpose of Considering Matters of Inter-colonial Importance in Connection with Diseases of Stock, The Government Printing and Stationery Office, Pretoria (1909)

1914, International Veterinary Congress, The Tenth International Veterinary Congress vol 1 and 3, London, August 3–8, 1914, John Bale, Sons and Danielsson Ltd, London (1915)

1923, Colony and Protectorate of Kenya, Report of the Proceedings of the Fifth Pan-African Veterinary Conference Held at Nairobi from 6th to 14th April, 1923, Government Press, Nairobi (1924)

1930, Share-Jones J (1931) The relationship of the veterinary surgeon to animal husbandry. In: International Veterinary Congress, Eleventh International Veterinary Congress, vol 2 London, Central Hall, Westminster, August 4–9, 1930, John Bale, Sons and Danielsson Ltd, London, p 234

1938, Babes V (1938) Recent experiences of the infection of man by animal diseases with especial regard to different occupations. In: Thirteenth International Veterinary Congress Zurich-Interlaken , vol 1 Zurich, August 21–27, 1938. Berne, Suisse, Impr. dr. G. Grunau, 1939, p 444

1938 Du Toit, PJ (1938) General conditions governing the origin and development of infectious diseases. In: Recent experiences of the infection of man by animal diseases with especial regard to different occupations. In: Thirteenth International Veterinary Congress Zurich-Interlaken 1938, vol 1 Zurich, August 21–27, 1938. Berne, Suisse, Impr. dr. G. Grunau, p 174

1940, United States Livestock Sanitary Association, Proceedings of the 44th Annual Meeting of the United States Livestock Sanitary Association, Hotel Morrison, Chicago, Ill, December 4–6, 1940, United States Livestock Sanitary Association, Reese Press, Baltimore (1941)

1950, New Zealand Society of Animal Production, Proceedings of the Tenth Annual Conference held in the Lecture Room, Dominion Museum, Wellington on May 3rd, 4th and 5th 1950, Volume 10 of the Proceedings of the New Zealand Society of Animal Production (1950)

5 Internationalisation of Disease and the Trade in Animals

Abstract

The evolution of international instruments that dealt with issues of quarantine, matters of cross-border trade of animals as well as disease prevention shaped the notion of animal wellbeing in an international context. These instruments began with Treaties of Friendship, Commerce and Navigation, progressed to broader bilateral instruments dealing with the national exercise of quarantine power (including over animal disease) and culminated in the use of standards and supervision by international organisations, such as the World Organisation for Animal Health (OIE). These instruments tackled how far national authority could be legitimately maintained over quarantine regulation, where the objective was to safeguard international boundaries against the transmission of disease. In this respect, international markets followed the pathway already set at the national level, harnessing advances from veterinary science to support trade. As with national jurisdictions, the internationalisation of disease control resulted in animal diseases being seen as a trade issue, but this time, control was equated with the "global good" deriving from international trade. Thus, the focus on disease-free shipments contributed greatly to commodification because animals were seen in terms of bulk shipments, rather than individual living beings, in circumstances where anti-cruelty provisions existing at the national level largely did not apply.

Keywords

Animal commodification · Animal disease · Internationalisation of disease · Quarantine treaties · Treaties of Friendship, Commerce and Navigation

S. Riley, *The Commodification of Farm Animals*, Animal Welfare 21,
https://doi.org/10.1007/978-3-030-85870-4_5

5.1 Introduction

This chapter discusses how the evolution of international instruments dealing with trade in animals, quarantine and disease prevention shaped the notion of animal wellbeing in a commercial context. As discussed in Chaps. 2 and 4 of this book, by the nineteenth century, international trade in farm animals and their products had become more globalised, requiring safeguards against the transmission of disease across international boundaries.[1] This led to the livestock sector harnessing advances in the field of veterinary science to support both domestic livestock production and international markets.[2] In the latter case, veterinary science became part of the international agenda when trading partners negotiated treaties requiring inspections and certification of shipments as part of quarantine regulation.

The role of quarantine evolved in three overlapping stages, which among other things tackled how far national authority could be legitimately maintained: first, early treaties of Friendship, Commerce and/or Navigation, which assumed that quarantine was a matter for determination solely by importing countries; second, a raft of bilateral instruments, which either reserved primary grants of quarantine power to the importing country or dealt with animal disease, but nevertheless gave much latitude to domestic quarantine; and, third, the reining-in of domestic quarantine regimes by the use of international standards and supervision by international organisations, such as the World Organisation for Animal Health (OIE).[3]

The second and third stages also evince a trajectory towards internationalisation of disease control, a movement that emerged after the cattle plagues of the 1860s when countries realised that effective quarantine required cooperation among trading partners. However, this standpoint took decades to gain traction, as apprehension persisted that quarantine could be a disguised restriction on international trade. Nevertheless, the regime which emerged made increasing use of objective material, such as inspections and certifications, to verify the disease status of animals. In this way, animal disease was seen as a trade issue and became equated with international trade as part of the "global good".[4] Ultimately, it was a development that greatly contributed to farm animal commodification.

The discussion commences with a brief account of disease control, in the context of international trade, before examining the three phases of quarantine. To undertake this task, the chapter evaluates selected international instruments, arguing that while

[1] R M Hartwell, *the Industrial Revolution and Economic Growth*, Methuen and Co Ltd. London (1971), 207–210.

[2] Ibid, 218.

[3] *International Agreement for the Creation at Paris of an International Office for Dealing with Contagious Diseases of Animals and Annex*, opened for signature 25 January 1924, [1925] *ATS* 15, (entered into force 12 January 1925). The organisation has 182 members. The original name of the OIE was the Office International des Épizooties. However, in May 2003 the name was changed to the World Organisation for Animal Health, while keeping the historical acronym, OIE.

[4] Mark Harrison, "A Global Perspective: Reframing the History of Health, Medicine, and Disease", (2015) 89 (4) *Bulletin of the History of Medicine*, 639, 642.

the putative focus of these instruments centred on animal disease, the most significant issue was the impact of disease on domestic production and international trade. Accordingly, animal diseases, and by implication animal health, were treated as trade issues, resulting in the animals themselves being subsumed as commodities within trade.

Before commencing the discussion, the use of terminology warrants a short explanation. The chapter refers to quarantine, a process that in some jurisdictions has now been incorporated within the concept of biosecurity.[5] However, the term quarantine is more consistent with the terminology used during the periods under discussion; hence it is adopted for this chapter. In addition, the discussion occasionally uses the phrase "international quarantine". This phrase is not defined in international instruments, although it has been used by institutions such as the World Health Organization.[6] For the purposes of this book, international quarantine refers to the collection of instruments dealing with animal health and protection that constitute a quarantine system based on international collaboration, cooperation and harmonisation, designed to prevent the spread of disease across international boundaries. Although quarantine laws are domestic laws, international law became increasingly relevant to establishing what these laws should be. Consequently, the term "international quarantine" encompasses the entirety of international and domestic instruments, laws and measures.

5.2 International Trade and Disease Control

Chapter 2 of this book detailed how, after 1843 and the progressive repeal of the Corn Laws, markets in the United Kingdom (UK) for meat and animal products acquired a robust international dimension. This nascent globalisation allowed UK merchants to make profits of up to 100%, as occurred with the Russian trade and the

[5]Biosecurity Act 2015 (Cth), (available from http://www.austlii.edu.au/cgi-bin/viewdb/au/legis/cth/consol_act/ba2015156/) replacing the Quarantine Act 1908 (Cth) (available from https://www.legislation.gov.au/Series/C1908A00003); Biosecurity Act 1993 (NZ) (available from https://www.legislation.govt.nz/act/public/1993/0095/latest/DLM314623.html) replacing a series of statutes and regulations, that are set out in schedules 3–7 inclusive, which had dealt with quarantine. For discussion on quarantine generally, Opi Outhwaite, "The International Framework for Biosecurity and the Challenges Ahead", (2010) 19 (2) *Review of European Community and International Environmental Law*, 207.

[6]The World Health Organization (WHO) was established as a United Nations Agency. Its objective is the attainment of the highest level of health by all peoples. WHO is set up by the Constitution of the World Health Organization, adopted on 22 July 1946, [1948] ATS 7, (entered into force 7 April 1948). WHO has 194 members. WHO has not defined the term "international quarantine", however, Chap. 18 of *The First Ten Years, of the World Health Organization* is titled "International Quarantine" and the chapter discusses international treaties and regulations on communicable diseases, WHO Geneva (1958) 264; See also generally the material and commentary in WHO, *Global Crises – Global Solutions* WHO Geneva (2002).

transport of cattle via the Baltic states.[7] Global trade, however, was also responsible for introducing disease and pathogens.[8] Although this was hardly a nineteenth-century phenomenon, it differed from earlier times, because of the frequency and severity of epidemics, as well as the diverse points of contagion.[9]

In the case of human health, regulators had long implemented a range of measures based on segregation and isolation. As far back as biblical times, communities isolated those who were ill or diseased,[10] while in 532, the Byzantine Emperor Justinian ordered the segregation of travellers arriving in Constantinople who had journeyed from known areas of disease or infection.[11] In 1377, during a Black Death epidemic, the city of Ragusa (now Dubrovnik) implemented border controls, detaining and segregating people who were ill, or suspected of being ill, initially for 30 days, but then for 40 days, which became the standard.[12] The word "quarantine" derives from the Latin word *quarantina*, which means "40 days", a reference to the period of quarantine initiated by Ragusa.[13]

In a strict sense, segregation of those who are already ill is more aptly described as "isolation", while quarantine is a preventative measure to stop the introduction or spread of disease by those who are suspected of being ill.[14] However, the two systems are analogous as they both involve a process of segregation aimed at preventing the introduction and spread of disease; hence the term "quarantine" is used in this extended sense in this book.

In England, the practice of quarantine commenced in the sixteenth century and over the years covered goods as well as people.[15] In some jurisdictions, the length of quarantine could be increased to 60, or even 90, days depending on travel routes and the goods in question.[16] This is not to say that quarantine was well-accepted. As discussed in Chap. 4 of this book, until the end of the nineteenth century, there was little consensus among the scientific community regarding the cause of disease, and even after vindication and acceptance of contagion theory, quarantine was not

[7] C A Spinage, *Cattle Plague, A History*, Kluwer Academic, New York (2003), 231.

[8] Mark Harrison, "A Global Perspective: Reframing the History of Health, Medicine, and Disease", above 4, 651.

[9] Ibid, 653–654.

[10] William H McNeill *Plagues and Peoples, Plagues and Peoples*, Anchor Books, New York (1989 reprint originally published 1976), 96–97.

[11] O Schepin and W Yermakov *International Quarantine*, International Universities Press Inc. Connecticut (1991). 11.

[12] William H McNeill *Plagues and Peoples, Plagues and Peoples*, above 10, 181.

[13] G Gensubu, M Yacoub, A Conti "The Concept of Quarantine in History: from Plague to SARS", (2004) Volume 49 *Journal of Infection* 257, 258; P. Sehdev, "The Origin of Quarantine" *Clinical Infectious Diseases*, volume 35 (2002), 1071.

[14] Ibid, 257, 257–258.

[15] John Booker, *Maritime Quarantine, the British Experience, c 1650–1900*, Ashgate, Aldershot (2007), 1, 385.

[16] Ibid, 385.

necessarily effective, so that the regime remained contentious and unpopular.[17] Travellers found measures intrusive and inconvenient, while merchants found quarantine tiresome and expensive.[18]

Critiques invariably targeted the cost and tardiness of measures and were particularly pronounced for trade that passed through the "Levant" or the eastern Mediterranean.[19] By the mid-nineteenth century, Turkey had set up 60 inland quarantines and 102 quarantines at seaports, which in the words of two commentators:

> [demonstrated a] . . .fanaticism. . . [which] put an insurmountable barrier in the way of trade. This aroused the grave discontent of the main trading powers, which sought the introduction of more liberal and rational quarantine measures and actual control over their operation.[20]

Yet, border controls to this day remain a country's first line of defence and invariably involve some type of trade restraint.[21] However, given that debates surrounding theories of disease were not resolved until the latter part of the nineteenth century, it is not surprising that by the middle of that century, discussion in Western Europe crystallised into two distinct approaches. One group wanted quarantine laws "relaxed" so that trade could be liberalised,[22] while the other group was concerned that relaxing quarantine would expose Europe to threats of epidemics from the east.[23] Depending on one's point of view, quarantine regulation was a necessary step towards disease control and the protection of a country's territory, or alternately it was "an instrument of foreign policy to be used aggressively in furtherance of national interests".[24]

Criticisms of quarantine were underscored by the fact that border controls were initially implemented unilaterally, resting solely within the national domain.[25] This criticism became more pronounced as trade and transportation advances exposed

---

[17] Ibid, 545, 548.

[18] Ibid, 195, 385; William Novak, *The People's Welfare,* The University of North Carolina Press Chapel Hill and London (1996), 208–211.

[19] M Harrison "Disease, Diplomacy and International Commerce: the Origins of International Sanitary Regulation in the Nineteenth Century" [2006] *Journal of Global History* 197 at 199.

[20] O Schepin and W Yermakov *International Quarantine*, above 11, 52.

[21] John Booker, *Maritime Quarantine, the British Experience, c 1650–1900*, above 15, 367; restrictions in place throughout 2020 during the COVID-19 pandemic provides a more recent example.

[22] For example, trade between Europe and Egypt in cotton. Mark Harrison "Disease, Diplomacy and International Commerce: the Origins of International Sanitary Regulation in the Nineteenth Century" [2006] *Journal of Global History*, above 19, 204–205.

[23] Mark Harrison "Disease, Diplomacy and International Commerce: the Origins of International Sanitary Regulation in the Nineteenth Century, above 19, 205.

[24] Ibid, 198.

[25] S Eystein and W Yngvild "Ecological Consequences of the Spreading of Pathogens and Genes Through an Increasing Trade in Foods" in *Proceedings of Norway/UN Conference on Alien Species* 141 Trondheim 1–5 July Directorate for Nature Management and Norwegian Institute for Nature Research Norway 1996.

border controls to political murmurings regarding trade liberalisation and the restrictive nature of quarantine.[26] Notwithstanding these misgivings, the matter did not receive formal attention until the early part of the twentieth century.[27] Yet, even before that time, governments were aware that they needed to grapple with two interrelated issues: the relationship between trade restraints and disease control in international trade and the effectiveness of an international regime based on a series of disconnected national laws. The first issue challenged regulators in balancing the complexities of protecting national health and territory, without unduly hindering international trade, while the second issue questioned how far international supervision could extend without compromising the ability of a country to retain control over its borders.

In the animal product sector, an outright prohibition on the importation of livestock and risky goods, such as skins and fibre, would have considerably reduced the risk of introducing disease. However, these types of trade bans would also have contradicted free trade policies adopted after the mid-nineteenth century, especially in countries such as the UK. Even less stringent measures, which allowed restricted landing of foreign animals, could be contentious.[28] When cattle plague recurred in the UK in 1872 and 1877, suggestions that all farm animals be killed on arrival met with resistance because live trade generated employment and provided food for the working classes.[29] Moreover, slaughter upon landing tended to encourage the export of diseased animals, who may have been strong enough to survive a sea voyage, but carried infection.[30]

These types of concerns beleaguered quarantine into the twentieth century, where commentators noted that trade restraints provided a temporary halt to the introduction and spread of disease but were inevitably at the cost of disrupted and destabilised international trade.[31] In particular, countries could introduce trade bans without warning, as well as impose stricter regulations than had thus far applied, all of which contributed to confusion and mistrust.[32] The ability of countries to introduce national measures at will is also relevant to the second issue, the effectiveness of a multi-jurisdictional regime based on discrete national laws.[33]

A difficult problem stemmed from the fact that as long as quarantine regulation derived from unilateral national initiatives, measures would be anchored to local

[26] Mark Harrison "Disease, Diplomacy and International Commerce: the Origins of International Sanitary Regulation in the Nineteenth Century", above 19, 213.

[27] Discussion in Sect. 5.3.3 of this Chapter.

[28] C A Spinage, *Cattle Plague, A History*, above 7, 304–305.

[29] Ibid, 304–305.

[30] Ibid, 300.

[31] The Royal Economic Society, "The Economic Effects of Cattle Disease Legislation", (1905) 15 (58) *The Economic Journal*, 156, 159–160.

[32] Ibid, 159–160.

[33] Ibid.

conditions and remain non-standardised among trading partners.[34] This could lead to substandard quarantine regimes in one country facilitating the spread and introduction of disease in other jurisdictions. In the early eighteenth century, the Netherlands rejected wholesale slaughter of infected animals, instead relying on treatments and cures that proved ineffective.[35] As a result, the region remained a source of infection for decades to come.[36] In an analogous example, Spain had managed to avoid cattle plague for the greater part of the nineteenth century, primarily because it did not import cattle.[37] However, when farmers in the border region of the Pyrenees acquired foreign animals, they also unknowingly imported cattle plague, which subsequently spread throughout the whole country.[38]

Unsurprisingly, lax border controls and regulatory deficiencies allowed sick animals to be traded, facilitating the repeated occurrence of infection.[39] When the cattle plague returned to Europe in the early 1870s, European governments belatedly acknowledged that an international solution was needed. This led to an intergovernmental conference held in Vienna in 1871, to discuss the formulation and adoption of standardised measures.[40] Although veterinarians had recognised the need for uniform measures from at least the First International Veterinary Conference held in 1863, governments still tended to adopt measures in response to local conditions.[41] Consequently, an international solution to the problem of disease control remained elusive. Part of the hesitancy stemmed from the fact that domestic regulation followed pathways predetermined by centuries of international commercial agreements, which assumed border controls and quarantine lay solely within the domain of national governments. Accordingly, attempts to subject domestic quarantine measures to the auspices of international law took time to become accepted.

---

[34] O Schepin and W Yermakov *International Quarantine*, above 11, 63.

[35] Lise Wilkinson, *Animals and Disease, An introduction to the History of Comparative Medicine*, Cambridge University Press, Cambridge (1992), 53.

[36] Ibid, 53.

[37] George Fleming, *Animal Plagues: Their History, Nature, and Prevention*, Chapman and Hall, London (1871), 479–480. This copy is a reprint.

[38] Ibid, 479–480.

[39] C A Spinage, *Cattle Plague, A History*, above 7, 265–266.

[40] Joanna Swabe, *Animals, Disease and Human Society, Human-animal Relations and the Rise of Veterinary Medicine*, Routledge, London (1999), 105; "Principles for an International Regulation for the Extinction of the Cattle Plague", and "Translation of the Appendix to the Debates of the International Conference at Vienna, Exposé by the English Delegate," in the *Report of the Veterinary Department for 1872,* London, Eyre and Spottiswoode, 1873, Appendix, 23–29, in *Parliamentary Papers,* the House of Commons and Command, 26 as cited, Amanda Kay McVety, *The Rinderpest Campaigns: A Virus, Its Vaccines, and Global Development in the Twentieth Century*, Cambridge University Press (2018), 22.

[41] Discussion in Sect. 4.5.1 of this Book.

5.3 International Trade and Quarantine Regulation

As foreshadowed in the introduction to this chapter, international quarantine regulation progressed through three overlapping phases, starting with early Treaties of Friendship, Commerce and Navigation, which gave way to treaties reserving primary grants of quarantine power to the importing country. This second phase also included treaties dealing with animal disease, where quarantine still largely remained the purview of domestic regimes. The third phase represents attempts to rein in quarantine, by transforming it from a regime based on unilateral national initiatives to one increasingly shaped by international prerogatives. A selection of treaties and conferences, up to the mid-twentieth century, is set out in Table 5.1 and forms the basis of the discussion which follows.

5.3.1 Treaties of Friendship, Commerce and Navigation

Treaties of Friendship, Commerce and Navigation largely encompassed bilateral treaties, dealing with an eclectic range of matters. Unlike treaties of the twentieth century that saw the negotiation of specific agreements on topics ranging from racial discrimination to the Law of the Sea,[42] Treaties of Friendship, Commerce and Navigation were catch-all treaties covering matters such as strategic alliances,[43] human rights and immigration.[44] One of the earliest, the *1373 Treaty between England and Portugal*,[45] stipulated that:

> There shall be between the respective kings and their successors, their realms, lands, dominions, provinces, vassals, and subjects... true, faithful, constant, mutual, and perpetual

[42] *Convention on the Elimination of all Forms of Racial Discrimination,* [1975] *ATS* 40, opened for signature by the United Nations General Assembly on 21 December 1965, entering into force on 4 January1969. The Convention has 182 parties; *The United Nations Law of the Sea Convention,* [78] *ATS* 31, opened for signature 10 December 1982, entering into force on 16 November 1994. The Convention has 168 parties.

[43] Treaty of Peace and Commerce between Great Britain and Tunis 1662 Volume 17 *CTS* 244, Articles I and II.

[44] J Leyser, "Treaties of Friendship, Commerce and Navigation", (1953) 7 (2) *Australian Outlook*, 132, 132; John F Coyle, "The Treaty of Friendship, Commerce and Navigation in the Modern Era", (2013) 51 *Columbia Journal of Transnational Law* 302, 302, 304, 311–312.

[45] Governments of the Kingdom of England and the Kingdom of Portugal, 16 June 1373, in *British Documents on foreign affairs—reports from the Foreign Office confidential Print. Part II. From the First to the Second World War. Series F, Europe*. Vol. 24., ed. Anthony Adamthwaite. (Frederick, MD: University Publications of America, 1991), 101, cited footnote 33 by Matthew Winslett, *The Nadir of Alliance: The British Ultimatum of 1890 and its Place in Anglo-Portuguese Relations, 1147–1945*, Presented to the Faculty of the Graduate School of The University of Texas at Arlington in Partial Fulfillment of the Requirements for the Degree of Master of Arts in History the University of Texas at Arlington August 2008, available from https://rc.library.uta.edu/uta-ir/handle/10106/1093. A copy of the treaty is also available from the Natonal Archives, UK http://discovery.nationalarchives.gov.uk/details/r/C3624900

Table 5.1 Treaties and intergovernmental conferences relating to quarantine

Instrument	Article
1373 Treaty between England and Portugal	No reference to quarantine
1662 The Treaty of Peace and Commerce between Great Britain and Tunis	No reference to quarantine
1734 Treaty of Commerce between Great Britain and Russia	No reference to quarantine
1815 Convention of Commerce between the United Kingdom of Great Britain and Ireland and the United States of America, and Declaration by United Kingdom Regarding St Helena	No reference to quarantine
1852 Sanitary Convention between France, Portugal, Sardinia, Tuscany and Turkey	An Appendix to the convention contained 137 Articles for dealing with cholera Did not enter into force
1871 Intergovernmental Conference held in Vienna	Adopted Six "Principles for an International Regulation for the Extinction of the Cattle Plague"
1882 Convention between France and Great Britain for the Regulation of Commercial and Maritime Relations and Declaration Prolonging the Treaties in Force	Article II stated that although the transit of goods was to be free of transit duties, the parties reserved for themselves to decide on prohibitions or temporary restrictions of import, export or transit, for sanitary reasons, to prevent the spread of cattle disease, or the destruction of crops
1887 Convention Designed to Remove the Danger of Epizootic Diseases in the Territories of the Two Countries	Article 2 deals with epizootic diseases of farm animals. Required certifications that the animal had come from a disease-free area Article 2 other quarantine measures included inspection and destruction of infected animals
1922 Commercial Treaty between the Esthonian Republic and the Kingdom of Hungary	Article 19(2) allowed exceptions in respect of free trade, for "considerations of public health or protection against infectious diseases of animals or plants"
1923 Treaty of Commerce and Navigation made between Denmark and Finland	Article VIII (3) reserved to each contracting party the right to enact laws to protect against infectious diseases of plants or animal
1923 Convention Relating to the Simplification of Customs Formalities	Article 17 specified that the contracting states could introduce "exceptional measures. . .to ensure the health of human beings, animals or plants"
1924 Treaty of Commerce and Navigation Between Italy and Albania	Article 7 permitted exemptions for the protection of useful animals and plants against disease and harmful parasites
1925 Convention of Commerce and Navigation with Final Protocol between Sweden and Czechoslovakia	Article 4 stated that measures should be made in "conformity with the universally recognized international regulations"
1927 Convention for the Abolition of Import and Export Prohibitions and Restrictions	Article 4 allowed eight exceptions to the general goal of abolition of trade restrictions, including prohibitions or restrictions that were

(continued)

Table 5.1 (continued)

Instrument	Article
	permitted to protect animals or plants against disease, insects and harmful parasites
1928 Treaty of Commerce and Navigation between Bulgaria and Turkey	Article 5(3) permitted exceptions to free trade that were applied for reasons connected with public health or to protect animals and useful plants against disease and noxious insects and parasites, in conformity with the international principles adopted with regard to this matter
1928 Treaty of Commerce and Navigation with Final Protocol and Protocol concerning the Import and Export Régime in Italy Between Hungary and Italy	Article 14 stipulated that quarantine regulation had to be applicable to all countries or to countries in similar circumstance
1928 Treaty of Commerce and Navigation with Protocol Between Union of South Africa and Germany	Article 10 allowed exceptions to free trade in consideration of public health or for protection of animals or useful plants against diseases, insects and harmful parasites
1930 Convention of Commerce and Navigation Between Poland and Roumania, Annex I, Veterinary Convention	Article 1 of the annexed convention stated that each country would comply with veterinary and commercial requirements of the other country Article 4 of the annexed convention stated that animals to be accompanied by a veterinary certificate of health and inspected at the border
1935 International Convention for the Campaign Against Contagious Diseases of Animals	Preamble states that the objective is to facilitate international trade in livestock and their products, noting that it is necessary to tighten quarantine and veterinary controls Article 5 refers to veterinary health bulletins
1935 International Convention Concerning the Transit of Animals, Meat and Other Products of Animal Origin	Article 1 transit of farm animals permitted for parties to the 1935 International Convention for the Campaign Against Contagious Diseases of Animals. Subject to prior notification and could be refused in cases of outbreak of disease Article 2 mandated the use of veterinary certificates and model certificate of health and origin of animals annexed Article 5 mandated veterinary inspections
1935 International Convention Concerning the Export and Import of Animals, Meat and Other Products of Animal Origin	Article 3 mandated the use of veterinary certificates and model certificate of health and origin of animals annexed
1952 Agreement Concerning Epizootic Diseases Between the Kingdom of Greece and The Federal people's Republic of Yugoslavia	Preamble indicates that the treaty is primarily aimed at ensuring the health of animals and animal products to facilitate trade Article 1 provides for animals to undergo veterinary inspection

(continued)

Table 5.1 (continued)

Instrument	Article
1959 Agreement Concerning Co-Operation in the Field of Veterinary Science	Preamble notes the objectives include the control of animal disease and developing economic and trade relations Article 1 provides for coordination of measures and exchange of information Article III stipulates that the parties shall draft coordinated bilateral regulations

> friendships, unions, alliances and... shall assist, maintain, and uphold each other mutually, by sea and by land...[46]

This treaty, acknowledged as the oldest still in force,[47] was invoked by Winston Churchill during World War II to assist the UK in protecting merchant shipping plying the Atlantic.[48]

In 1662, the *Treaty of Peace and Commerce between Great Britain and Tunis*[49] covered a typical assortment of issues, including the rights of free passage for citizens of each party, diplomatic immunity and payment for runaway slaves.[50] Yet, the treaty was silent on matters of quarantine. Some 72 years later, the 1734 *Treaty of Commerce between Great Britain and Russia* dealt with passports, imposition of duties and customs as well as giving the subjects of each contracting party "liberty of navigation and commerce", but again did not broach quarantine.[51] Similarly, the 1815 *Convention of Commerce between the United Kingdom of Great Britain and Ireland and the United States of America, and Declaration by United Kingdom Regarding St Helena* (1815 Convention between the United Kingdom and the United States of America) covered issues relating to reciprocal liberty of citizens, commerce and shipping, as well as the appointment of consuls, but not quarantine.[52] As Steve Charnovitz pointed out, while Treaties of Friendship,

[46] J J Munro, "England's Treaties of Guarantee", (1881) CCXL May, *Law Magazine and Review*, 215, 226.

[47] Lesyer 132; Winston 14.

[48] Winston Churchill, Agreement with Portugal, 12 October, 1943, House of Commons Debate, Volume 392, Hansard (1943) column 716–719, available from https://api.parliament.uk/historic-hansard/commons/1943/oct/12/agreement-with-portugal

[49] Treaty of Peace and Commerce between Great Britain and Tunis, 5 October, 1662, Vol 7 *CTS* 241.

[50] *Treaty of Peace and Commerce between Great Britain and Tunis*, Articles I, VII, VIII and XII.

[51] *Treaty of Commerce between Great Britain and Russia*, 2 December 1734, Vol 34 *CTS* 211, Articles 2, 4, 5 and 17.

[52] *Convention of Commerce between the United Kingdom of Great Britain and Ireland and the United States of America, and Declaration by United Kingdom Regarding St Helena*, 3rd July 1815, Articles 1, 2 and 4, 1901 *Australian Treaty Series*, 107, (electronic), available from http://www.austlii.edu.au/au/other/dfat/treaties/1901/107.html

Commerce and Navigation had particularly broad remits, up to the later part of the nineteenth century, they omitted references to quarantine and disease control.[53]

This was a significant omission, given that from 1377, 40 days' quarantine had become an increasingly established practice, and quarantine had been implemented in the UK from the sixteenth century.[54] One likely explanation is that states themselves tacitly accepted the legitimacy of national sovereignty over borders, hence there was no need to reference this in their treaties. Another explanation stems from the fact that as long as theories on the origin and control of disease remained unsettled, it was difficult to achieve the type of consensus which could have supported binding treaties and conventions.[55] Whatever the reason, by the nineteenth century, countries began to reserve powers over health, safety and disease control to themselves.

5.3.2 Reservations, Exemptions and Quarantine

Agreements reserving quarantine power to the importing nation are best understood in the context of international trade relations of the time, particularly the use of most favoured nation (MFN) clauses. MFN provisions were incorporated into English treaties from the twelfth century, eventually becoming a standard provision in trade instruments across the globe,[56] including within the World Trade Organization (WTO).[57] Although the exact wording of these clauses varied, the intent was consistent, in that all trading partners would receive privileges and concessions equivalent to those bestowed on the most favoured trading partner.[58]

The 1815 Convention between the United Kingdom and the United States of America provided that the citizens of the United States of America (US) would pay no higher duties or charges on cargo than "shall be payable on the same articles when imported or exported in the vessels of the most favoured European nations".[59] In a

[53] S Charnovitz "Exploring the Environmental Exceptions in GATT Article XX", (1991) 25 (5) *Journal of World Trade Law* 37, 40; J Jackson *The World Trading System – Law and Policy of International Economic Relations* MIT Press, Cambridge MA and London (2002), 35.

[54] William H McNeill *Plagues and Peoples, Plagues and Peoples*, above 10, 181; John Booker, *Maritime Quarantine, the British Experience, c 1650–1900*, above 15, 1, 385.

[55] Discussion in Chap. 4, Sect. 4.3 of this book.

[56] Georg Schwarzenberger, "The Most-Favored-Nation Standard in British State Practice", (1945) 22 *British Year Book of International Law*, 96, 97–98.

[57] *Marrakesh Agreement Establishing the World Trade Organization,* opened for signature 15 April 1994, 1867 UNTS 3, entering into force on 1 January 1995; *Agreement on the Application of Sanitary and Phytosanitary Measures*, 1867 UNTS 3 annex 1A; The WTO has 164 members; *General Agreement on Tariffs and Trade* (GATT 1994) [1995] *ATS* No 8 Article 1. GATT 1947 has been subsumed into the 1994 GATT; Members of the WTO are members of GATT 1994.

[58] Generally, Georg Schwarzenberger, "The Most-Favored-Nation Standard in British State Practice," above 56, 96.

[59] *Convention of Commerce between the United Kingdom of Great Britain and Ireland and the United States of America, and Declaration by United Kingdom Regarding St Helena*, Article 3.

similar way, Article 1 of the 1882 *Convention between France and Great Britain for the Regulation of Commercial and Maritime Relations and Declaration Prolonging the Treaties in Force* (1882 Convention between France and Great Britain) gave each party "the benefit of every favour, immunity or privilege. . . conceded. . . to any third nation whatsoever".[60] Outside of the UK, the 1922 *Commercial Treaty between the Esthonian Republic and the Kingdom of Hungary*[61] provided a typical MFN clause of the twentieth century, stipulating that the nationals of each party would be afforded treatment equivalent to the nationals of the most favoured nation.[62] By the end of the twentieth century, Article 1 of GATT 1994 required that:

> any advantage, favour, privilege or immunity granted by any contracting party to any product originating in or destined for any other country shall be accorded immediately and unconditionally to the like product originating in or destined for the territories of all other contracting parties.

It is important to keep in mind that beyond equivalence of treatment, these provisions did not impose obligations to comply with specific treaty provisions. Instead, the articles granted rights that were "merely the counterpart of the rights granted by the promisor to third States".[63] In this way, MFN clauses aimed at eliminating discrimination among trading partners and as a result standardised some elements of international trade. Additionally, the clauses allowed trade agreements to evolve and adapt to change, promoting free trade by an incremental increase of privileges and concessions.[64] By way of contrast, reservations and exercise of quarantine power would tend to have a limiting effect on trade, because measures inevitably imposed trade restraints or trade embargoes. As well, to the extent that quarantine regulation imposed trade restrictions against some countries but not others, the laws were discriminatory, technically infringing the free trade principles that underpinned MFN provisions.

Accordingly, from the late nineteenth century, nations drafted quarantine powers as an exemption to free trade arrangements. Article II of the 1882 Convention between France and Great Britain acknowledged that while the transit of goods was to be free of duties, the parties nevertheless reserved to themselves the power to make "such prohibitions or temporary restrictions of import, export or transit which they may think necessary to enforce for sanitary reasons, to prevent the spread of

[60] *Convention between France and Great Britain for the Regulation of Commercial and Maritime Relations and Declaration Prolonging the Treaties in Force,* 28 February 1882, (1882) 160 *CTS* 143.

[61] *Commercial Treaty between the Esthonian Republic and the Kingdom of Hungary*, 19, October 1922, Vol 30 LNTS, 347, available from https://treaties.un.org/doc/Publication/UNTS/LON/Volume%2030/v30.pdf

[62] Ibid, Article 1.

[63] Georg Schwarzenberger, "The Most-Favored-Nation Standard in British State Practice," above 56, 96.

[64] Ibid, 99–100.

cattle disease, or the destruction of crops". Analogously, Article 19(2) of the 1922 Treaty between Estonia and Hungary allowed exceptions for "considerations of public health or protection against infectious diseases of animals or plants". Similarly, Article VIII (3)(2) of the 1923 *Treaty of Commerce and Navigation made between Denmark and Finland* included exemptions from free trade provisions, for laws enacted to protect against infectious diseases of plants or animals.[65] Likewise, Article 7 of the 1924 *Treaty of Commerce and Navigation Between Italy and Albania* permitted exemptions for reasons of public health or the protection of useful animals and plants against disease, insects and harmful parasites.[66] A similar and further example stems from Article 10 of the 1928 *Treaty of Commerce and Navigation with Protocol Between the Union of South Africa and Germany*, which also allowed exemptions to protect public health, animals or useful plants.[67]

These types of reservations persisted throughout the twentieth century and permitted countries to retain extensive powers over their borders.[68] It is also significant that exemptions gave trading partners the ability to enact laws, but did not oblige them to do so. Consequently, as long as disease, safety and quarantine were reserved to the national domain, international regulation remained dependent on unilateral national initiatives, leading to regulation that was not standardised among trading partners.

For these reasons, countries also negotiated specific treaties dealing with animal disease. These provided frameworks for establishing national systems to deal with epizootics, assisting trading partners in having equivalent measures, so that substandard regulation in one jurisdiction would not foster the introduction and spread of disease into other jurisdictions. One of the earliest of these treaties, the 1887 *Convention Designed to Remove the Danger of Epizootic Diseases in the Territories of the Two Countries*, was negotiated between Austria-Hungary and Italy.[69] The objective, set out in Article 1, was to stop animal diseases spreading across the border between the two countries, emphasising that measures were to be based on

---

[65] *Treaty of Commerce and Navigation made between Denmark and Finland*, 3 August, 1923, Vol 21 LNTS 269, available from https://treaties.un.org/doc/Publication/UNTS/LON/Volume%2021/v21.pdf

[66] *Treaty of Commerce and Navigation between Italy and Albania*, 20 January 1924, Vol 44 *LNTS*, 359, available from https://treaties.un.org/doc/Publication/UNTS/LON/Volume%2044/v44.pdf

[67] *Treaty of Commerce and navigation with Protocol between the Union of South Africa and Germany*, 1 September 1928, 95 *LNTS*, 289, available from https://treaties.un.org/doc/Publication/UNTS/LON/Volume%2095/v95.pdf

[68] Levent Sabanogullari, *General Exception Clauses in International Investment Law: The Recalibration of Investment Agreements* via *WTO-Based Flexibilities*, Nomos Verlagsgesellschaft, Germany (2018), 118, footnote 294 that sets out more than 35 bilateral treaties of the twentieth century with health and sanitary exemptions.

[69] *Convention Designed to Remove the Danger of Epizootic Diseases in the Territories of the Two Countries*, 7 December, 1887, reprinted in Bernd Rüster and Bruno Simma (eds) Vol IV *International Protection of the Environment: Treaties and Related Documents*, Dobbs Ferry: Oceana Publications Inc., New York (1975) 1586.

veterinary requirements and be non-discriminatory in the sense that they applied equally to both parties. However, the right to halt trade was maintained by Article 2.3, where disease spread across borders, even if animals or their products came from non-infected areas.[70]

The use of veterinary knowledge became a common feature, typified by the 1930 *Convention of Commerce and Navigation Between Poland and Roumania*.[71] Annex 1 of the convention contained a veterinary convention, which stipulated that animals were to be accompanied by a certificate of health and inspected at the border.[72] In addition, the customs office of each country needed to comply with the veterinary and commercial requirements of the other country.[73] Similarly, the 1952 *Agreement Concerning Epizootic Diseases Between the Kingdom of Greece and The Federal people's Republic of Yugoslavia* provided that animals needed to undergo veterinary inspection prior to their importation.[74] Article 7(5) also specified that if there was an outbreak of foot and mouth disease, the importing country could prohibit trade in animal products, but if the importing country did not do so, the export country had "an obligation to suspend exports".

Although these types of treaties made national regimes subject to veterinary supervision, and evinced an intention that each party should undertake equivalent regulation, measures were still based on individual national action, so that quarantine depended on the law and policy of the importing country. This approach did little to address either the role of globalisation in the introduction and spread of disease or the place of international law in domestic quarantine regimes. Moreover, complaints about quarantine persisted. By its nature, quarantine imposed trade restraints, could be discriminatory and continued to be perceived as a disguised restriction on international trade.[75]

---

[70] Similar bilateral treaties were concluded by Austria-Hungary with Serbia, Switzerland and Germany, Austria-Hungary Serbia, 24 April, 1881 Vol 8 *Nouveau Recueil Général, Deuxième Série*, 352; Austria-Hungary Switzerland 31 March 1883 Vol 8 *Nouveau Recueil Général, Deuxième Série*, 588; Austria-Hungary Germany 6 December 1891 Vol 17 *Nouveau Recueil Général, Deuxième Série*, 923.

[71] *Convention of Commerce and Navigation Between Poland and Roumania*, 23 June 1930, Annex I, reprinted in Bernd Rüster and Bruno Simma (eds) Vol IV *International Protection of the Environment: Treaties and Related Documents*, Dobbs Ferry: Oceana Publications Inc., New York (1975) 1682.

[72] 1930 *Convention of Commerce and Navigation Between Poland and Roumania*, 23 June 1930, Annex I, Article 4.

[73] Ibid, Article 1.

[74] *Agreement Concerning Epizootic Diseases Between the Kingdom of Greece and The Federal people's Republic of Yugoslavia* 2 February 1952, reprinted in Bernd Rüster and Bruno Simma (eds) Vol IV *International Protection of the Environment: Treaties and Related Documents*, Dobbs Ferry: Oceana Publications Inc., New York (1975), 1833.

[75] League of Nations, *General Report of the Sub-Committee of Experts on Veterinary Questions*, Official No 0.633.M.252. Economic and Financial 1930. II. 49. Geneva, 4 November 1930, 4, available from https://biblio-archive.unog.ch/Dateien/CouncilMSD/C-633-M-252-1930-II_EN.pdf

All the same, the international community was aware of the limitations inherent in a regime that dealt with regional and/or global issues but was based on autonomous domestic measures. In the field of human health, nations tempered their assumptions that quarantine was solely a domestic matter and from 1851 European governments initiated a series of International Sanitary Conferences aimed at developing standardised measures to prevent the spread of cholera, yellow fever and plague.[76] These initiatives paved the way for multilateral agreements where quarantine regulation was brought under an international umbrella.

5.3.3 Quarantine and International Supervision

The third stage in the evolution of quarantine regulation encompassed attempts to rein in domestic quarantine regulation, either by reference to international standards or by establishing international organisations with a supervisory role.

In the field of human health, the Annex to the 1852 *Sanitary Convention between France, Portugal, Sardinia, Tuscany and Turkey, with Annex*[77] contained 137 articles, setting out the basis for uniform measures to prevent the spread of cholera. However, Tuscany and Turkey did not ratify the convention, while Portugal and Sardinia, who had initially ratified the convention, withdrew, so that the convention never entered into force.[78] Indeed, the idea of subordinating national measures to international supervision was so contentious that it was not until 1892, and the seventh International Sanitary Conference, held in Venice, that states reached agreement on a binding treaty.[79]

With respect to animal disease, the 1871 Intergovernmental Conference of Vienna saw representatives from Austria, Belgium, France, Germany, Hungary Italy, Romania, Russia, Serbia, Switzerland and Turkey convene to negotiate standardised measures for dealing with the cattle plague.[80] The conference, which

[76] Norman Howard-Jones, *The Scientific Background of the International Sanitary Conferences*, World Health Organization, Geneva (1975), available from https://apps.who.int/iris/bitstream/handle/10665/62873/14549_eng.pdf;jsessionid=2B113F7596689E48F313A2187319269E?sequence=1

[77] The *1852 Sanitary Convention between France, Portugal, Sardinia, Tuscany and Turkey*, 3 February 1852, Vol 107 *CTS* 345; The convention was negotiated after a series of meetings in Paris in 1851; Ministère des Affaires Étrangères, *Procès-Verbeaux, de la Conférence Sanitaire Internationale*, Tome I, Paris Imprimerie Internationale MDCCCLII, available from Harvard Law School Library, https://iiif.lib.harvard.edu/manifests/view/drs:7309389$9i

[78] Discussion in V Huber "The Unification of the Globe by Disease? The International Sanitary Conferences on Cholera 1851–1894", (2006) Volume 49 No 2 *The Historical Journal* 453 at 461.

[79] *International Sanitary Convention between Austria-Hungary, Belgium, Denmark, France, Germany, Great Britain, Italy, the Netherlands, Portugal, Russia, Spain, Sweden-Norway, and Turkey, with Annexes I to V*, signed 30 January 1892, Vol 176 *CTS* 395. The convention had 13 parties; discussion in Oleg Schepin and Waldemar Yermakov, *International Quarantine*, above 11, 70–134.

[80] C A Spinage, *Cattle Plague, A History*, above 7, 299.

post-dated the international veterinary conferences by some 8 years, developed six "Principles for an International Regulation for the Extinction of the Cattle Plague", including obligations relating to communication of outbreaks of disease, the adoption of veterinary strategies to provide a timely response to the threat of disease, disinfection of conveyances used to transport animals, preparedness to deploy measures to eradicate cattle plague, compensation for owners where animals were destroyed and contingency provisions with respect to international trade.[81] These principles were based on measures already successfully implemented throughout the UK, but as with the 1852 Sanitary Convention, did not lead to a binding treaty. Nevertheless, parties entered into numerous bilateral treaties, similar to the 1877 Convention between Austria-Hungary and Italy.[82]

The fact that neither the 1852 Sanitary Convention nor the 1871 Vienna Conference resulted in binding arrangements illustrated the difficulties involved in reaching an international agreement. This led to a quarantine-and-trade paradox. Without effective quarantine, diseases would continue to be introduced and spread among trading partners, so that trade restraints, including trade embargoes, would eventually need to be imposed. The goal, therefore, was to implement effective quarantine but in a way that neither unduly impinged on domestic border controls nor unduly hindered international trade. Efforts, therefore, turned to standardising measures, on the basis that this would provide much-needed regulation, but applied in a non-discriminatory way, which would lessen unnecessary trade restrictions.[83]

Accordingly, treaties from the 1920s included references to quarantine in terms of "recognised" or "adopted" international principles. Article 4 of the 1925 *Convention of Commerce and Navigation with Final Protocol between Sweden and Czechoslovakia* stipulated that sanitary measures should be made in "conformity with the universally recognized international regulations".[84] The 1928 *Treaty of Commerce and Navigation between Bulgaria and Turkey* specified that measures needed to be non-discriminatory, to the extent that they were "applicable to all countries or to countries in the like circumstances" and had to be made in conformity with adopted international principles.[85] In an almost identical manner, the 1928 *Treaty of Commerce and Navigation with Final Protocol and Protocol concerning*

[81] Ibid, 299–300.

[82] Amanda Kay McVety, *The Rinderpest Campaigns: A Virus, Its Vaccines, and Global Development in the Twentieth Century*, above n 40, 22; for other treaties, Levent Sabanogullari, *General Exception Clauses in International Investment Law: The Recalibration of Investment Agreements via WTO-Based Flexibilities*, above 68.

[83] S Charnovitz "Exploring the Environmental Exceptions in GATT Article XX", above 53, 40–44.

[84] *Convention of Commerce and Navigation with Final Protocol Sweden and Czechoslovakia*, 36 *LNTS* 289, available from https://treaties.un.org/doc/Publication/UNTS/LON/Volume%2036/v36.pdf

[85] *Treaty of Commerce and Navigation between Bulgaria and Turkey*, 12 February 1928, 81 *LNTS* 383, Article 5, 5(3), available from https://treaties.un.org/doc/Publication/UNTS/LON/Volume%2081/v81.pdf

the Import and Export Régime in Italy Between Hungary and Italy provided that quarantine regulation had to apply to all countries or to countries in a similar circumstance.[86]

Notwithstanding treaty references to international regulations or accepted principles, these simply did not exist.[87] In the case of animal disease, delegates to international veterinary conferences had called for standardised measures in international trade from at least the 1860s, yet to no avail.[88] Even after the cattle plagues of the 1860s and the Vienna Conference in 1871, governments were reluctant to adopt multilateral treaties to oversee domestic quarantine. Attempts at multilateral standardisation started again in earnest from the 1920s, in the form of two customs and tariff conventions.

The first, the 1923 *International Convention Relating to the Simplification of Customs* (Customs Convention), aimed at relieving "international commerce from the burden of unnecessary, excessive, or arbitrary customs or other similar formalities".[89] Articles 1 and 2 obliged the contracting parties to promote equitable treatment of commerce by revising national law, including technical requirements, such as health and sanitary regulation.[90] In particular, where commerce depended on the sanitary condition of imported products, the parties were expected to negotiate agreements that facilitated reliance on certifications issued by the exporting country. However, Article 17 emphasised that the contracting states could introduce "exceptional measures…to ensure the health of human beings, animals or plants". While this convention did not prescribe standardised measures, it evinced an expectation that parties would negotiate agreements that moved towards standardisation.

The second instrument, the 1927 Convention for the Abolition of Import and Export Prohibitions and Restrictions, was sponsored by the League of Nations.[91] As with the Customs Convention, it focussed on the elimination of trade restraints, and

[86] *Treaty of Commerce and Navigation with Final Protocol and Protocol concerning the Import and Export Régime in Italy Between Hungary and Italy*, 4 July 1928, 92 *LNTS* 117, Article 14, available from https://treaties.un.org/doc/Publication/UNTS/LON/Volume%2092/v92.pdf

[87] S Charnovitz "Exploring the Environmental Exceptions in GATT Article XX" above 53, 41.

[88] Discussion in Chap. 4, Sect. 4.5.1 of this Book.

[89] *International Convention Relating to the Simplification of Customs Formalities, Preamble*. The Convention opened for signature on 3 November 1923, Geneva, and entered into force on 27 November 1924, in accordance with article 26 when it was ratified by 5 powers, (1924–1925) 30 LNTS 37. The convention had 36 ratifications by 1934.

[90] 1923 *International Convention Relating to the Simplification of Customs*, Article 13.

[91] 1927 *International Convention for the Abolition of Import and Export Prohibitions and Restrictions*, 97 LNTS 391, opened for signature on 8 November 1927, but did not enter into force, even though as provided by Article 17, seven parties agreed to observe the convention among themselves; available from https://treaties.un.org/doc/Publication/UNTS/LON/Volume%2097/v97.pdf. Discussion "Entry into force of the Convention. Situation on July I, 1930", 97 LNTS 391, 397. This arrangement collapsed when the United States and Great Britain withdrew in 1933; Office of the Historian, *Foreign Relations of the United States, Diplomatic Papers, General, Vol 1*, Rogers P Churchill, George V Blue and Shirley F Landau (ed) "Withdrawal of the United States from the International Convention for the Abolition of Import and Export Prohibitions and Restrictions",

somewhat ambitiously, Article 2 indicated that within 6 months of the convention coming into force, the parties were to abolish any "prohibitions and restrictions" that impeded free trade. Nevertheless, Article 4 envisaged eight exceptions to this far-reaching goal, including the protection of animals or plants, against disease, insects and harmful parasites. The proviso to Article 4 was that measures would not be applied in a manner that constituted arbitrary discrimination between countries where the same conditions prevailed nor were measures to amount to a disguised restriction against international trade. The 1927 Convention for Abolition of Restrictions never fully entered into force, collapsing when the US and Great Britain withdrew from the agreement in 1933.

It took the resurgence of the cattle plague (by that time known as *rinderpest*), in 1920, in Belgium, to act as the catalyst for initiating cooperative arrangements.[92] It was perhaps surprising that the move towards cooperation had taken so long, given that from at least the end of the nineteenth century, governments had realised that strong quarantine, based on veterinary advice and timely communication, was the key to infection control.[93] In any event, the devastation that the outbreak in Belgium caused galvanised European nations, and in 1924 28 states came together to sign the 1924 International Agreement for the Creation at Paris of an International Office for Dealing with Contagious Diseases of Animals, and Annex, creating the OIE. The Annex consists of a document known as the "Organic Statutes of the International Office for Dealing with Contagious Diseases of Animals" (Organic Statutes), which sets out mechanisms for collecting and disseminating information on outbreaks of disease and providing members with guidance on how best to maintain animal health and safety.[94]

Undoubtedly, intrusions into national sovereignty were at the forefront of OIE negotiations; thus in its early days, the OIE did not set standards for animal health, relying instead on advances developed by national veterinary services. Indeed, Article 2 of the Organic Statutes provides that "The Office may not interfere in any way with the administration of the different States. It is independent of the authorities of the country in which it is situated". Accordingly, at its inception, the role of the OIE involved facilitating international cooperation by disseminating information and supporting veterinary research. With respect to the latter, the OIE confirmed the importance of veterinary science, resolving at its first meeting in 1928 that "only sanitary documents emanating from nations with correctly organised

(1933), documents 587–593, available from https://history.state.gov/historicaldocuments/frus1933v01/comp10

92 Amanda Kay McVety, *The Rinderpest Campaigns: A Virus, Its Vaccines, and Global Development in the Twentieth Century*, above 40, 33.

93 Ibid, 30.

94 *International Agreement for the Creation at Paris of an International Office for Dealing with Contagious Diseases of Animals, Annex, Organic Statutes*, Articles 4, 5, 9, 10.

veterinary services can be considered as providing importers with sufficient guarantees".[95]

Over the twentieth century, the role of the OIE evolved, assuming increasing significance in the development of standards by way of animal health codes and manuals.[96] The adoption of the Agreement on the Application of Sanitary and Phytosanitary Measures (SPS Agreement), in 1994, with its emphasis on sound science as justification for trade restrictions, meant that OIE codes and manuals became recognised as international standards within the WTO.[97]

This is not to say, however, that prior to the WTO, countries ignored the potential of the OIE to become a unifying force in international quarantine. In 1930, the League of Nations noted that at meetings of the 1927 Convention for Abolition of Restrictions, delegates pointed to the success of coordinated measures in the field of plant quarantine and called on the League of Nations to convene an international conference, to prevent the introduction and spread of both plant *and* animal diseases by joint and coordinated action.[98] Yet, as just indicated, an official role for standard-setting bodies in international trade did not start in earnest until the commencement of the WTO.[99]

Following the establishment of the OIE, three agreements designed to provide international supervision for dealing with animal diseases were drafted by the Economic Committee of the League of Nations: the 1935 International Convention for the Campaign Against Contagious Diseases of Animals;[100] the 1935 International Convention Concerning the Transit of Animals, Meat and Other Products of

[95] OIE, *History*, History, available from https://www.oie.int/about-us/history/

[96] Although the OIE first published its *Terrestrial Animal Health Code* in 1968, it was not until 1995 and the start of the WTO, that its role as a standard-setting organisation was formalised; OIE, *International Standards*; OIE, Standard Setting, *Aquatic Animal Health Code, Aquatic Manual, Terrestrial Animal Health Code, Terrestrial Manual*, both references available from https://www.oie.int/standard-setting/overview/

[97] *Agreement on the Application of Sanitary and Phytosanitary Measures*, 1867 UNTS 3, Article 5(1), Annex A 3(b); The WTO has 164 members. The Agreement on the Application of Sanitary and Phytosanitary Measures is a specific agreement that provides a primary source of rights and obligations for WTO members with respect to quarantine, now more commonly referred to as biosecurity. For discussion, generally, Christophe Bonneuil and Les Levidow, "How Does the World Trade Organization Know? The Mobilization and Staging of Scientific Expertise in the GMO Trade Dispute", (2012) 42 (1) *Social Studies of Science, 75.*

[98] League of Nations, *General Report of the Sub-Committee of Experts on Veterinary Questions*, above 75, 2.

[99] *Agreement on the Application of Sanitary and Phytosanitary Measures*, Article 3.4, Annex A, Article 3.

[100] *International Convention for the Campaign Against Contagious Diseases of Animals*, 20 February, 1935, 186 *LNTS* 173, entered into force March 23rd, 1938, signed by nine parties and entered into force with five ratifications. Available from https://treaties.un.org/doc/Publication/UNTS/LON/Volume%20186/v186.pdf

Animal Origin;[101] and the 1935 International Convention Concerning the Export and Import of Animals, Meat and Other Products of Animal Origin.[102] These instruments confirmed the importance of veterinary science to international trade, setting up a system of bulletins, notifications of outbreaks of disease and certifications of animal health.[103] The desirability of international instruments to underpin a coordinated and global regime was not lost on governments, and countries of the former Soviet bloc similarly negotiated the 1959 Agreement Concerning Co-Operation in the Field of Veterinary Science.[104] This agreement provided for coordinated bilateral regulation, focussing on the import, export and transit of animals and their products.[105]

Overall, these instruments displayed a trend towards acceptance of international supervision of national measures, particularly by international organisations, shifting disease control from the national to the international arena. In the latter case, the purpose of the treaties rested squarely on preventing animals from introducing and spreading infection.

5.4 Purposes of International Instruments

Treaties evolved dual purposes, focussing on stopping the introduction and spread of animal diseases while promoting international trade. Stopping disease was a core objective, consistently spelt out in instruments, even though parties came to link quarantine to the facilitation of trade.[106] The 1935 Convention Against Contagious Diseases of Animals noted that to boost international trade, it was necessary to

[101] *International Convention Concerning the Transit of Animals, Meat and Other Products of Animal Origin*, 20 February, 1935, 193 *LNTS* 37, entered into force 6 Dec 1938, signed by nine parties and entered into force with five ratifications. Available from https://treaties.un.org/doc/Publication/UNTS/LON/Volume%20193/v193.pdf

[102] *International Convention Concerning the Export and Import of Animals, Meat and Other Products of Animal Origin*, 20 February 1935, 193 *LNTS* 59, entered into force 6 Dec 1938, signed by nine parties and entered into force with five ratifications. Available from https://treaties.un.org/doc/Publication/UNTS/LON/Volume%20193/v193.pdf

[103] 1935 *International Convention for the Campaign Against Contagious Diseases of Animals*, Articles 2, 5 and 6; 1935 *International Convention Concerning the Transit of Animals, Meat and Other Products of Animal Origin*, Articles 1, 2, 5 and Model Certificates annexed; 1935 *International Convention Concerning the Export and Import of Animals, Meat and Other Products of Animal Origin*, Article 3 and Model Certificates annexed.

[104] *Agreement Concerning Co-Operation in the Field of Veterinary Science*, 14 December 1959, 422 *UNTS* 59, The Agreement entered into force 12 September 1960 and at the end of 1960 the treaty had ten parties.

[105] *Agreement Concerning Co-Operation in the Field of Veterinary Science*, Article III.

[106] 1887 *Convention Designed to Remove the Danger of Epizootic Diseases in the Territories of the Two Countries*, Article 1; 1930 *Convention of Commerce and Navigation Between Poland and Roumania*, preamble; 1935 *International Convention for the Campaign Against Contagious Diseases of Animals*, preamble and Article 1; 1952 *Agreement Concerning Epizootic Diseases Between the Kingdom of Greece and The Federal people's Republic of Yugoslavia*, Preamble.

tighten veterinary controls, including fostering closer international cooperation.[107] Likewise, the Preamble to the 1959 Agreement on Veterinary Science highlighted the importance of mutual protection of the parties' territories from epizootic outbreaks while simultaneously "developing economic and trade relations".

However, if quarantine were to facilitate international trade, then it needed not to hinder trade. On this point, the League of Nations said that:

> health considerations and commercial policy are not always kept absolutely separate, and that prohibitions on sanitary grounds are sometimes applied in such a way as to constitute a disguised restriction on international trade.[108]

As discussed, unsuccessful attempts at standardising quarantine regulation were made in the 1920s. However, prior to that time, trading partners had on occasion turned to, other, familiar techniques, such as specifying that goods among trading partners would be treated in equivalent ways. Accordingly, Article 1 of the 1877 Convention between Austria-Hungary and Italy stipulated that measures should be applied equally and in a non-discriminatory manner, as well as be based on veterinary requirements.

Certifications and inspections could also be used analogously, because, among trading partners, shipments were expected to be treated equivalently, based on certificates issued by the exporting country.[109] This process also led to a degree of standardisation, particularly where treaties had model certificates annexed.[110] By the 1930s, the benefits of certification were reflected in Article 2 of the 1935 International Convention Concerning the Transit of Animals, which specified that model certificates should be "drawn up in accordance with the principles laid down by the International Office for Contagious Diseases of Animals". This is not to say that the OIE put forward a model certificate, but the OIE Organic Statutes envisaged sharing knowledge, including how "veterinary police measures" were implemented, laying a

---

[107] 1935 *International Convention for the Campaign against Contagious Diseases of Animals*, preamble.

[108] League of Nations, *General Report of the Sub-Committee of Experts on Veterinary Questions*, above 75, 4.

[109] 1887 *Convention Designed to Remove the Danger of Epizootic Diseases in the Territories of the Two Countries*, Article 2, no model certificate attached; 1930 *Convention of Commerce and Navigation Between Poland and Roumania*, Article 4 – model certificate attached; 1935 *International Convention for the Campaign Against Contagious Diseases of Animals*, Article 2, no model certificate; 1935 *International Convention Concerning the Transit of Animals, Meat and Other Products of Animal Origin*, Article 2 and annexes containing one certificate for health and another for the origin of the shipment; 1935 *International Convention Concerning the Export and Import of Animals, Meat and Other Products of Animal Origin*, annex with one model certificate dealing both with health and origin.

[110] 1935 *International Convention Concerning the Transit of Animals, Meat and Other Products of Animal Origin*, Article 2 and annexes containing one certificate for health and another for the origin of the shipment; 1935 *International Convention Concerning the Export and Import of Animals, Meat and Other Products of Animal Origin*, annex with one model certificate dealing both with health and origin.

foundation for cooperative arrangements.[111] Indeed, the call for model certificates was well-accepted in other treaty systems where the threat of introducing pests and diseases could similarly hinder international trade.[112]

Given the importance of these approaches, it almost went without saying that the regime needed to be underpinned by appropriate domestic regulation, including national legislation, appropriate administrative procedures, a means for communicating outbreaks of disease, the creation of an official veterinary organisation, and an enhanced role for veterinary science, with veterinarians acting as certifiers.

Effective domestic legislation was critical, as it provided the legal framework for inspections, notifications and penalties for breaches of the law. This had been recognised from the earliest of treaties that specifically required the parties to implement measures that were "necessary" to prevent the movement of diseased animals or animals suspected of harbouring disease.[113] By 1935, these calls were supplemented by the obligation to establish administrative procedures that could foster joint action against animal diseases.[114] The 1935 International Convention for the Campaign Against Contagious Diseases of Animals contained some of the most detailed obligations, which as well as calling for joint action, targeted production across the supply chain and included penalties for infringement.[115] The desirability of joint action, or cooperative arrangements, is also evident in the 1959 Agreement on Veterinary Science, which favoured such arrangements to achieve harmonised regulation.[116] At this time, regional and bilateral harmonisation, by way of individual treaties, was important, because as already indicated, the role of the OIE did not extend to setting formal standards.[117]

Over the late nineteenth and early twentieth centuries, commonalities in state practices emerged, particularly concerning communication and the use of veterinary science.[118] The OIE played a critical role, acting as a clearinghouse for information

[111] *International Agreement for the Creation at Paris of an International Office for Dealing with Contagious Diseases of Animals, Annex, Organic Statutes* Article 4.

[112] *International Plant Protection Convention 1951* adopted on 6 December 1951, [1952] ATS No 5 (entered into force 3 April 1952), Articles V and VI; 127 governments adhered to this convention. It was replaced by the *International Plant Protection Convention 1997*, adopted 17 November 1997, [2005] ATS No 23 (entered into force 2 October 2005). This convention has 183 parties.

[113] 1887 *Convention Designed to Remove the Danger of Epizootic Diseases in the Territories of the Two Countries*, Article 1.

[114] 1935 *International Convention for the Campaign Against Contagious Diseases of Animals*, Article 1.

[115] Ibid.

[116] 1959 *Agreement on Veterinary Science*, Article III.

[117] *International Agreement for the Creation at Paris of an International Office for Dealing with Contagious Diseases of Animals and Annex*, Article 2.

[118] Examples include: 1887 *Convention Designed to Remove the Danger of Epizootic Diseases in the Territories of the Two Countries*, Articles 1, 2; 1930 *Convention of Commerce and Navigation Between Poland and Roumania*, Articles 4, 8, 11; 1952 *Agreement Concerning Epizootic Diseases Between the Kingdom of Greece and The Federal people's Republic of Yugoslavia*, Articles 3, 4, 5, 7, 8.

as well as coordinating research on animal diseases.[119] Significantly, the OIE's notifications and correspondence went direct to the veterinary department of its member countries, enabling responses in real time.[120] Otherwise if communication was made through diplomatic channels, there could be delays in information reaching the appropriate institution, highlighting the importance of national veterinary organisations.[121] To overcome these barriers, some treaties mandated notification to veterinary departments,[122] while others, such as the 1935 Convention on Transit of Animal Products, also stipulated that exporting countries needed to send prior notification of consignments to countries along transit routes.[123] In each case, obligations underscored the importance of veterinary organisations and veterinarians as gatekeepers for preventing the transmission of animal disease.

These developments all loosely stemmed from the move towards internationalisation of animal health, where inspections and certifications elevated veterinary science to a pivotal role. This left a legacy that not only provided the foundations for quarantine regulation but also the treatment of farm animals, into the twenty-first century.

5.5 Internationalisation of Disease and Animal Commodification

Internationalisation of animal disease was closely linked to the operation of quarantine in international trade, which led to disease being regarded as a trade issue, with flow-on effects for the treatment of farm animals.

Governments generally took one of two approaches: the "closed door" system, where domestic law and policy prohibited the importation of animals and their products, unless a permit was obtained, which would only be granted in extraordinary circumstances, or the "open door" system, which ostensibly adhered to principles of free trade, but could be very restrictive, due to quarantine regulation.[124] In either case, the impediment to free trade led the League of Nations to assert that

[119] *International Agreement for the Creation at Paris of an International Office for Dealing with Contagious Diseases of Animals, Annex, Organic Statutes* Article 4.

[120] Ibid, Article 2.

[121] League of Nations, *General Report of the Sub-Committee of Experts on Veterinary Questions*, above 75, 7.

[122] 1930 *Convention of Commerce and Navigation Between Poland and Roumania*, Articles 11, 13; 1935 *International Convention for the Campaign Against Contagious Diseases of Animals*, Article 4; 1935 *International Convention Concerning the Transit of Animals, Meat and Other Products of Animal Origin*, article 1; 1952 *Agreement Concerning Epizootic Diseases Between the Kingdom of Greece and The Federal people's Republic of Yugoslavia*, Article 3.

[123] 1935 *International Convention Concerning the Transit of Animals, Meat and Other Products of Animal Origin*, article 1.

[124] League of Nations, *General Report of the Sub-Committee of Experts on Veterinary Questions*, above 75, 10.

import permits should be freely granted and refused only "for conclusive veterinary reasons".[125]

This statement was indicative of a move, throughout the twentieth century, of using science to assess and manage quarantine risks, rather than relying on traditional discrimination-based approaches. The system of inspections and certifications initiated in early quarantine treaties formed the backbone of the regime and as already noted resulted in the adoption of the SPS Agreement, with its emphasis on sound science and nomination of the OIE as a standard-setting body.

In the 1930s, the League of Nations observed that certifications provided countries with an imperative to adhere to "the essential rules of veterinary hygiene and veterinary inspection, the importance of which has been endorsed by science".[126] This resulted in a practice that came to be known as "pushing back the border".[127] In practical terms, it means that pests and diseases are initially managed in the jurisdiction of the exporting country, reducing the chances of accidental introductions into the territory of importing countries.[128] Such mechanisms have clear international underpinnings and are consistent with acknowledging the need for international solutions to the problem of disease in international trade. They are also consistent with the move to place disease control under the auspices of international bodies, such as the OIE.[129] This move progressed slowly because members still clung to the right to design and implement their own regimes, even while accepting that disease control required international consensus.[130]

Notwithstanding this hesitancy, border controls and quarantine were increasingly drawn into the international spotlight, and disease control became a legitimate concern at the international level. This sealed the fate of farm animals because their health status became identified with the "global good", a concept with many facets, but above all linked to promoting international trade.[131]

In one sense, the global good could be seen in terms of stopping the spread and introduction of disease; on another level, it could be identified with reducing the

[125] Ibid, 11.

[126] League of Nations, *General Report of the Sub-Committee of Experts on Veterinary Questions*, above 75, 4.

[127] M E Nairn, P G Allen, A R Inglis and C Tanner, *Australian Quarantine: A Shared Responsibility* Department of Primary Industries and Energy, Canberra (1996) paragraph 79, available from https://www.agriculture.gov.au/sites/default/files/sitecollectiondocuments/aqis/about/corporate-docs/nairn/nairn_report.pdf

[128] Carolyn Tanner and Mike Nunn, "Australian Quarantine Post the Nairn Review", (1998) 42 - *Australian Journal of Agricultural and Resource Economics* 451, 452.

[129] Richard Perren, *Taste, Trade and Technology: The Development of the International Meat Industry Since 1840,* (2006 Ashgate) Routledge, Abingdon, Oxfordshire (2017), 226.

[130] Ibid, 226; Istvan Hont, *Jealousy of Trade, International Competition and the Nation-State in Historical Perspective*, Belknap Press of Harvard University Press, Cambridge MA and London (2005), 5.

[131] Mark Harrison, "A Global Perspective: Reframing the History of Health, Medicine, and Disease", above 4, 642.

disruption of quarantine to international trade; and in yet another sense, it encompassed concern that diminished agricultural production at the domestic level should not obstruct international trade. Hence, the global good encompassed strong economic elements that synthesised national and international commerce. Trading partners anticipated that a combination of science and effective quarantine would stop the spread and introduction of disease, averting potentially devastating consequences for the long-term viability of international trade. In this way, the global good was equated with the free flow of international trade, focussing on the health of animals, but not necessarily their wellbeing.

Treaty provisions commonly required exporting countries to verify that animals were not infected and that consignments had not been shipped from infected areas.[132] Otherwise, the importing country could reject shipments and/or destroy live animals.[133] These types of constraints were clearly aimed at protecting the territory of the importing country.[134] In the 1930 Treaty between Poland and Roumania, the sole objective was to ensure "the sanitary protection of native animals". The reference to native animals might be mistaken for an allusion to indigenous wild animals, but the term actually related to horses, asses, cattle, sheep, goats, swine and poultry, or in other words, domestic animals of importance to food production and cross-border trade.[135] Likewise, the 1952 Agreement Between Greece and Yugoslavia consistently referred to protecting the territory of the contracting parties, a reference to domestic herds and domestic production.[136] Analogous prohibitions were also imposed on the transit of animals and their products if there was a chance of introducing disease and infection.[137] Significantly, the wording of these instruments indicates that the harm addressed was not so much the ill health of the animals themselves, as the impact of that ill health on the

---

[132] 1887 *Convention Designed to Remove the Danger of Epizootic Diseases in the Territories of the Two Countries*, Articles 2.1 and 2.3; 1930 Treaty between Poland and Roumania, Article 4; 1935 *International Convention for the Campaign Against Contagious Diseases of Animals*, Article 2; 1952 *Agreement Concerning Epizootic Diseases Between the Kingdom of Greece and The Federal people's Republic of Yugoslavia*, Article 1; 1959 Agreement on Veterinary Science, Article III.

[133] 1887 *Convention Designed to Remove the Danger of Epizootic Diseases in the Territories of the Two Countries*, Articles 2.1 and 2.3; 1930 *Convention of Commerce and Navigation Between Poland and Roumania*, Article 8; 1952 *Agreement Concerning Epizootic Diseases Between the Kingdom of Greece and The Federal people's Republic of Yugoslavia*, Article 8.

[134] Steve Charnovitz, "Trade Measures and the Design of International Regimes", (1966) 5 (2) *Journal of Environment and Development*, 168, 179.

[135] 1930 *Convention of Commerce and Navigation Between Poland and Roumania*, Annexed Veterinary Convention, Article 2.

[136] 1952 *Agreement Concerning Epizootic Diseases Between the Kingdom of Greece and The Federal people's Republic of Yugoslavia*, Article 7.

[137] 1935 *International Convention Concerning the Transit of Animals, Meat and Other Products of Animal Origin*, Article 1.3; *International Convention Concerning the Export and Import of Animals, Meat and Other Products of Animal Origin* Article 2.

importing country, including, as just noted, on domestic herds, domestic animal production and flow-on effects to international trade.

Animal disease was therefore seen in economic terms, where the emphasis lay on commerce and production. This mirrored developments during veterinary conferences of the nineteenth century, where delegates concluded that the health of farm animals was critical to economic growth and prosperity.[138] By implication, this also meant that animal health was primarily relevant as far as it related to merchantability in the context of trade. Accordingly, animals were treated like any other commodity, but without reference to sentience or the ability to suffer, which differentiates them from inanimate objects.

Where trading partners were obliged to implement measures that were "necessary", this did not include consideration of animal cruelty or animal suffering,[139] nor was animal suffering relevant where trading partners were permitted to reject and destroy sick animals.[140] Indeed, there were few attempts to identify whether practices in international trade could amount to cruelty, or far less, whether animal suffering could be legally or morally justified. Consequently, the animal, as a living being, became subsumed into its economic value, where the financial and international interests involved in the livestock sector rationalised ignoring animal sentience.[141] This standpoint also had a somewhat perverse outcome, because, in at least one treaty, the parties were obliged to exterminate stray dogs, wild boars, wolves and other wild animals in case they transmitted disease to domestic livestock.[142] The economic imperative thus extended to all wild and domestic animals capable of diminishing production and trade.

One exception to these generalisations stemmed from the 1935 Convention on Transit of Animal Products. Article 3 specified "animals may only be transported in wagons constructed to prevent the escape or dissemination of excreta and other materials liable to transmit infection". In addition, the Annex to Article 5 prohibited overloading, specified that floors had to be covered with proper litter and if animals were not shipped in boxes or cages, they were to be accompanied by a person to care for them, who should not have more animals in his charge than was possible to look after. The Article further stipulated that animals should be transported by the

[138] Leonard Pearson, "A Review of the Field of Veterinary Medicine", Proceedings of the First General Meeting of the National Veterinary Association of 1897, 126, 131, 136–137, NVA, Kansas City (1897); Discussion, Chap. 4, Sect. 4.5.1 of this book.

[139] 1887 *Convention Designed to Remove the Danger of Epizootic Diseases in the Territories of the Two Countries*, Article 1.

[140] Ibid, Articles 2.1 and 2.3; 1930 *Convention of Commerce and Navigation between Poland and Roumania*, Article 8; 1952 *Agreement Concerning Epizootic Diseases between the Kingdom of Greece and The Federal people's Republic of Yugoslavia*, Article 8.

[141] Michael J Bowman, "The Protection of Animals under International Law", (1989) 4 *Connecticut Journal of International Law*, 487, 489.

[142] 1952 *Agreement Concerning Epizootic Diseases between the Kingdom of Greece and The Federal people's Republic of Yugoslavia*, Article 6.

quickest route, including the use of "specially accelerated goods trains". These provisions clearly targeted animal wellbeing, an objective underscored by the additional stipulation in Article 5, that exporting countries should take all steps, including proper loading and feeding, to "avoid unnecessary suffering". Bowman has concluded that this agreement was the first to approach an aspect of international trade "from a true animal welfare standpoint".[143]

This attitude, however, was not the norm. Usual practices mirrored domestic processes that remained fixated on controlling imports by rejection and/or destruction of diseased animals.[144] As discussed in Chap. 4 of this book, during the cattle plagues of the 1860s the UK initiated government policy based on stamping out disease, which invariably involved slaughtering diseased animals, diverting attention from finding cures that could have improved animal wellbeing.[145] This differed from human diseases for which there was greater knowledge and understanding, including advances in vaccination.[146] Some areas of Europe that were disease-free undertook experimental trials with vaccination, but as a scientific curiosity, rather than an alternative to wholesale slaughter.[147]

In general, officials charged with enforcement of veterinary regulations retained existing methods including "border controls and using legal regulations and penalties".[148] This not only meant that the regime failed to be proactive in disease prevention, but it also failed to engage with broader notions of animal suffering and animal wellbeing. Largely, the latter required improved veterinary competency and a detailed understanding of treatments and cures, something that, in reality, only became possible after World War I.[149] From that time, treaties occasionally referred to vaccinations as a preventative measure, but not so much for the health of the animal as for the viability of trade.[150] Ultimately, the legacy of international instruments discussed in this chapter was that they reinforced the perception of farm animals as a product in the marketplace, crystallising their commodification in international trade.

---

[143] Michael J Bowman, "The Protection of Animals under International Law", above 141, 489.

[144] C A Spinage, *Cattle Plague, A History*, above 7, 329.

[145] Michael Worboys, "Germ Theories of Disease and British Veterinary Medicine, 1860–1890", (1991) 35 *Medical History*, 308, 310.

[146] Ibid. 310.

[147] C A Spinage, *Cattle Plague, A History*, above 7, 329.

[148] Ibid, 329.

[149] Richard Waller, "'Clean' and 'Dirty': Cattle Disease and Control Policy in Colonial Kenya, 1900–40", (2004) 45 (1) *The Journal of African History*, 45, 57.

[150] 1952 *Agreement Concerning Epizootic Diseases Between the Kingdom of Greece and The Federal people's Republic of Yugoslavia*, Articles 6, 7.

5.6 Conclusion

International trade was, and is, primarily an economic activity; however, by at least the eighteenth century, it had acquired a political dimension that merged private contractual arrangements with the national interest.[151] Accordingly, countries pursued economic goals, such as profits and increased volumes of trade, but as demonstrated by early treaties of commerce and navigation, international trade also incorporated components of international relations. In the animal product sector, these became important, because as trade grew, so too did the increased likelihood of introducing and spreading disease. Therefore, preventative efforts saw the national interest overlaid with conceptions of the global good. In this way, the legitimacy of unilateral quarantine measures slowly gave way to calls for a more harmonised and standardised regime, based on veterinary science and international oversight.

While veterinary science was critical to the health of animals, the fact that animal disease, and by implication animal wellbeing, was categorised as an issue of international trade, the regime contributed greatly to the commodification of farm animals. Animal health came to be regarded less as a matter of animal wellbeing and more as an issue of protecting national territory against diseased shipments. Only a handful of instruments dealing with trade and animal disease considered that animal suffering was part of their remit. Yet, as discussed in Chap. 3 of this book, by the 1930s the UK had passed several iterations of anti-cruelty laws. So too, had other European countries and the USA.[152] Consequently, among trading partners, recognition of animal suffering was not a foreign concept. Still, the notion did not reach the international arena.

In one sense, animal suffering and animal cruelty are moral issues that fall within "the purview of personal ethics".[153] However, in the case of farm animals, ethical considerations were first addressed as part of the law at the national level and then virtually ignored at the international level. In the latter context, animal suffering became conflated with disease prevention, commerce and trade. Consequently, in the same way that animal cruelty in domestic regimes had become production and marketing issues, animal disease in international law had become a trade and commodification issue. The next chapter evaluates why this occurred, asking the question: Whither Ethics?

[151] Istvan Hont, *Jealousy of Trade, International Competition and the Nation-State in Historical Perspective*, above 130, 4–6.

[152] For the influence of English law and policy, David Favre and Vivien Tsang, "The Development of Anti-Cruelty Laws during the 1800's". (1993) Spring (1) *Detroit College of Law Review*; Lois Laimene Lelanchon "Detailed Discussion of Anti-Maltreatment Laws in France and Spain", (2013) *Animal Legal & Historical Center*, Michigan State University College of Law, available from https://www.animallaw.info/article/detailed-discussion-anti-maltreatment-laws-france-and-spain

[153] Bernard E Rollin, "Putting the Horse before the Descartes: My Life's Work on Behalf of Animals", Temple University Press, Philadelphia (2011), 31.

Bibliography

Bonneuil C, Levidow L (2012) How does the World Trade Organization know? The mobilization and staging of scientific expertise in the GMO trade dispute. Soc Stud Sci 42(1):75

Booker J (2007) Maritime quarantine, the British experience, c 1650–1900. Ashgate, Aldershot

Bowman MJ (1989) The protection of animals under international law. Conn J Int Law 4:487

Charnovitz S (1991) Exploring the environmental exceptions in GATT article XX. J World Trade Law 25(5):37

Charnovitz S (1996) Trade measures and the Design of International Regimes. J Environ Dev 5(2): 168

Coyle JF (2013) The treaty of friendship, commerce and navigation in the modern era. C J Transnatl Law 51:302

Eystein S, Yngvild W (1996) Ecological consequences of the spreading of pathogens and genes through an increasing trade in foods. In: Proceedings of Norway/UN Conference on Alien Species, Trondheim, Directorate for Nature Management and Norwegian Institute for Nature Research Norway, 1–5 July

Favre D, Tsang V (1993) The development of anti-cruelty Laws during the 1800's. Detroit Coll Law Rev 1:1

Fleming G (1871) Animal plagues: their history, nature, and prevention. Chapman and Hall, London, pp 479–480. This copy is a reprint

Gensubu G, Yacoub M, Conti A (2004) The concept of quarantine in history: from plague to SARS. J Infect 49:257

Harrison M (2006) Disease, diplomacy and international commerce: the origins of international sanitary regulation in the nineteenth century. J Global Hist:197

Harrison M (2015) A global perspective: reframing the history of health, medicine, and disease. Bull Hist Med 89(4):639

Hartwell RM (1971) The industrial revolution and economic growth. Methuen and Co Ltd., London

Hont I (2005) Jealousy of trade, international competition and the nation-state in historical perspective. Belknap Press of Harvard University Press, Cambridge, MA/London

Howard-Jones N (1975) The scientific background of the international sanitary conferences. World Health Organization, Geneva. https://apps.who.int/iris/bitstream/handle/10665/62873/14549_eng.pdf;jsessionid=2B113F7596689E48F313A2187319269E?sequence=1

Huber V (2006) The unification of the globe by disease? The international sanitary conferences on cholera 1851-1894. Hist J 49(2):453

Jackson J (2002) The world trading system—law and policy of international economic relations. MIT Press, Cambridge, MA/London

League of Nations, General Report of the Sub-Committee of Experts on Veterinary Questions, Official No 0.633.M.252. Economic and Financial 1930. II. 49. Geneva, November 1930. https://biblio-archive.unog.ch/Dateien/CouncilMSD/C-633-M-252-1930-II_EN.pdf

Lelanchon LL (2013) Detailed discussion of anti-maltreatment laws in France and Spain. Animal Legal & Historical Center, Michigan State University College of Law. https://www.animallaw.info/article/detailed-discussion-anti-maltreatment-laws-france-and-spain

Leyser J (1953) Treaties of friendship, commerce and navigation. Aust Outlook 7(2):132

McNeill WH (1989) Plagues and peoples, plagues and peoples. Anchor Books (1989 reprint originally published 1976). New York

McVety AK (2018) The rinderpest campaigns: a virus, its vaccines, and global development in the twentieth century. Cambridge University Press

Munro JJ (1881) England's treaties of guarantee. Law Mag Rev CCXL:215

Nairn ME, Allen PG, Inglis AR, Tanner C (1996) Australian quarantine: a shared responsibility. Department of Primary Industries and Energy, Canberra. https://www.agriculture.gov.au/sites/default/files/sitecollectiondocuments/aqis/about/corporate-docs/nairn/nairn_report.pdf

Novak W (1996) The People's welfare. In: The University of North Carolina Press Chapel Hill and London

Office of the Historian (1933) Foreign relations of the United States, diplomatic papers, general, vol 1. In: Churchill RP, Blue GV, Landau SF (eds) Withdrawal of the United States from the International convention for the abolition of import and export prohibitions and restrictions, documents 587–593. https://history.state.gov/historicaldocuments/frus1933v01/comp10
OIE History, https://www.oie.int/about-us/history/
OIE International Standards; OIE, Standard Setting, Aquatic Animal Health Code, Aquatic Manual, Terrestrial Animal Health Code, Terrestrial Manual, both references available from https://www.oie.int/standard-setting/overview/
Outhwaite O (2010) The international framework for biosecurity and the challenges ahead. Rev Eur Commun Int Environ Law 19(2):207
Pearson L (1897) A review of the field of veterinary medicine. In: Proceedings of the First General Meeting of the National Veterinary Association of 1897, 126, NVA, Kansas City
Perren R (2017) Taste, trade and technology: the development of the international meat industry since 1840 (2006 Ashgate). Routledge, Abingdon
Rollin BE (2011) Putting the horse before the descartes: my life's work on behalf of animals. Temple University Press, Philadelphia
Sabanogullari L (2018) General exception clauses in international investment law: the recalibration of investment agreements via WTO-based flexibilities. Nomos Verlagsgesellschaft, Germany
Schepin O, Yermakov W (1991) International Quarantine. International Universities Press Inc, Connecticut
Schwarzenberger G (1945) The Most-Favored-nation standard in British state practice. Br Year Book Int Law 22:96
Sehdev P (2002) The origin of quarantine. Clin Infect Dis 35:1071
Spinage CA (2003) Cattle plague, a history. Kluwer Academic, New York
Swabe J (1999) Animals, disease and human society, human-animal relations and the rise of veterinary medicine. Routledge, London
Tanner C, Nunn M (1998) Australian quarantine post the Nairn review. Aust J Agric Resource Econ 42:451
The Royal Economic Society (1905) The economic effects of cattle disease legislation. Econ J 15(58):156
Waller R (2004) 'Clean' and 'Dirty': cattle disease and control policy in colonial Kenya, 1900-40. J Afric Hist 45(1):45
WHO (2002) Global crises—global solutions. WHO, Geneva
Wilkinson L (1992) Animals and disease, an introduction to the history of comparative medicine. Cambridge University Press, Cambridge
Winslett M (2008) The Nadir of Alliance: The British Ultimatum of 1890 and its place in Anglo-Portuguese Relations, 1147–1945. Presented to the Faculty of the Graduate School of The University of Texas at Arlington in Partial Fulfillment of the Requirements for the Degree of Master of Arts in History the University of Texas at Arlington August. https://rc.library.uta.edu/uta-ir/handle/10106/1093
Worboys M (1991) Germ theories of disease and British veterinary medicine, 1860-1890. Med Hist 35:308

Legislation, Regulations, Gazettes and Hansard

1943, Winston Churchill, Agreement with Portugal, 12 October, 1943, House of Commons Debate, Volume 392, *Hansard* (1943) column 716–719. https://api.parliament.uk/historic-hansard/commons/1943/oct/12/agreement-with-portugal

International Instruments

1373, Governments of the Kingdom of England and the Kingdom of Portugal, 16 June 1373, in *British Documents on foreign affairs—reports from the Foreign Office confidential Print. Part II. From the First to the Second World War. Series F, Europe*. Vol. 24., ed. Anthony Adamthwaite. (Frederick, MD: University Publications of America, 1991). A copy of the treaty is also available from the National Archives, UK. http://discovery.nationalarchives.gov.uk/details/r/C3624900

1662, *Treaty of Peace and Commerce between Great Britain and Tunis* 1662 Volume 17 *CTS* 244

1734, *Treaty of Commerce between Great Britain and Russia*, 2 December 1734, Vol 34 *CTS* 211

1815, *Convention of Commerce between the United Kingdom of Great Britain and Ireland and the United States of America, and Declaration by United Kingdom Regarding St Helena*, 3rd July 1815, 1901 *Australian Treaty Series*, 107, (electronic). http://www.austlii.edu.au/au/other/dfat/treaties/1901/107.html

1851, Ministère des Affaires Étrangères, *Procès-Verbeaux, de la Conférence Sanitaire Internationale*, Tome I, Paris Imprimerie Internationale MDCCCLII, available from Harvard Law School Library, https://iiif.lib.harvard.edu/manifests/view/drs:7309389$9i

1852, *Sanitary Convention between France, Portugal, Sardinia, Tuscany and Turkey*, 3 February 1852, Vol 107 *CTS* 345

1882, *Convention between France and Great Britain for the Regulation of Commercial and Maritime Relations and Declaration Prolonging the Treaties in Force,* 28 February 1882, (1882) 160 *CTS* 143

1887, *Convention Designed to Remove the Danger of Epizootic Diseases in the Territories of the Two Countries*, 7 December, 1887, reprinted in Rüster B and Simma B (eds) Vol IV *International Protection of the Environment: Treaties and Related Documents*, Dobbs Ferry: Oceana Publications Inc, New York (1975) 1586

1892, *International Sanitary Convention between Austria-Hungary, Belgium, Denmark, France, Germany, Great Britain, Italy, the Netherlands, Portugal, Russia, Spain, Sweden-Norway, and Turkey, with Annexes I to V*, signed 30 January 1892, Vol 176 *CTS* 395. The convention had 13 parties

1922, *Commercial Treaty between the Esthonian Republic and the Kingdom of Hungary*, 19, October 1922, Vol 30 LNTS, 347. https://treaties.un.org/doc/Publication/UNTS/LON/Volume%2030/v30.pdf

1923, *Treaty of Commerce and Navigation made between Denmark and Finland*, 3 August, 1923, Vol 21 LNTS 269. https://treaties.un.org/doc/Publication/UNTS/LON/Volume%2021/v21.pdf

1923, *International Convention Relating to the Simplification of Customs Formalities*, 30 LNTS 37, opened for signature on 3 November 1923, Geneva, and entered into force on 27 November 1924, in accordance with article 26 when it was ratified by 5 powers. The convention had 36 ratifications by 1934

1924, International Agreement for the Creation at Paris of an International Office for Dealing with Contagious Diseases of Animals and Annex, opened for signature 25 January 1924, [1925] ATS 15, (entered into force 12 January 1925). The organisation has 182 members

1924, Treaty of Commerce and Navigation Between Italy and Albania, 20 January, 44 *LNTS*, 359. https://treaties.un.org/doc/Publication/UNTS/LON/Volume%2044/v44.pdf

1925, *Convention of Commerce and Navigation with Final Protocol Sweden and Czechoslovakia*, 36 *LNTS* 289. https://treaties.un.org/doc/Publication/UNTS/LON/Volume%2036/v36.pdf

1927, *International Convention for the Abolition of Import and Export Prohibitions and Restrictions*, 97 LNTS 391, opened for signature on 8 November 1927, but did not enter into force. https://treaties.un.org/doc/Publication/UNTS/LON/Volume%2097/v97.pdf

1928, *Treaty of Commerce and Navigation between Bulgaria and Turkey*, 12 February 1928, 81 *LNTS* 383. https://treaties.un.org/doc/Publication/UNTS/LON/Volume%2081/v81.pdf

1928, *Treaty of Commerce and Navigation with Final Protocol and Protocol concerning the Import and Export Régime in Italy Between Hungary and Italy*, 4 July 1928, 92 *LNTS* 117. https://treaties.un.org/doc/Publication/UNTS/LON/Volume%2092/v92.pdf

1928, Treaty of Commerce and navigation with Protocol Between the Union of South Africa and Germany, 1 September 1928, 95 *LNTS*, 289. https://treaties.un.org/doc/Publication/UNTS/LON/Volume%2095/v95.pdf

1930, *Convention of Commerce and Navigation Between Poland and Roumania*, 23 June 1930, Annex I, reprinted in Bernd Rüster and Bruno Simma (eds) Vol IV *International Protection of the Environment: Treaties and Related Documents*, Dobbs Ferry: Oceana Publications Inc, New York, (1975) 1682

1935, *International Convention for the Campaign Against Contagious Diseases of Animals*, opened for signature, 20 February, 1935, 186 *LNTS* 173, entered into force March 23rd, 1938a, signed by 9 parties and entered into force with 5 ratifications. https://treaties.un.org/doc/Publication/UNTS/LON/Volume%20186/v186.pdf

1935, *International Convention Concerning the Transit of Animals, Meat and Other Products of Animal Origin*, opened for signature, 20 February, 1935, 193 *LNTS* 37, entered into force 6 Dec 1938b, signed by 9 parties and entered into force with 5 ratifications. https://treaties.un.org/doc/Publication/UNTS/LON/Volume%20193/v193.pdf

1935, *International Convention Concerning the Export and Import of Animals, Meat and Other Products of Animal Origin*, opened for signature, 20 February 1935, 193 *LNTS* 59, entered into force 6 Dec 1938, signed by 9 parties and entered into force with 5 ratifications. Available from https://treaties.un.org/doc/Publication/UNTS/LON/Volume%20193/v193.pdf

1946, *Constitution of the World Health Organization*, adopted on 22 July 1946, [1948] *ATS* 7, (entered into force 7 April 1948). WHO has 194 members

1951, *International Plant Protection Convention 1951* adopted on 6 December 1951, [1952] *ATS* No 5 (entered into force 3 April 1952), 127 governments adhered to this convention

1959, *Agreement Concerning Co-Operation in the Field of Veterinary Science*, opened for signature, 14 December 1959, 422 *UNTS* 59, The Agreement entered into force 12 September 1960 and at the end of 1960 the treaty had 10 parties

1975, *Convention on the Elimination of all Forms of Racial Discrimination,* [1975] *ATS* 40, opened for signature by the United Nations General Assembly on 21 December 1965, entering into force on 4 January 1969. The Convention has 182 parties

1952, *Agreement Concerning Epizootic Diseases Between the Kingdom of Greece and The Federal people's Republic of Yugoslavia* 2 February 1952, reprinted in Bernd Rüster and Bruno Simma (eds) Vol IV *International Protection of the Environment: Treaties and Related Documents*, Dobbs Ferry: Oceana Publications Inc, New York, (1975), 1833

1982, *The United Nations Law of the Sea Convention,* [1994] *ATS* 31, opened for signature 10 December 1982, entering into force on 16 November 1994. The Convention has 168 parties

1994, *Agreement on the Application of Sanitary and Phytosanitary Measures*, [1995] *ATS* No 8, 65. *Marrakesh Agreement Establishing the World Trade Organization,* opened for signature 15 April 1994, [1995] *ATS* No 8, 1, entering into force on 1 January 1995; The WTO has 164 members

1994, *General Agreement on Tariffs and Trade* (GATT 1994) [1995] *ATS* No 8, 14. *Marrakesh Agreement Establishing the World Trade Organization,* opened for signature 15 April 1994, [1995] *ATS* No 8, 1, entering into force on 1 January 1995; The WTO has 164 members

1997, *International Plant Protection Convention 1997*, adopted 17 November 1997, [2005] *ATS* No 23 (entered into force 2 October 2005). This convention has 183 parties

6 Whither Ethics?

Abstract

From the nineteenth century, market growth in animal agriculture was accompanied by anti-cruelty legislation, an increasing number of animal protection organisations, as well as general critiques of the sector. The regulatory dilemma centred on how to balance these competing ideologies, to support animal production and trade while simultaneously dealing with cruelty and animal disease. Ultimately, neither anti-cruelty regulation of the nineteenth century nor animal welfare regulation of the twentieth century, led to the optimum treatment of farm animals. Instead, the individual animal became invisible, subsumed into the merchantability of the whole, receiving what little protection the market could spare. By 1964, problems in the sector culminated in the publication of Ruth Harrison's seminal tome, *Animal Machines*, detailing institutionalised cruelty in British livestock production. This resulted in the establishment of a committee to investigate these allegations and draft what came to be known as the Brambell Report. Part of the problem derived from the fact that farm animals were primarily treated as articles of trade, or commodities; their wellbeing shaped by commercial values ascribed to them (as explicated by Karl Marx) and delineated by the whims of the marketplace.

Keywords

Animal commodification · Animal welfare · Anti-cruelty legislation · Brambell Report · Karl Marx · Ruth Harrison

6.1 Introduction

This chapter draws together the threads of arguments presented thus far, addressing the question why anti-cruelty regulation of the nineteenth century, and later animal welfare regulation of the twentieth century, did not lead to greater consideration of

S. Riley, *The Commodification of Farm Animals*, Animal Welfare 21,
https://doi.org/10.1007/978-3-030-85870-4_6

the ethical treatment of farm animals. Given that from the nineteenth century, market growth in the sector was accompanied by a larger number of animal protection organisations,[1] the question "Whither Ethics?" is an important one. To address this issue, it helps keep in mind that stakeholders, such as producers, veterinarians and government, had differing points of view regarding how farm animals should be treated. Nevertheless, at a basic level, each evaluated the same question: how to support production and trade while simultaneously dealing with cruelty and animal disease? Against this backdrop, animal protection organisations faced an uphill battle.

As discussed in Chap. 3 of this book, judicial interpretation of anti-cruelty regulation made it difficult for prosecutors to secure convictions in circumstances where husbandry methods were accepted practices. In consequence, the scope of anti-cruelty regulation became restricted in commercial settings. Further, as discussed in Chaps. 4 and 5 of this book, the cattle plagues of the 1860s led to an expanded role for the veterinary profession, which acted as a gatekeeper at the behest of government. Veterinarians were also important in international trade, as regulators used veterinary expertise to protect national territory. Throughout these developments, little attention was paid to the ethical dimensions of animal production, beyond half-hearted attempts at dealing with the worst cases of animal cruelty.[2] By 1964, the problems in the sector had culminated in the publication of Ruth Harrison's (1920–2000) seminal tome, *Animal Machines*, where she detailed abhorrent institutionalised cruelty.[3]

Harrison's findings are complex, but the issues she raised begin with the preference given to commercial interests during the formative years of the livestock sector so that the lives, transportation and slaughter conditions of animals were of less importance than the profitability of production and trade. Among the many problems that needed resolving was the variety of ways that animal cruelty became entangled with animal death so that in many respects animal mortality became a metaphor for welfare. At one extreme, death could be ignored where it was seen as a private property matter that did not otherwise impinge on profitability, human health or food security. At the other extreme, animal mortality could morph into a measure of animal wellbeing within specific industries.

Other reasons for the failure of regulation to prevent institutionalised cruelty stem from animal commodification. This theme wove its way through the supply chain, ensuring that farm animals were treated primarily as articles of trade. As such, their wellbeing was subject to the whims of the marketplace, where they were regarded as

---

[1] Kimberley Smith, *Governing Animals Animal Welfare and the Liberal State*, Oxford University Press, New York, (2012), 28.

[2] Peter Sandøe, Stine B Christiansen and Björn Forkman, "Animal Welfare: What is the Role of Science?" in Jacky Turner and Joyce D'Silva (eds), *Animals, Ethics and Trade*, 41, 45, Earthscan, UK (2006).

[3] Ruth Harrison, *Animal Machines*, first published 1964 by Vincent Stuart Publishers, 2013 Edition J Harrison and J Wilson, CABI Oxfordshire (2013).

DIAGRAM ONE

COMMON AIMS OF LAW, POLICY AND MARKET FORCES

LAW AND POLICY			COMMERCE AND ECONOMICS	
Anti-Cruelty and Later, Animal Welfare	Health Issues		Values and Commodification	
	Veterinary Practice	International Instruments	Transformation of animals into exchange values	
• Cruelty could be justified by commercial practices • What level of cruelty could society accept? • Individual animals subsumed into commercial wholes and regulated as food products	• Health issues equated to maintaining disease-free status of animals as a whole • Animals equated with commodities in trade	• Aimed at preventing the introduction and spread of disease across international boundaries • Animals treated as bulk commodities in trade	• Use value determined by consumer needs • Exchange value determined by marketplace • Surplus value realised on exchange	• Animals and their products treated as consumer goods, shaped by rules and standards of the marketplace, which aimed at enhancing revenue garnered from animals. • Exchange value treated animals as a form of capital
The primary focus was on the profitability and/or durability of the animal product sector				

Fig. 6.1 Common aims of law, policy and market forces

a bulk commodity or, in economic terms, a form of capital.[4] Within these systems, the individual animal became invisible, subsumed into the merchantability of the whole. This turned the concepts of anti-cruelty regulation and, later, animal welfare on their heads because both depend on animal sentience, a notion based on individual pain and suffering. By favouring the collective over the individual, the regime reinforced the status of farm animals as objects of trade, receiving what little protection the market could spare. Figure 6.1 sets out how law and policy converged with developments in trade and commerce, leading to an overriding concern for the profitability of the sector.

These matters are elaborated in the discussion that follows, identifying how animal welfare evolved from anti-cruelty regulation and, in the process, retained several drawbacks of the latter. Following this, Sect. 6.3 of the chapter examines how regulation for animal welfare became conflated with animal mortality, particularly in transportation by sea with respect to live exports, where welfare outcomes are measured against the percentage of animals who die in transit. Developments

[4] F Bailey Norwood and Jayson L Lusk, *Compassion, by the Pound: The Economics of Farm Animal Welfare*, Oxford University Press, New York (2011) 258.

such as these contributed to the failure of the sector to come to grips with challenges presented by institutionalised cruelty. This leads to an evaluation of commodification, where the discussion draws on the writings of Karl Marx (1818–1883) in *Capital*, particularly his explanation of different forms of value.[5] This material is extrapolated to farm animals, with the chapter concluding that classifying farm animals as commodities has facilitated and entrenched their systemic exploitation into the twenty-first century.

6.2 Anti-cruelty and Animal Welfare

The reader may have noticed that thus far, apart from the introduction to this book, the discussion has referred to "animal welfare" in detail only once, when quoting Michael Bowman in Chap. 5 (Sect. 5.5). This omission is deliberate, as during the time under discussion the phrase "animal welfare" was not generally in use. This is not to say that animal suffering was ignored. As discussed in Chap. 3 of this book, the evolution of anti-cruelty regulation developed as a means of transforming ideals of animal sentience into a legal concept, which occurred by creating offences for ill-treatment or abuse of animals.

These proscriptions, however, were generally not proactive in the sense that they required actions to ensure or improve animal wellbeing.[6] Instead, regulation was reactive to negative states in animals, resulting from human-inflicted cruelty.[7] Hence, the Cruelty to Animals Act 1835 (UK) and Cruelty to Animals Act 1849 (UK) stipulated that failing to feed and water animals confined for sale or slaughter

[5] Karl Marx, "*Capital, A Critique of Political Economy*", Volume I Book One: The Process of Production of Capital, Progress Publishers, Moscow, USSR, first published in German in 1867, English edition first published in 1887, available from https://www.marxists.org/archive/marx/works/download/pdf/Capital-Volume-I.pdf

[6] 1850, *An Act for the More Effectual Prevention of Cruelty to Animals in Scotland 1850* (UK) 13 & 14 Vic c 59, *Public General Statutes*, s II, Thirteenth and Fourteenth Year, Queen Victoria, 1850, Printed by George Edward Eyre and William Spottiswoode and published by W Benning and Co, London (1850), 801, available from https://play.google.com/books/reader?id=Z6pKAAAAMAAJ&hl=en_GB&pg=GBS.PA801

[7] Examples include: 1849, *An Act for the more Effectual Prevention of Cruelty to Animals*, 1849, 12 &13 Vic c 92, *Public General Statutes*, Twelfth and Thirteenth Year, Queen Victoria, 1849, Printed by George Edward Eyre and William Spottiswoode and published by W Benning and Co, London (1849), 592, available from https://play.google.com/books/reader?id=RKJKAAAAMAAJ&hl=en_GB&pg=GBS.PA592;

1850, *An Act for the More Effectual Prevention of Cruelty to Animals in Scotland 1850* (UK) 13 & 14 Vic c 59, *Public General Statutes*, s II, Thirteenth and Fourteenth Year, Queen Victoria, 1850, Printed by George Edward Eyre and William Spottiswoode and published by W Benning and Co, London (1850), 801, available from https://play.google.com/books/reader?id=Z6pKAAAAMAAJ&hl=en_GB&pg=GBS.PA801;

1876, *Cruelty to Animals Act 1876* (UK), 39 & 40 Vic c 77, available from, The Statutes Project, http://statutes.org.uk/site/the-statutes/nineteenth-century/1876-39-40-victoria-c77-cruelty-to-animals-act/

was a breach of the legislation, as was transporting animals in a cruel manner.[8] These provisions operated after the event and did not mandate standards of care. Even legislation, such as the *Protection of Animals Act 1911* (UK), which aimed at "protection", still adhered to regulatory pathways whose objectives outlawed the most egregious cases of cruelty, with the addition of offences for infuriating or terrifying an animal.[9] A small number of exceptions to this general proposition were found in statutes that regulated road transport, rail transport and slaughter, which included provisions, such as providing food and water for animals, considered necessary for animal wellbeing.[10]

Judicial interpretation of anti-cruelty legislation evinced an economic bias that led to animal cruelty becoming subordinated to production and trade.[11] As such, while developments in anti-cruelty law and policy were ground-breaking for their day, they could not be equated with the type of broader animal wellbeing which became part of the regulatory agenda towards the end of the twentieth century. Indeed, it was largely from the 1960s that stakeholders embraced the term "animal welfare" and building on this, from the 1990s, emphasised the importance of positive states in animals.[12]

Prior to the 1960s, the literature contains only a handful of references to animal welfare. One of the earliest of these was made by a veterinarian, George Fleming (1833–1901), who referred to animal welfare in the context of domestication of animals when he observed that "pastoral people derive their subsistence... [from farm animals, and in] return, the welfare of these animals is carefully studied, their increase in number and in individual value is a matter of social as well as political importance".[13] This statement clearly regarded animal welfare as the equivalent of animal health and linked it to societal benefits.

Just over 20 years later, Henry Salt in his 1892 treatise, *Animals' Rights: Considered in Relation to Social Progress*, argued in favour of positive obligations,

[8] Cruelty to Animals Act 1835 (UK), ss 1, 2, 4; Cruelty to Animals Act 1849 (UK) ss 8, 12.

[9] 1911, *Protection of Animals Act 1911* (UK), 1 & 2 Geo 5 c 271, s 1, available from http://www.bailii.org/uk/legis/num_act/1911/1069356.html

[10] 1835, *An Act to Consolidate and Amend the Several Laws Relating to the Cruel and Improper Treatment of Animals and the Mischiefs Arising from the Driving of Cattle, and to Make Other Provisions in Regard Thereto,* 1835, 5 & 6 William 4 c59, Cruelty to Animals Act, available from The Statutes Project, http://statutes.org.uk/site/the-statutes/nineteenth-century/1835-5-6-william-4-c-59-cruelty-to-animals-act/.Cruelty to Animals Act 1835 (UK) s 8; Cruelty to Animals Act 1849 (UK) s 8, 12.

[11] Discussion, Chap. 3, Sect. 3.5 of this book.

[12] David J Mellor and Cam S W Reid, "Concepts of Animal Well-Being and Predicting the Impact of Procedures on Experimental Animals", in *Improving the Well-Being of Animals in the Research Environment*, R M Baker, G Jenkin and D J Mellor (eds) 3, 11, Australian and New Zealand Council for the Care of Animals in Research and Teaching (ANZCCART), Glen Osmond, SA, Australia (1994), available from https://pdfs.semanticscholar.org/6890/adfa3127be363a7c3b03925fbf22363f4489.pdf

[13] George Fleming, *Animal Plagues: Their History, Nature, and Prevention*, Chapman and Hall (1871), London XV. This copy is a reprint.

to maintain the "welfare" of captive wild animals by providing them with necessities.[14] Salt likened these obligations to animal "rights", which in his view comprised an essential element of social reform or as he termed it "social progress". To Salt, human and animal wellbeing were not separate social movements, but merely facets of the one social agenda.[15] Moreover, while social progress initially benefitted humans, it also created a sufficiently high degree of social cohesion to provide moral space for animals.[16] This further facilitated society valuing animals as living beings in their own right rather than products to be used at will. By way of a telling illustration, Salt noted that commercial activities frequently pushed humans to produce animal products in the quickest and cheapest ways.[17] These methods may not have been designed to inflict gratuitous cruelty, yet economic imperatives could lead to animal abuse, exemplified by slaughterhouses skinning sheep alive to save time and money.[18] Salt anticipated that social reform would help alleviate these practices.

In addition, Salt's ideologies acknowledged affinity between humans and animals, because both were commensurate parts of nature, allowing "the lower races [to be admitted] within the pale of human sympathy".[19] In this context, sympathy was not intended to be a maudlin reaction to animal cruelty but was based on compassion towards animals, activating a form of animal "rights", which included obligations not to kill animals unless absolutely necessary.[20] Sympathy also compelled humans to be honest and transparent about the severity and extent of suffering they inflicted on animals.[21] When Salt wrote his treatise, the notion of sympathy towards animals was well-accepted.[22] It was seen as an indication of heightened moral sense, frequently forming the basis of arguments against vivisection.[23] In this context, apathy towards suffering denoted a "callous and insensitive" personality, common in the lower classes, but "undesirable in a medical practitioner or scientist".[24]

Major Charles Westley Hume (1886–1981), a World War I veteran and animal activist, who used the term animal welfare, similarly espoused the concept of

---

[14] Henry Salt, *Animals' Rights: Considered in Relation to Social Progress*, (1894), revised edition G Bell and Sons Ltd. (1922), 26, available from https://ia600901.us.archive.org/32/items/cu31924030305332/cu31924030305332.pdf

[15] Ibid, 115.

[16] Ibid, 13, 115.

[17] Ibid, 115.

[18] Ibid.

[19] Ibid, 15, 115.

[20] Ibid, 15, 22.

[21] Ibid, 15–16, 22.

[22] A W H Bates, *Anti-Vivisection and the Profession of Medicine in Britain, a Social History* Palgrave Macmillan Animal Ethics Series (2017), 29 (eBook) DOI https://doi.org/10.1057/978-1-137-55,697-4

[23] Ibid, 29.

[24] Ibid.

sympathy. In 1926, he established the University of London Animal Welfare Society (ULAWS) to tackle animal cruelty, particularly in the field of scientific research.[25] ULAWS adopted a policy platform of "maximum... sympathy but a minimum of sentimentality".[26] Presumably, for this reason, the Society did not take a position on whether humans should refrain from using animals in scientific experiments, but called upon researchers to avoid using animals in situations that involved "serious suffering".[27] In 1938, the Society changed its name to UFAW, the Universities Federation for Animal Welfare,[28] and its predominantly academic base continued their existing mandate to reduce animal pain and suffering.[29] UFAW was involved in campaigns that today would be regarded as advancing animal welfare, and the federation was instrumental in championing cruelty-free cosmetics, humane methods of animal trapping as well as improved living conditions for laboratory animals.[30]

Notwithstanding these initiatives and the existence of anti-cruelty regulation, atrocious conditions in farm animal production led to the publication of Ruth Harrison's seminal tome, *Animal Machines* in 1964.[31] Harrison did not use the term "animal welfare" in the book and likely first used the phrase at a conference in 1969.[32] The book criticised institutionalised cruelty in the British livestock and poultry sectors, leading to an outcry that was instrumental in establishing a committee of inquiry to investigate intensive animal production in Britain.[33] In 1965, the committee handed down its report, under the leadership of Professor F W Rogers Brambell, the *Report of the Technical Committee to Enquire into the Welfare of Animals kept under Intensive Livestock Husbandry Systems* (often referred to as the Brambell Report), ushering the term "animal welfare" into everyday use.[34] The report found that animal welfare comprised:

---

[25] S M Wickens, (compiler and editor) *Science in the Service of Animal Welfare, A Chronicle of Eighty Years of* UFAW, UFAW, Wheathampstead, (2007), 6, 7.

[26] Ibid, 6, 7.

[27] Richard P Haynes, *Animal Welfare, Competing Conceptions and their Ethical Implications*, Springer NL (2010), 7.

[28] S M Wickens, (compiler and editor) *Science in the Service of Animal Welfare, A Chronicle of Eighty Years of* UFAW, above 25, 7.

[29] Ibid, 7.

[30] Ibid 6, 10.

[31] Ruth Harrison, *Animal Machines*, above 3.

[32] Donald M Broom, "Ruth Harrison's Later Writings and Animal Welfare Work", in Ruth Harrison, *Animal Machines*, 21, 25, (cited as an endnote R Harrison, 1969(a) Intensive Livestock Farming and Health Welfare. *Proceedings of Health Congress, Eastbourne, 28 April to 2 May 1969*. Royal Society of Health, London) Vincent Stuart Publishers, 2013 Edition J Harrison and J Wilson, CABI Oxfordshire (2013).

[33] Abigail Woods, "From Cruelty to Welfare: The Emergence of Farm Animal Welfare in Britain, 1964–71", (2011) 36 (1) *Endeavour*, 14, 18.

[34] F W Rogers Brambell, *Report of the Technical Committee to Inquire into the Welfare of Animals Kept under Intensive Livestock Husbandry Systems*, Her Majesty's Stationery Office, (1965, Reprinted 1967), available from https://edepot.wur.nl/134379

> both the physical and mental well-being of the animal. Any attempt to evaluate welfare therefore must take into account the scientific evidence available concerning the feelings of animals that can be derived from their structure and functions and also from their behaviour.[35]

To its credit, the Brambell Report rejected economic benchmarks, such as rates of meat or egg production, as the primary arbiter for determining animal welfare.[36] The report further noted that while production criteria could furnish information on whether animals were given sufficient food and water, these criteria did not capture animal emotions and reactions to "stress-causing circumstances", including the inability to express instinctive behaviours.[37] However, a gap in knowledge at the time of the Brambell Report meant that scientific observations were not sufficiently advanced to make determinations on the feelings of animals.[38] Moreover, farmers and producers who were closest to animals, and who were better placed to make these determinations, had significant economic stakes in maintaining animal welfare as a production issue, to the exclusion of animal emotions.[39] This meant that translating animal emotion and reactions into a workable concept would be challenging.

Nevertheless, Chap. 4 of the Brambell Report discussed animal welfare in detail, including benchmarks such as freedom from disease, injury, hunger and thirst, provision of sufficient space for animals to move and groom themselves, and nurturing the ability of animals to express innate behaviour.[40] These benchmarks formed the foundation of what later came to be known as the Five Freedoms, a concept that was influential in characterising animal welfare during the last three decades of the twentieth century:[41]

- Freedom from hunger and thirst
- Freedom from discomfort
- Freedom from pain, injury and disease
- Freedom to express normal behaviours
- Freedom from fear and distress

---

[35] Ibid, 25.

[36] Ibid, 10–11.

[37] F W Rogers Brambell, *Report of the Technical Committee to Inquire into the Welfare of Animals Kept under Intensive Livestock Husbandry Systems*, above 34, 10, 73.

[38] Ibid, 9; Abigail Woods, "From Cruelty to Welfare: The Emergence of Farm Animal Welfare in Britain, 1964–71", above 33, 19.

[39] Abigail Woods, "From Cruelty to Welfare: The Emergence of Farm Animal Welfare in Britain, 1964–71", above 33, 19.

[40] Steven P McCulloch, "A Critique of FAWC's Five Freedoms as a Framework for the Analysis of Animal Welfare", (2013) 26 (5) *Journal of Agricultural and Environmental Ethics*, 959, 960–961, DOI https://doi.org/10.1007/s10806-012-9434-7

[41] Ibid, 961.

These criteria have been incorporated into the definition of animal welfare adopted by the World Organisation for Animal Health (OIE):[42]

> the physical and mental state of an animal in relation to the conditions in which it lives and dies. An animal experiences good welfare if the animal is healthy, comfortable, well nourished, safe, is not suffering from unpleasant states such as pain, fear and distress and is able to express behaviours that are important for its physical and mental state.[43]

Two comments may be made about the totality of definitions and descriptions proffered by the OIE, the Five Freedoms and the Brambell Report: first, the definitions refer to individual animals, and, second, the definitions and descriptions differ in many respects from anti-cruelty provisions of the nineteenth and early twentieth centuries. The first point is relevant to the application of welfare principles to animals en masse and is discussed later in this chapter, as part of the analysis on commodification. The second point represents something of a conundrum because notwithstanding the differences between anti-cruelty regulation and animal welfare, both share common features, notably their failings in the degree of protection proffered to animals. This is partly explained by the fact that animal welfare regulation evolved from concepts of anti-cruelty,[44] and hence both represent attempts to determine what level of cruelty or welfare is "acceptable to society".[45] A further explanation derives from the fact that both neglected to address, at least to a meaningful extent, threshold issues regarding the legitimacy of using animals.

With respect to the evolution of animal welfare, the Brambell Report simply assumed the existence of "animal welfare legislation" based on prohibitions against cruelty in statutes such as the *Protection of Animals Act 1911* (UK) Act.[46] In interpreting animal welfare in this way, the report deemed the concept to be a legal construct (although dependent on science), allowing humans "to subjugate animals", but not authorising unnecessary cruelty or unwarranted suffering.[47] Accordingly, animal welfare was framed within recognised parameters of allowable use. This premise, which is consistent with anti-cruelty regulation, remains

---

[42] International Agreement for the Creation at Paris of an International Office for Dealing with Contagious Diseases of Animals and Annex, opened for signature 25 January 1924, [1925] *ATS* 15, (entered into force 12 January 1925). As at May 2020 the organisation had 182 members.

[43] OIE, *Terrestrial Animal Health Code*, Article 7.1.1, 28th edition (2019) available from https://www.oie.int/index.php?id=169&L=0&htmfile=chapitre_aw_introduction.htm

[44] John McEldowney, Wyn Grant, Graham Medley et al, *The Regulation of animal Health and Welfare*, Routledge, Taylor & Francis Group Abindgon, Oxfordshire (2013), 17.

[45] F Ohl and F J Van der Staay, "Animal Welfare: At the Interface Between Science and Society", (2012) 192 *The Veterinary Journal*, 13, 13.

[46] F W Rogers Brambell, *Report of the Technical Committee to Inquire into the Welfare of Animals Kept under Intensive Livestock Husbandry Systems*, above 34, 7; Abigail Woods, "From Cruelty to Welfare: The Emergence of Farm Animal Welfare in Britain, 1964–71", above 33, 18.

[47] F W Rogers Brambell, *Report of the Technical Committee to Inquire into the Welfare of Animals Kept under Intensive Livestock Husbandry Systems*, above 34, 7.

unchanged by the OIE definition, notwithstanding the organisation's inclusion of animals' emotional states and the Five Freedoms.[48]

Indeed, achieving the aspirational aims of the Five Freedoms is problematic in a practical sense,[49] because in common with anti-cruelty regulation, animal welfare represents an inequitable "trade-off between human and animal interests".[50] Consequently, one of the most pointed critiques aimed at the Five Freedoms derives from lack of meaningful engagement with ethical principles, particularly the fundamental question of whether humans should use animals in the first place.[51]

Ruth Harrison made this point, noting that intensive farming "pushed animals outside their effective biological functioning",[52] resulting in production methods that paid scant regards to animals' physical and psychological needs.[53] These deficiencies were particularly conspicuous in the case of farm animals who provide food and economic benefits and whose treatment entrenched husbandry practices, which invariably avoided classification as "cruel", even when accompanied by a high degree of harm.[54] Consequently, Harrison concluded that existing regulation was ineffective in dealing with institutionalised cruelty.[55] At the same time, Harrison's concerns went deeper than settling the contours of allowable use, and she reasoned that factory farming was part of a broader ethical problem concerning humanity's relationship with animals.[56] The failure of anti-cruelty regulation to address this issue carried over to animal welfare regulation.

Both anti-cruelty and animal welfare are based on utilitarianism, an ethic that allows the use of animals provided humans do not cause animals to suffer unnecessarily. In the twentieth century, Peter Singer qualified this approach by the concept of "equal consideration", so that humans are obliged to consider animal interests equally with human interests.[57] In *Animal Liberation*, he concluded that:

> If a being suffers there can be no moral justification for refusing to take that suffering into consideration. No matter what the nature of the being, the principle of equality requires that

---

[48] Discussion Sect. 3.4.2 of this book.

[49] Steven P McCulloch, "A Critique of FAWC's Five Freedoms as a Framework for the Analysis of Animal Welfare", above 40, 959, 974.

[50] Steven White, "Animals and the Law: A New Legal Frontier", (2005) 29 Melbourne University Law Review, 298, 301.

[51] Steven P McCulloch, "A Critique of FAWC's Five Freedoms as a Framework for the Analysis of Animal Welfare", above 40, 959, 974.

[52] Donald M Broom, "Ruth Harrison's Later Writings and Animal Welfare Work", above 32, 21.

[53] Ruth Harrison, *Animal Machines*, above 3, 176–192.

[54] Ibid, 173, 175.

[55] Abigail Woods, "From Cruelty to Welfare: The Emergence of Farm Animal Welfare in Britain, 1964–71", above 33, 19.

[56] Ibid, 19.

[57] Peter Singer, *Animal Liberation*, fourth Edition Harper Perennial New York (2009), 6. The first edition was published as Peter Singer, *Animal Liberation*, Harper Collins, 1st Edition, New York (1975).

> its suffering be counted equally with the like suffering—insofar as rough comparisons can be made—of any other being.[58]

Although the phrase "no moral justification" appears to provide an overriding veto in favour of animal sentience, neither anti-cruelty regulation nor, in a practical sense, animal welfare operates in this way. Primarily this is because utilitarianism is also a consequentialist ethic, which depends on the parties' interests and the harmonisation of human and animal interests.[59] Indeed, had the principle of equal consideration been adopted by the animal product sector, it should have led to the demise of intensive production methods.[60] However, in the decades following the Brambell Report, systems based on animal welfare science bolstered "merely limited reform" in the same way that anti-cruelty regulation had done before it.[61] In economic terms, this meant that reforms were predominantly accepted where they were consistent with preserving profitability within the sector. Hence, the quality of life of farm animals, including their innate feelings, was not generally acknowledged, with the focus resting on production. Moreover, while animal suffering was ostensibly taken into account, apart from laws regarding transportation, animal welfare was largely seen in terms of unplanned deaths and where possible ameliorating pain and suffering at the point of death. In this sense, animal mortality became a metaphor for welfare.

6.3 Mortality as a Metaphor for Welfare

The Oxford English dictionary defines a metaphor as a term which is "a figure of speech... regarded as representative or symbolic of something else...".[62] This part of the chapter argues that in the evolution of intensive agriculture, the quality of life of the animal took a subordinate role to managing its death, a trend that was replicated in regulation. In the case of animals raised for meat, animal death was inevitable as animals were destined for slaughter, so this planned death was intended to be as painless as possible, notwithstanding the fact that this outcome was not always achieved. In addition, unplanned deaths were accepted as a corollary of farming, whether through disease (mis)management or death in trade and transportation. Indeed, in some industries, such as the live export trade by sea, percentages of animal mortality currently provide benchmarks for acceptable animal welfare.

---

[58] Ibid, 8.

[59] Gaverick Matheny, "Utilitarianism and Animals", in Peter Singer (ed) *In Defense of Animals the Second Wave*, 13, 15, Blackwell Publishing, New Jersey (2006).

[60] Johannes Kniess, "Bentham on Animal Welfare", (2019) 27 (3) *British Journal for the History of Philosophy*, 556, 566, DOI: https://doi.org/10.1080/09608788.2018.1524746

[61] Richard P Haynes, *Animal Welfare, Competing Conceptions and their Ethical Implications*, above 27, ix, 151.

[62] Angus Stevenson (ed), *Oxford Dictionary of English*, Oxford University Press, third edition, (2015), online by subscription.

To start with, farmers who raise animals have long been accustomed to animal suffering and premature deaths as a normal part of farming.[63] In 1871, George Fleming pointed to the utilitarian underpinnings of farm management, where maintaining the health of animals was linked to their economic value.[64] Consequently, when farm animals became ill, the decision whether to spend money attempting a cure depended on the cost of the remedy compared to the cost of the animal.[65] Although some death and suffering could have been averted, this came "at a significant cost… [potentially making businesses] unprofitable".[66] Premature death was therefore treated as a routine occurrence that farmers had to learn to live with.

This line of reasoning unfolded on a larger scale during the epizootics of the later part of the nineteenth century.[67] Veterinarians who would have been well placed to research cures instead acted as inspectors to stop trade and commerce from introducing and spreading disease.[68] As discussed in Chap. 4 of this book, until vaccines became commonly available, government saw the role of veterinarians as gatekeepers, implementing regulations based on border controls and containment of disease by slaughtering infected animals.[69] Accordingly, animal health became subsumed into disease control, inevitably leading to the death of thousands of animals in state-sanctioned wholesale slaughter.[70] Indeed, this approach persists to the present day[71] and, in accordance with animal welfare principles, is not necessarily cruel as long as the method of slaughter is humane. Nevertheless, this does not change the point to the argument made here, which is that law and policy predominantly centred on animal death.

This argument is reinforced by the fact that animal death was a common and economically acceptable feature of transportation of animals by sea. Regulation of sea transport differed from overland transport, with the latter being better controlled. By way of illustration, Section 12 of the 1849 Cruelty to Animals Act 1849 (UK) created an offence for conveying animals in ways that caused them to suffer unnecessarily, providing a broad-based criminal offence for animal cruelty. Sea transport, however, markedly differed, with the oceanic trade of the nineteenth century following substandard frameworks set by slave ships many decades

[63]F Bailey Norwood and Jayson L Lusk, *Compassion, by the Pound: The Economics of Farm Animal Welfare*, above 4, 112.

[64]George Fleming, *Animal Plagues: Their History, Nature, and Prevention*, Chapman and Hall (1871) xxx. This copy is a reprint.

[65]Ibid, xxx.

[66]F Bailey Norwood and Jayson L Lusk, *Compassion, by the Pound: The Economics of Farm Animal Welfare*, above 4, 112.

[67]Discussion, Chap. 4, Sect. 4.4.1 of this book.

[68]C A Spinage, *Cattle Plague, A History*, Kluwer Academc, New York (2003), 329.

[69]Ibid, 329.

[70]Discussion, Chap. 4, Sect. 4.4.1 of this book.

[71]*Animal Health and Welfare (Scotland) Act 2006* (UK), ss1–3, available from https://www.legislation.gov.uk/asp/2006/11/contents

earlier.[72] Notwithstanding the fact that from a commercial perspective, it was in the parties' interests to ensure animals survived sea voyages in good health, this was not always the case.[73] Conditions were often dank and unhygienic with many animals succumbing to disease because the ships were not built to deal with the live trade:

> Animals were...given scanty amounts of food and water, and worked continually with prods, sticks and clubs to keep them awake and alive....Hot paraffin was poured down their ear canal, hay stuffed inside their ears and set afire, and tails twisted and broken—all to incite near-dead animals to move. [The ships]... were difficult to manoeuvre in stormy weather... [sinking] and taking up to a thousand animals and crew to a watery death.[74] ...

Samuel Plimsoll (1824–1898) wrote on the deficiencies of this trade, providing numerous illustrations of abject cruelty. One example describes the steamship "Iowa" that carried 150 cattle on deck and 300 in the deck-hold. On route from New York to the UK, the ship encountered a storm that raged for 2 days.[75] The animals on deck were washed away, and those in the hold trampled and gored each other as the ship rolled and listed in the storm.[76] To save the ship, dead and dying cattle were thrown overboard, resulting in the drowning of animals still alive when jettisoned.[77] In another example, the steamship "Santiago" caught fire, causing the passengers and crew to abandon ship. Many animals were burned alive, although some ended up in the sea and swam towards the lifeboats, where they were "beaten off with hatchets and oars".[78]

These animals were clearly regarded as expendable, even though they were sentient beings, to whom anti-cruelty regulation should have applied. Moreover, storms and bad weather were a foreseeable and regular occurrence, yet regulation did not extend to contingency planning, such as avoiding the storm season or reducing the number of animals carried on board. Indeed, throwing unwanted animals over the side of a ship appeared to be a well-accepted practice. During the cattle plagues of the 1860s, it was common for ships docking in the UK to throw sick animals overboard, rather than declare them to quarantine authorities.[79] Apart from cruelty

---

[72] Laurie Winn Carlson, *Cattle, An Informal Social History*, Ivan R. Dee, Chicago (2001), 125.

[73] Richard Perren, "The North American Beef and Cattle Trade, with Great Britain, 1870–1914", (1971) 24 (3) *Economic History Review*, 430, 436; Richard Perren, *Taste, Trade and Technology: the Development of the International Meat Industry Since 1840,* (2006 Ashgate) Routledge, Abingdon, Oxfordshire (2017), 53.

[74] Laurie Winn Carlson, *Cattle, An Informal Social History*, above 72, 125.

[75] S Plimsoll, *Cattle Ships, Being the Fifth Chapter of Mr Plimsoll's Second Appeal for Our Seaman*, Kegan Paul, Trench, Trübner, & Co Ltd., London (1890) 54–55, republished by Forgotten Books www.ForgottenBooks.com

[76] Ibid, 54–55.

[77] Ibid.

[78] Ibid, 56–58.

[79] C A Spinage, *Cattle Plague, A History*, above 68, 288.

issues, this also meant that remaining animals could have been infected, increasing the likelihood of introducing and spreading disease.[80]

Even when animals were transported by purpose-built cattle ships, the aim was to achieve low mortality rates, rather than provide adequate (let alone optimum) shipping conditions, regardless that the two objectives would have overlapped.[81] Some of the highest losses occurred in the trade between Ireland and the UK, which was less regulated than transatlantic shipping because it was classified as part of the domestic trade.[82] Even when veterinary inspectors were appointed, their task was to uncover disease, rather than identify or deal with animal cruelty.[83] It was, therefore, not surprising that animal protection organisations regularly denounced the sector.[84] Criticisms continued into the twentieth century leading the UFAW to undertake a series of inspections in 1948 along shipping routes from Ulster to Perth, Arwnwick and Glasgow.[85] As a result, UFAW became proactive in advising the Ministry of Agriculture (UK) on how to improve transportation conditions for cattle.[86]

In countries that still engage in live export, such as Australia, animal welfare is often measured by mortality rates, which addresses the worst aspects of the trade, without necessarily improving shipping conditions. Australia has the world's largest sheep export trade and the world's fourth largest cattle trade, which is regulated by law and policy including the Australian Standards for the Export of Livestock and the Australian Meat and Live-stock Industry (Standards) Order 2005.[87] These regulations mandate reporting of deaths which exceed percentages, ranging from 0.5% for cattle and buffalo on a voyage of fewer than 10 days to 2% for goats and sheep.[88] The total animals involved in the live export trade for sheep, cattle, goats

[80]Ibid, 288.

[81]Richard Perren, "The North American Beef and Cattle Trade, with Great Britain, 1870–1914", above 73, 436.

[82]Ibid, 436.

[83]Richard Perren, *The Meat Trade in Britain 1840–1914*, Routledge and Kegan Paul Ltd. London (1978), 98.

[84]Richard Perren, "The North American Beef and Cattle Trade, with Great Britain, 1870–1914", above 73, 434.

[85]S M Wickens, (compiler and editor) *Science in the Service of Animal Welfare, A Chronicle of Eighty Years of* UFAW, above 25, 11.

[86]Ibid, 11.

[87]Parliament of Australia, *Live Animal Exports*, (undated), available from https://www.aph.gov.au/About_Parliament/Parliamentary_Departments/Parliamentary_Library/pubs/BriefingBook44p/AnimalExports; Australian Standards for the Export of Livestock (Version 2.3) 2011 and Australian Position Statement on the Export of Livestock, paragraphs 5.5, 6.5, available from https://www.agriculture.gov.au/sites/default/files/sitecollectiondocuments/animal-plant/animal-welfare/standards/version2-3/australian-standards-v2.3.pdf; Australian Meat and Live-stock Industry (Standards) Order 2005 made under the Australian Meat and Live-stock Industry Act 1997, Compilation 8 (4 June 2020), section 5, available from https://www.legislation.gov.au/Details/F2020C00475/Download

[88]Australian Standards for the Export of Livestock (Version 2.3) 2011 and Australian Position Statement on the Export of Livestock, paragraphs 5.5, 6.5; Australian Meat and Live-stock Industry

Table 6.1 Live exports, cattle, sheep, goats and buffalo

Financial year	Sheep	Cattle	Goats	Buffalo
2015–2016	1,937,429	1,238,789	80,144	4671
2016–2017	1,826,240	909,065	29,208	7692
2017–2018	1,934,904	958,905	17,316	10,325
2018–2019	1,009,795	1,242,100	18,324	7996
2019–2020	996,690	1,290,667	12,627	10,132

Source: Australian Government, Department of Agriculture, Water and the Environment

and buffalo are set out in Table 6.1, with the numbers being largest for sheep (996,690–1,937,429) and cattle (909,065–1,290,667).[89] This would equate to between 4545 and 6453 cattle deaths and 19,933–25,813 sheep deaths per annum, before reporting requirements become mandatory.

Although from a production perspective, these figures might be acceptable, as will be discussed in Sect. 6.4, sentience which is the basis of the welfare paradigm is experienced at an individual level. Therefore, using animal mortality percentages as a basis for animal welfare does not adequately consider the pain and suffering of some 24,000–32,000 sheep and cattle per year. This conclusion is reinforced by the fact that the type of criticism made against substandard conditions aboard live export ships has scarcely changed from the nineteenth century.[90]

In 2018, footage taken aboard a live export ship travelling from Perth, Australia, to the Middle East was reviewed by Dr Lynn Simpson, a veterinarian who has travelled on some 57 live export voyages. Dr Simpson concluded that extreme heat endured by the animals was reflected in the condition of their carcasses, leading her to conclude that the animals had been cooked alive.[91] The focus on percentages of animal deaths means that from a regulatory perspective, matters such as conditions during the voyage and the way the animals die are consigned to a status of lesser significance.

With respect to planned animal death, human consumption of meat necessarily requires animals to be killed specifically for that purpose, because animals who die

(Standards) Order 2005 made under the Australian Meat and Live-stock Industry Act 1997, Compilation 8 (4 June 2020), section 5.

[89] Australian Government, Department of Agriculture, Water and the Environment, *All Live Exports*, 14 October 2020, available from https://www.agriculture.gov.au/export/controlled-goods/live-animals/live-animal-export-statistics/livestock-exports-by-market

[90] Report of the Australian Senate, released in 1985, inferences drawn from the history of live exports in Australia, The Senate Select Committee on Animal Welfare, *The Export of Live Sheep from Australia*, Commonwealth of Australia (1985), 100–106, available from https://www.aph.gov.au/Parliamentary_Business/Committees/Senate/Significant_Reports/animalwelfarectte/exportlivesheep/index; generally, Bidda Jones and Julian Davies, *Backlash*, Finlay Lloyd Publishers, Braidwood (Aus) (2016).

[91] Latika Bourke, "'Boiled Alive': New Footage Shows Full Scale of Live Exports Horror" Sydney Morning Herald, 5 May 2018, available from https://www.smh.com.au/politics/federal/boiled-alive-new-footage-shows-full-scale-of-live-exports-horror-20180503-p4zd9q.html

of illness or natural causes are deemed unfit for human consumption.[92] Economically speaking, the slaughter of an animal is a "value-producing moment", "naturalised" to transform a living being into commodities for trade.[93] Not surprisingly, slaughter also carries the potential for animal cruelty, as occurred in the UK during the nineteenth century when the demand for meat and other animal products put pressure on urban slaughterhouses. This led to illegal and questionable practices, such as increased sales of diseased meat and slaughter methods that compelled animals to watch others being killed before they themselves were slaughtered.[94] The latter practice still occurs in some jurisdictions, detailed in an observational study conducted by Dr Bidda Jones, the chief scientist for the Royal Society for the Protection of Animals (Australia), on the slaughter of Australian cattle in Indonesia.[95] In that study, animals were slaughtered within view of other animals. Dr Jones concluded that animals awaiting slaughter were plainly distressed, noting that: "Still visibly trembling, Steer 3 looks behind him at Steer 2 which is being hosed and continues to bleed out. Steer 3 vocalises. . . . . . Steer 2 and 4 are being skinned. Steer 1 is on his back with his legs in the air. Steer 3 is still standing in the box watching—trembling, ears twitching".[96]

Notwithstanding the harshness of these types of practices, the UK government in the nineteenth century hesitated to interfere, and when it did, the emphasis remained on safeguarding public health and assisting trade.[97] This regulatory approach was common throughout the animal product sector, as evinced by reforms in the dairy industry.[98] Conditions in urban dairies were particularly poor, as cows were

---

[92] Jan Dutkiewicz, "'Postmodernism,' Politics, and Pigs", (2013) 8 (2) *Society for Existential and Phenomenological Theory and Culture*, 296, 302; Dinesh Joseph Wadiwel, "Like One Who is Bringing his Own Hide to Market", (2016) 21 (2) *Angelaki Journal of the Theoretical Humanities*, 65, 74, DOI: https://doi.org/10.1080/0969725X.2016.1182725; Noëlie Vialles, *Animal to Edible*, Cambridge University Press, Cambridge (1994), 5.

[93] Jan Dutkiewicz, "'Postmodernism,' Politics, and Pigs," above 92, 302; Dinesh Joseph Wadiwel, "Like One Who is Bringing his Own Hide to Market", above 92, 74.

[94] Gareth Shaw, "Changes in Consumer Demand and Food Supply in Nineteenth-Century British Cities", (1985) 11 (3) *Journal of Historical Geography*, 280, 288; Chris Otter, "The Vital City: Public Analysis, Dairies and Slaughterhouses in Nineteenth-century Britain, (2006) 13 *Cultural Geographies*, 517, 527; Chris Otter, "Civilizing Slaughter: The Development of the British Public Abattoir, 1850–1910", in *Meat, Modernity and the Rise of the Slaughterhouse* (Paula Young Lee ed), University of New Hampshire Press, Hanover (2008) 89, 92.

[95] Bidda Jones, *The Slaughter of Australian Cattle in Indonesia – An Observational Study*, (2011) RSPCA Report, 42. https://kb.rspca.org.au/wp-content/uploads/2019/03/The-slaughter-of-Australian-cattle-in-Indonesia-RSPCA-Research-Report-2011.pdf

[96] Ibid, 42–43.

[97] Richard Perren, *The Meat Trade in Britain 1840–1914*, above 83, 53; Chris Otter, "The Vital City: Public Analysis, Dairies and Slaughterhouses in Nineteenth-century Britain", above 94, 527; Matthew Hilton, *Consumerism in Twentieth-Century Britain: The Search for a Historical Movement*, Cambridge University Press, Cambridge (2003), 30.

[98] Gareth Shaw, "Changes in Consumer Demand and Food Supply in Nineteenth-Century British Cities", above 94, 288; Chris Otter, "The Vital City: Public Analysis, Dairies and Slaughterhouses in Nineteenth-century Britain", above 94, 524.

deliberately kept indoors in crowded and hot conditions, because the optimum temperature for stimulating milk production was between 65 and 75 °F.[99] This meant that "flatulent air, flakes of skin [and] cobwebs were easily absorbed into milk... making it positively dangerous".[100] Until the advent of pasteurisation, drinking contaminated milk could cause typhoid, diphtheria and scarlet fever.[101] Reforms targeted building standards, and later housing conditions for cows, yet improvements were largely introduced to protect human health.[102]

Reforms in the meat industry included licensing butchers who slaughtered horses and cattle,[103] replacing privately owned slaughterhouses with publicly owned abattoirs and developing national standards to prevent the sale of adulterated meat.[104] In common with political ideologies that applied to dairies, these reforms aimed at safeguarding humans, rather than protecting animals.[105] In any event, reform proceeded slowly, partly due to a backlog of public health matters and partly due to resistance from stakeholders, such as butchers who opposed government inspections and control of public abattoirs.[106] For these reasons, at the beginning of the twentieth century, the structure and operation of many slaughterhouses had changed little from the beginning of the nineteenth century.[107]

The sluggish rate of reform and the emphasis on public health also had an impact on the way act of slaughter was seen. In reality, it was framed in terms of economics and societal wellbeing, leading to animals being regulated as food products, rather than sentient beings. Accordingly, beyond safeguarding adequate sanitary conditions, regulation of slaughter tended to ignore matters such as animal cruelty, where it did not affect the public as consumers.[108] This oversight was consistent with

---

[99] Chris Otter, "The Vital City: Public Analysis, Dairies and Slaughterhouses in Nineteenth-century Britain", above 94, 524.

[100] Ibid.

[101] Richard Perren, *The Meat Trade in Britain 1840–1914*, above 83, 53; Chris Otter, "The Vital City: Public Analysis, Dairies and Slaughterhouses in Nineteenth-century Britain", above 94, 524.

[102] Chris Otter, "The Vital City: Public Analysis, Dairies and Slaughterhouses in Nineteenth-century Britain", above 94, 525.

[103] Cruelty to Animals Act 1849 (UK), ss 7–10.

[104] Richard Perren, *The Meat Trade in Britain 1840–1914*, above 83, 53; Chris Otter, "The Vital City: Public Analysis, Dairies and Slaughterhouses in Nineteenth-century Britain", above 94, 527; Matthew Hilton, *Consumerism in Twentieth-Century Britain: The Search for a Historical Movement*, above 97, 30.

[105] Richard Perren, *The Meat Trade in Britain 1840–1914*, above 83, 53; Chris Otter, "The Vital City: Public Analysis, Dairies and Slaughterhouses in Nineteenth-century Britain", above 94, 527; Matthew Hilton, *Consumerism in Twentieth-Century Britain: The Search for a Historical Movement*, above 97, 30.

[106] Richard Perren, "Filth and Profit, Disease and Health: Public and Private Impediments to Slaughterhouse Reform in Victorian Britain", in *Meat, Modernity and the Rise of the Slaughterhouse* (Paula Young Lee ed), University of New Hampshire Press, Hanover (2008) 127, 128.

[107] Ibid, 127–128.

[108] Ibid, 128.

legal interpretations of anti-cruelty, which authorised harmful acts unless they were unnecessary or unjustifiable. Yet, then as now, production practices were not likely implemented for the purpose of inflicting cruelty.[109] Instead, these practices were shaped by producers and traders, reflecting a level of cruelty that was commercially acceptable.[110] This pattern conforms with laissez-faire trends evident in the UK in the nineteenth century where, epizootics aside, both factions of politics left the organisation and management of the sector to market forces.[111] The outcome, however, was an inequitable compromise that favoured human interests over animal wellbeing.[112]

Thus, the answer to the question "Whither Ethics?" commences with an understanding that both anti-cruelty and animal welfare regulation spotlight only a selection of issues, which depend on one's point of view. From a historical perspective, producers regarded animals as food, veterinarians saw their role as preventing the introduction and spread of animal disease and government regarded farm animals as objects of trade and commerce to be regulated for the national good.[113] Although a deeper understanding of animal welfare evolved from the mid-twentieth century and has challenged these views, the underlying premise has not altered: farm animals can be used by humans in a way that "increase[es] productivity or reduc[es] labor costs. . .[without recognising]. . . that animals have inherent value requiring that we respect their interests even when there is no benefit to us".[114] This approach follows what Tom Regan (1938–2017) has described as an "exploitation principle",[115] ensuring that animal cruelty is consistently shaped by what is useful to humans.[116]

In this context, ethical issues based on animal welfare can have perverse outcomes where animal death is under the spotlight because utilitarianism, which is the basis of animal welfare, frequently condenses welfare to questions involving the degree of pain and suffering inflicted. A critical point stems from the fact that animals are thought not to be self-aware or self-conscious.[117] If this indeed is the

---

[109] Aaron Garrett, "Francis Hutcheson and the Origin of Animal Rights", (2007) 45 (2) *Journal of the History of Philosophy*, 243, 262.

[110] Gary L. Francione, "Reflections on Animals, Property, and the Law and Rain Without Thunder", (2007) 70 (9) *Law and Contemporary Problems*, 9, 51–52; Taimie L Bryant, "Sacrificing the Sacrifice of Animals: Legal Personhood for Animals, the Status of Animals as Property, and the Presumed Primacy of Humans", (2008) 39 *Rutgers Law Journal* 247, 249.

[111] Richard Perren, Agriculture in Depression, 1870–1940, Cambridge University Press, Cambridge (1995), 26.

[112] Steven White, "Animals and the Law: A New Legal Frontier", above 50, 301.

[113] Discussion, Chap. 4, Sects. 4.5.1 and 4.5.2 and Chap. 5, Sect. 5.5 of this book.

[114] Gary L Francione, "Reflections on Animals, Property, and the Law and Rain Without Thunder", above 110, 13.

[115] Tom Regan, "Sentience and Rights" in Jacky Turner and Joyce D'Silva (eds), *Animals, Ethics and Trade*, Earthscan, UK (2006), 79, 92.

[116] Robert Garner, "Rawls, Animals and Justice: New Literature, Same Response", (2012) 18 *Res Publica*, 159, 163, DOI https://doi.org/10.1007/s11158-011-9173-z

[117] Johannes Kniess, "Bentham on Animal Welfare", above 60, 567.

case, then animals do not have an interest in staying alive, allowing humans to kill them if this is done painlessly.[118] At the very point of death, "welfare ceases and if consciousness is lost instantly and not regained before death, there is no welfare problem".[119] As indicated previously, however, this line of thought may sidestep the treatment animals receive before the point of death (as in the live export trade by sea) and also fails to address preliminary issues regarding whether, and how, humans should use animals.

These dilemmas also call into question whether a welfare paradigm is sufficiently robust to offset commercial biases, which have become increasingly ingrained. Indeed, animal protection organisations have found it challenging to stop or curtail economic pathways towards animal commodification, which in effect deconstruct animals into objects of the market place.[120]

6.4 Commodification, Anti-cruelty and Animal Welfare

The introduction to this book highlighted Karl Marx's work on commodification, as explicated in his seminal tome, *Capital*.[121] Marx starts by defining a commodity as an object which "satisfies human wants of some sort or other", with commodification being the process that transforms products and materials into usable forms.[122]

Commodities have at least two functions. By definition, they are "objects of utility" to humans, and in addition, they are repositories of value, including use value, exchange value and surplus value.[123] Use value depends on the utility of the commodity, which directly flows from human needs or wants and in the marketplace becomes evident by consumption.[124] Exchange value, which is also a function of the market, links to use value because "[t]o become a commodity a product must be transferred to another, whom it will serve as a use value, by means of an exchange".[125] In other words, use value depends on what the commodity can be used for, and exchange value depends on what the commodity can be exchanged for. A key component of this process is the creation of objects whose net value is greater than the individual value of materials, goods and services used to produce the objects.[126] Marx called this greater value the "surplus value", a benefit that accrues

[118] Ibid, 567.

[119] D M Broom, "The Welfare of Vertebrate Pests in Relation to their Management", in *Advances in Vertebrate Pest Management*, P D Cowan and C J Feare (eds), 309, 309, Filander Verlag, Fürth (1999).

[120] Kimberley Smith, *Governing Animals Animal Welfare and the Liberal State*, above 1, 28.

[121] Karl Marx, "*Capital, A Critique of Political Economy*", above 5.

[122] Ibid, 27, 41, 47.

[123] Ibid, above 5, 33.

[124] Ibid, 27, 30.

[125] Ibid, 27, 30.

[126] Ibid, 131.

to owners of commodities when the commodity is traded.[127] Commodification thus happens through market dealings, which give products an exchange value that turns people's labour into goods and services that can be bought and sold.[128]

At the same time, an exchange value is not necessarily identical to the price of a commodity, although exchange values can be represented in monetary terms.[129] The difficulty is that the price may neither correlate with how the commodity was produced nor accurately represent the commodity's cost in social terms.[130] Marx explained that "social relations" underlie the exchange value of commodities, which derives from the (exploited) labour of workers.[131] In this respect, the use of labour conceals "a vast. . .network of productive forces and relationships" that contribute towards the making of commodities.[132] This leads to commodities being removed or alienated from peoples' labour.

In transposing these arguments to animals, commentators have used analogies of free labour and slavery to argue that animals provide their labour without charge and are otherwise treated as slaves.[133] Put simply (and probably rather simplistically), these arguments substitute animal labour for human labour, leading to animal commodification. This reasoning rests on the conceptual claim that animals are "mere means to the ends of another", an explanation that applies equally to the analogy of slaves or exploited labourers.[134]

However, as indicated in the introduction to this book, the arguments proposed by this study rest on less ambitious foundations, taking a legalistic approach based on the fact that the law classifies animals as personal property.[135] Commodification results from human labour, which deems animals to be goods in the marketplace. Christian Stache takes this approach, observing that animals are neither wage

[127] Ibid, 60.

[128] Christian Stache, "Conceptualising Animal Exploitation in Capitalism: Getting Terminology Straight", (2020) 44 (3) *Capital and Class*, 408.

[129] Dinesh Joseph Wadiwel, "Like One Who is Bringing his Own Hide to Market", above 92, 67–68.

[130] Karl Marx, "*Capital, A Critique of Political Economy*", above 5, 48; Discussion, Matthew Hilton, *Consumerism in Twentieth-Century Britain: The Search for a Historical Movement*, above 97, 35.

[131] Karl Marx, "*Capital, A Critique of Political Economy*", above 5, 47–48, 330, 359, 379.

[132] Bob Torres *Making a Killing: The Political Economy of Animal Rights*, Oakland CA: AK Press (2007) 39.

[133] Gary L Francione, *Rain Without Thunder, the Ideology of the Animal Rights Movement*, Temple University Press (1996), Philadelphia 2004 Reprint, 179–180; Bob Torres *Making a Killing: The Political Economy of Animal Rights*, above 132, 41–43, 56; Diana Stuart and Ryan Gunderson, "Nonhuman Animals as Fictitious Commodities: Exploitation and Consequences in Industrial Agriculture", (2020) 28 (3) *Society and Animals*, 291, 300–304, available from doi:https://doi.org/10.1163/15685306-12341507

[134] Bob Torres *Making a Killing: The Political Economy of Animal Rights*, above 132, 31.

[135] William Blackstone, *Blackstone's Commentaries on the Laws of England Book the Second—Chapter the First: Of Property in General* 1765–1769, 5, 14, available from http://avalon.law.yale.edu/18th_century/blackstone_bk2ch1.asp

labourers nor slaves because they do not produce commodities or value in the sense explained by Marx.[136] Instead, Stache contends that Marx is useful for analysing the way capitalist systems organise human relations because that signposts how human input creates pathways towards animal commodification and exploitation.[137]

These pathways have been discussed in previous chapters of this book and are summarised in Fig. 6.1 of this chapter. The starting point originates with increased demand for animals and their products, followed by the advent of anti-cruelty regulation and advances in veterinary science.[138] Notwithstanding the differing bearings of these developments, in combination, they engendered a strong economic bias that favoured production and trade. The confluence of these different strands found expression in the operations of the market, where the exchange value of animals and their products dominated, concealing underlying social relations, including animal pain and suffering. Although market forces require both the production of commodities and trade to create value,[139] exchange value is particularly important, because once this is realised it unlocks the surplus value invested in animals.[140] On a large scale, this leads to a profit-driven system, where animal interests are "subjugated—*en masse*—to the productive ends of agricultural capital".[141]

Marx himself critiqued the exploitation of farm animals several times in *Capital*, describing how animals were used as "raw material" for production and commodification.[142] It was a sentiment echoed almost 100 years later by Ruth Harrison who similarly criticised society for disregarding animals' physical and psychological needs.[143] This critique was hardly surprising, as by the time *Animal Machines* was published, capital markets had made significant progress towards perfecting pathways that harnessed scientific and technological advances to intensify animals' surplus value.[144] While the formative years of animal agriculture had provided opportunities for anti-cruelty regulation and veterinary science to rein in unbridled economic objectives, both were subordinated by a commodification pathway that resulted in animals being regarded as objects of trade, their individualism ignored

[136] Christian Stache, "Conceptualising Animal Exploitation in Capitalism: Getting Terminology Straight", above 128, 401, 402.

[137] Ibid, 406–407.

[138] Discussion in Chaps. 3 and 4 of this book.

[139] Christian Stache, "Conceptualising Animal Exploitation in Capitalism: Getting Terminology Straight", above 128, 401, 409.

[140] Karl Marx, "*Capital, A Critique of Political Economy*", above 5, 122.

[141] Bob Torres *Making a Killing: The Political Economy of Animal Rights*, above 132, 56.

[142] Karl Marx, "*Capital, A Critique of Political Economy*", above 5, 128, 129.

[143] Ruth Harrison, *Animal Machines*, above 3, 176–192.

[144] John Bellamy Foster and Brett Clark, "Marx and Alienated Speciesism", (2018) 70 (7) *Monthly Review*, 1, 14, DOI https://doi.org/10.14452/MR-070-07-2018-11_1

and their value determined by the market. By at least the end of the nineteenth century, this had already led to "systemic exploitation".[145]

The reasons for this development cannot be entirely laid at the door of science and technology. Rather, the catalysts are found in consumer demand, the growth of markets and lack of appropriate ethical engagement. Science and technology contributed, but it was the drive to attain profits at an ever-increasing rate that provided the clincher.[146] This not only meant that animals were regulated as commodities in trade, but their exchange value was also seen as a basis for profit making, in effect a form of capital.[147] As a result, animals were "treated as consumer goods" with the aim of extracting as much revenue as possible from them.[148]

A contributing factor also stemmed from overriding economic and political power wielded by humans. This led to animals receiving only as much attention as stakeholders with power decided to provide,[149] a situation exacerbated by the fact that political power was distributed unequally among humans, favouring those with economic interests in animal production.[150] Thus, utilitarianism, in the form of anti-cruelty regulation, became skewed towards market gains and failed to address unresolved conflicts between animal and human interests in an even-handed manner.[151] In consequence, the regime bolstered "anthropocentric assumptions and practices" which have persisted to the twenty-first century.[152]

One of the biggest obstacles to meaningful engagement with animal wellbeing remains the invisibility of individual animals. This is a bewildering development given that both anti-cruelty and animal welfare address harm as experienced and felt at the individual level. Nevertheless, it is a development that is consistent with the propensity of economic ideologies to locate value in collectives or wholes so that in animal production, productivity has been regarded as a "measure of the entire operation".[153] Accordingly, the paradigm that has emerged perpetuates the economic status quo, which forges human-animal relationships in terms of economic aggregates. In practical terms, this also reduces the non-economic worth of individual animals at every level of production, a circumstance that is reflected in large-

---

[145]Steven McMullen, *Animals and the Economy*, Palgrave Macmillan, Basingstoke (2016), 17, building on the work of David Nibert *Animal Rights/Human Rights* Lanham, MD; Rowman and Littlefield (2002), 11; Bob Torres *Making a Killing: The Political Economy of Animal* Rights, above 132, 2007.

[146]John Bellamy Foster and Brett Clark, "Marx and Alienated Speciesism", above 144, 14.

[147]Dinesh Joseph Wadiwel, "Like One Who is Bringing his Own Hide to Market", above 92, 69.

[148]Kimberley Smith, *Governing Animals Animal Welfare and the Liberal State*, above 1, 29, 87.

[149]Steven McMullen, *Animals and the Economy*, above 145, 29.

[150]Ibid, 17.

[151]Ibid, 17.

[152]Ibid, 18.

[153]Bernard E Rollin, "Putting the Horse before the Descartes: My Life's Work on Behalf of Animals", Temple University Press, Philadelphia (2011), 204.

scale farming where individual welfare can be ignored because it is insignificant compared to profits achieved from a large number of animals used.[154]

The focus on wholes and aggregates is also consistent with utilitarian philosophies where the measure of pain and pleasure is quantified and aggregated, without considering individual preferences.[155] The question, nonetheless, remains how society can account for individual animals in the great vastness of animal production.[156] This challenge goes to the heart of intensive agriculture where the focus on collectives has created many points of contention.[157] Chapter 3 of this book, for example, discussed arguments surrounding dishorning of cattle, which stemmed from this very issue. Evidence provided in early court cases indicated that dishorning cattle made them easier to handle, stopped cattle from injuring each other and facilitated their transportation; all of these points were influential in deciding cases in favour of production, notwithstanding the pain and suffering inflicted on animals.[158]

The focus on collectives was also one of the reasons that veterinary practitioners acted as the gatekeepers of animal health during the epizootics of the 1860s. Although animal health is an issue of individual animal wellbeing, it also carries economic consequences, understood in terms of disease-free shipments. Hence, regulators used scientific and veterinary advances to grapple with the commercial problem of how to maintain trade in thousands of animals while preventing the introduction of pests and diseases. This reasoning persists to the present, with Rollins concluding that by the twentieth century the emphasis on aggregates had led to animal suffering of epidemic proportions:

> If a nineteenth-century agriculturalist had tried to put 100,000 egg-laying hens in cages in a building, they all would have died of disease in a month; today, such systems dominate. . .causing new and unprecedented degrees of suffering.[159]

Rollins' arguments are even more compelling because while intensive practices had their origins in the processes of the nineteenth century, they continue to be magnified and entrenched on ever-larger scales.

---

[154] Marie Blosh, *The History of Animal Welfare and the Future of Animal Rights*, (2012) Electronic Thesis and Dissertation Repository, 803, 50, available from https://ir.lib.uwo.ca/etd/803

[155] Norman Wilde, "Kant's Relation to Utilitarianism", (1894) 3 (3) *The Philosophical Review*, 289, 290.

[156] Bernard E Rollin, "Putting the Horse before the Descartes: My Life's Work on Behalf of Animals", above, 153, 204.

[157] Kimberley Smith, *Governing Animals Animal Welfare and the Liberal State*, above 1, 90.

[158] *Callaghan v Mc'Evoy* 16 LR Ir 325, 332–333, 334; *The Queen v McDonagh* 28 LR Ir 204; *Todrick v Wilson* 18 RJ 41.

[159] Bernard Rollin, "Animal Ethics and the Law", (2008) 106 *Michigan Law Review First Impressions*, 143, 145.

6.5 Conclusion

Both anti-cruelty regulation and animal welfare were built on the foundations of utilitarianism, a concept that finds expression in animal sentience. Utilitarianism, however, balances animal interests against human interests, where the practical, if not theoretical, outcome has been an inequitable bias in favour of human interests. Indeed, human concern for farm animals has mostly been pragmatic, leading to their treatment as articles of trade and commerce. While this approach did not totally exclude farm animals from the sphere of human concern, government largely focussed on issues, such as public health and facilitating trade, rather than mitigating animal suffering. Moreover, stakeholders understood animal cruelty in their own way, so that regulation only covered a small part of the problem.

Overlaying these issues were consumer expectations regarding the supply and availability of animal products, expectations that were significant for two reasons. First, consumption is said to be the goal of production so it was important to satisfy consumer demands[160]; and, second, consumer pressure carries a political dimension capable of influencing law and policy.[161] The obstacle, however, lay with the degree of commodification wreaked on farm animals, coupled with the fact that by and large, government relinquished regulation to market forces.[162] Consequently, improvements to animal wellbeing depended on the vagaries of the market, which were dominated by industry stakeholders, placing the burden on consumers or animal welfare groups to initiate reforms.[163] This was an onerous expectation given that government determines law and policy and the regime sanctioned animal commodification. The next chapter examines the extent to which societal sway can counterbalance these forces.

Bibliography

Australian Government, Department of Agriculture, Water and the Environment, All Live Exports, 14 October 2020. https://www.agriculture.gov.au/export/controlled-goods/live-animals/live-animal-export-statistics/livestock-exports-by-market

Australian Parliament, Live Animal Exports, (undated). https://www.aph.gov.au/About_Parliament/Parliamentary_Departments/Parliamentary_Library/pubs/BriefingBook44p/AnimalExports

[160] Adam Smith, *An Inquiry into the Nature and Causes of the Wealth of Nations*, Digital Edition, ΜεταLibri (2007), 218, available from https://www.ibiblio.org/ml/libri/s/SmithA_WealthNations_p.pdf

[161] Matthew Hilton, *Consumerism in Twentieth-Century Britain: The Search for a Historical Movement*, above 97, 1.

[162] F Bailey Norwood and Jayson L Lusk, *Compassion, by the Pound: The Economics of Farm Animal Welfare*, above 4, 258.

[163] Ibid.

Australian Senate, The Senate Select Committee on Animal Welfare, The Export of Live Sheep from Australia, Commonwealth of Australia (1985). https://www.aph.gov.au/Parliamentary_Business/Committees/Senate/Significant_Reports/animalwelfarectte/exportlivesheep/index

Bates AWH (2017) Anti-vivisection and the profession of medicine in Britain, a social history. Palgrave Macmillan Animal Ethics Series, (eBook). https://doi.org/10.1057/978-1-137-55697-4

Blackstone W Blackstone's commentaries on the laws of England Book the second—Chapter the First: of property in general 1765–1769. http://avalon.law.yale.edu/18th_century/blackstone_bk2ch1.asp

Blosh M (2012) The history of animal welfare and the future of animal rights. Electronic Thesis and Dissertation Repository, 803. https://ir.lib.uwo.ca/etd/803

Bourke L 'Boiled Alive': new footage shows full scale of live exports horror. Sydney Morning Herald, 5 May 2018. https://www.smh.com.au/politics/federal/boiled-alive-new-footage-shows-full-scale-of-live-exports-horror-20180503-p4zd9q.html

Brambell FWR Report of the technical committee to inquire into the welfare of animals kept under intensive livestock husbandry systems. Her Majesty's Stationery Office (1965, Reprinted 1967). https://edepot.wur.nl/134379

Broom DM (1999) The welfare of vertebrate pests in relation to their management. In: Cowan PD, Feare CJ (eds) Advances in vertebrate pest management. Fürth, Filander Verlag, p 309

Broom DM (2013) Ruth Harrison's later writings and animal welfare work. In: Harrison R (ed) Animal machines, Edition J Harrison and J Wilson edn. Vincent Stuart Publishers, CABI Oxfordshire

Bryant TL (2008) Sacrificing the sacrifice of animals: legal personhood for animals, the status of animals as property, and the presumed primacy of humans. Rutg Law J 39:247

Carlson LW (2001) Cattle, an informal social history. Ivan R. Dee, Chicago

Dutkiewicz J (2013) 'Postmodernism,' politics, and pigs. Soc Exist Phenomenol Theory Culture 8(2):296

Fleming G (1871) Animal plagues: their history, nature, and prevention. Chapman and Hall, London

Foster JB, Clark B (2018) Marx and alienated speciesism. Mon Rev 70(7):1. https://doi.org/10.14452/MR-070-07-2018-11_1

Francione GL (1996) Rain without thunder, the ideology of the animal rights movement. Temple University Press, Philadelphia, 2004 Reprint

Francione GL (2007) Reflections on animals, property, and the law and rain without thunder. Law Contemp Probl 70(9):9

Garner R (2012) Rawls, animals and justice: new literature, same response. Res Publica 18:159. https://doi.org/10.1007/s11158-011-9173-z

Garrett A (2007) Francis Hutcheson and the origin of animal rights. J Hist Philos 45(2):243

Harrison R (2013) Animal machines, first published 1964 by Vincent Stuart Publishers, 2013 Edition J Harrison and J Wilson, CABI Oxfordshire

Haynes RP (2010) Animal welfare, competing conceptions and their ethical implications. Springer NL

Hilton M (2003) Consumerism in twentieth-century Britain: the search for a historical movement. Cambridge University Press, Cambridge

Jones B, Davies J (2016) Backlash. Finlay Lloyd Publishers, Braidwood (Aus)

Kniess J (2019) Bentham on animal welfare. Br J Hist Philos 27(3):556. https://doi.org/10.1080/09608788.2018.1524746

Marx K Capital, a critique of political economy, Volume I Book One: The process of production of capital. Progress Publishers, Moscow, USSR, first published in German in 1867, English edition first published in 1887. https://www.marxists.org/archive/marx/works/download/pdf/Capital-Volume-I.pdf

Matheny G (2006) Utilitarianism and animals. In: Singer P (ed) In defense of animals the second wave. Blackwell Publishing, New Jersey

McCulloch SP (2013) A critique of FAWC's five freedoms as a framework for the analysis of animal welfare. J Agric Environ Ethics 26(5):960–961. https://doi.org/10.1007/s10806-012-9434-7

McEldowney J, Grant W, Medley G et al (2013) The regulation of animal health and welfare. Routledge/Taylor & Francis Group, Abingdon

McMullen S (2016) Animals and the economy. Palgrave Macmillan, Basingstoke

Mellor DJ, Reid CSW (1994) Concepts of animal well-being and predicting the impact of procedures on experimental animals. In: Baker RM, Jenkin G, Mellor DJ (eds) Improving the well-being of animals in the research environment. Australian and New Zealand Council for the Care of Animals in Research and Teaching (ANZCCART), Glen Osmond, SA, Australia, p 3. https://pdfs.semanticscholar.org/6890/adfa3127be363a7c3b03925fbf22363f4489.pdf

Norwood FB, Lusk JL (2011) Compassion, by the pound: the economics of farm animal welfare. Oxford University Press, New York

Ohl F, Van der Staay FJ (2012) Animal welfare: at the interface between science and society. Vet J 192:13

OIE (2019) Terrestrial animal health code, 28th edn. https://www.oie.int/index.php?id=169&L=0&htmfile=chapitre_aw_introduction.htm

Otter C (2006) The Vital City: public analysis, dairies and slaughterhouses in nineteenth-century Britain. Cult Geogr 13:517

Otter C (2008) Civilizing slaughter: the development of the British public abattoir, 1850–1910. In: Lee PY (ed) Meat, modernity and the rise of the slaughterhouse. University of New Hampshire Press, Hanover, p 13

Perren R (1971) The North American beef and cattle trade, with Great Britain, 1870-1914. Econ Hist Rev 24(3):430

Perren R (1978) The meat trade in Britain 1840–1914. Routledge and Kegan Paul Ltd., London

Perren R (1995) Agriculture in depression, 1870–1940. Cambridge University Press, Cambridge

Perren R (2008) Filth and profit, disease and health: public and private impediments to slaughter-house reform in Victorian Britain. In: Lee PY (ed) Meat, modernity and the rise of the slaughterhouse. University of New Hampshire Press, Hanover, p 127

Perren R (2017) Taste, trade and technology: the development of the international meat industry since 1840 (2006 Ashgate). Routledge, Abingdon

Plimsoll S (1890) Cattle ships, being the fifth chapter of Mr Plimsoll's second appeal for our seaman. Kegan Paul, Trench, Trübner, & Co Ltd, London, republished by Forgotten Books. www.ForgottenBooks.com

Regan T (2006) Sentience and rights. In: Turner J, D'Silva J (eds) Animals, ethics and trade. Earthscan, UK, p 79

Rollin B (2008) Animal ethics and the law. Mich Law Rev First Impress 106:143

Rollin BE (2011) Putting the horse before the Descartes: my life's work on behalf of animals. Temple University Press, Philadelphia

Salt H (1922) Animals' rights: considered in relation to social progress (1894), revised edition G Bell and Sons Ltd. https://ia600901.us.archive.org/32/items/cu31924030305332/cu31924030305332.pdf

Sandøe P, Christiansen SB, Forkman B (2006) Animal welfare: what is the role of science? In: Turner J, D'Silva J (eds) Animals, ethics and trade. Earthscan, UK, p 41

Shaw G (1985) Changes in consumer demand and food supply in nineteenth-century British cities. J Hist Geogr 11(3):280

Singer P (2009) Animal liberation, 4th edn. Harper Perennial, New York. The first edition was published as Peter Singer, Animal Liberation, Harper Collins, 1st Edition, New York (1975)

Smith A (2007) An inquiry into the nature and causes of the wealth of nations. Digital Edition, ΜεταLibri. https://www.ibiblio.org/ml/libri/s/SmithA_WealthNations_p.pdf

Smith K (2012) Governing animals animal welfare and the Liberal state. Oxford University Press, New York

Spinage CA (2003) Cattle plague, a history. Kluwer Academic, New York

Stache C (2020) Conceptualising animal exploitation in capitalism: getting terminology straight. Capital Class 44(3):401. https://journals.sagepub.com/doi/abs/10.1177/0309816819884697

Stevenson A (ed) (2015) Oxford dictionary of English, 3rd edn. Oxford University Press, online by subscription

Stuart D, Gunderson R (2020) Nonhuman animals as fictitious commodities: exploitation and consequences in industrial agriculture. Soc Anim 28(3):291. https://doi.org/10.1163/15685306-12341507

Torres B (2007) Making a killing: the political economy of animal rights. AK Press, Oakland CA

Vialles N (1994) Animal to edible. Cambridge University Press, Cambridge

Wadiwel DJ (2016) Like one who is bringing his own hide to market. Angelaki J Theor Human 21(2):65. https://doi.org/10.1080/0969725X.2016.1182725

White S (2005) Animals and the law: a new legal frontier. Melbourne Univ Law Rev 29:298

Wickens SM (compiler and editor) (2007) Science in the service of animal welfare, a chronicle of eighty years of UFAW, UFAW, Wheathampstead

Wilde N (1894) Kant's relation to utilitarianism. Philos Rev 3(3):289

Woods A (2011) From cruelty to welfare: the emergence of farm animal welfare in Britain, 1964–71. Endeavour 36(1):14

Legislation, Regulations, Gazettes and Hansard

1835, An Act to Consolidate and Amend the Several Laws Relating to the Cruel and Improper Treatment of Animals and the Mischiefs Arising from the Driving of Cattle, and to Make Other Provisions in Regard Thereto, 1835, 5 & 6 William 4 c59, Cruelty to Animals Act, available from The Statutes Project, http://statutes.org.uk/site/the-statutes/nineteenth-century/1835-5-6-william-4-c-59-cruelty-to-animals-act/

1849, An Act for the more Effectual Prevention of Cruelty to Animals, 1849, 12 &13 Vic c 92, Public General Statutes, Twelfth and Thirteenth Year, Queen Victoria, 1849, Printed by George Edward Eyre and William Spottiswoode and published by W Benning and Co, London (1849), 592. https://play.google.com/books/reader?id=RKJKAAAAMAAJ&hl=en_GB&pg=GBS.PA592

1850, An Act for the More Effectual Prevention of Cruelty to Animals in Scotland 1850 (UK) 13 & 14 Vic c 59, Public General Statutes, s II, Thirteenth and Fourteenth Year, Queen Victoria, 1850, Printed by George Edward Eyre and William Spottiswoode and published by W Benning and Co, London (1850), 801. https://play.google.com/books/reader?id=Z6pKAAAAMAAJ&hl=en_GB&pg=GBS.PA801

1876, Cruelty to Animals Act 1876 (UK), 39 & 40 Vic c 77, available from, The Statutes Project, http://statutes.org.uk/site/the-statutes/nineteenth-century/1876-39-40-victoria-c77-cruelty-to-animals-act/

1911, Protection of Animals Act 1911 (UK), 1 & 2 Geo 5 c 271, s 1, available from http://www.bailii.org/uk/legis/num_act/1911/1069356.html

2005, Australian Meat and Live-stock Industry (Standards) Order 2005 made under the Australian Meat and Live-stock Industry Act 1997, Compilation 8 (4 June 2020), section 5, available from https://www.legislation.gov.au/Details/F2020C00475/Download

2006, Animal Health and Welfare (Scotland) Act 2006 (UK), ss1-3, available from https://www.legislation.gov.uk/asp/2006/11/contents

2011, Australian Standards for the Export of Livestock (Version 2.3) 2011 and Australian Position Statement on the Export of Livestock, paragraphs. https://www.agriculture.gov.au/sites/default/files/sitecollectiondocuments/animal-plant/animal-welfare/standards/version2-3/australian-standards-v2.3.pdf

Cases

Callaghan v Mc'Evoy 16 LR Ir 325
The Queen v McDonagh 28 LR Ir 204
Todrick v Wilson 18 RJ 41

International Treaties

International Agreement for the Creation at Paris of an International Office for Dealing with Contagious Diseases of Animals and Annex, opened for signature 25 January 1924, [1925] ATS 15, (entered into force 12 January 1925). As at May 2020 the organisation had 182 members

A Sufficient Level of Repugnance 7

Abstract

Norbert Elias's civilising process provided a glimpse of the future of animal commodification. He argued that Western European society, from the Middle Ages to the turn of the twentieth century, had steadily become less violent as a result of increasing disgust and repugnance at violence. Elias's work is relevant to farm animals because violence is an inherent part of animal production and is linked with society's prevailing stage of civilisation. A critical feature of Elias's theory is that the state holds a monopoly over violence, which it should use to pacify society and thus diminish individual and group violence. At the same time, the civilising process is subject to de-civilising forces that shape the progress of civilisation, including law and policy. The political power wielded by the animal product sector has emerged as a de-civilising force, evident by the anomaly of regulatory capture and the introduction of "Ag-Gag" laws, which limit the ability of whistle-blowers to inform regulators and the public of substandard production practices. These events have afforded a substantial degree of ascendancy to market forces, which will continue to exert de-civilising pressures if not counterbalanced.

Keywords

Civilising process · De-civilising process · Norbert Elias · Violence against animals

7.1 Introduction

Using the theoretical framework of Norbert Elias's civilising process, this chapter evaluates society's approaches to raising farm animals, against the backdrop of violence. As used in this chapter, violence refers to the intentional act of harming

S. Riley, *The Commodification of Farm Animals*, Animal Welfare 21,
https://doi.org/10.1007/978-3-030-85870-4_7

animals, whether or not procedures and processes are authorised by regulation.[1] It is a concept that is well-accepted in the literature and is discussed in more detail throughout the course of this chapter.[2] Elias's work is relevant to understanding how society arrived at this point and is also significant for the future direction of regimes, because it identifies weaknesses and gaps, underscoring where improvements might be made.

The civilising process concentrates on Western European society, predominantly from medieval times to the beginning of the twentieth century, analysing how manners and court etiquette influenced society to be less accepting of aggression, violence and cruelty.[3] Elias argues that over centuries, increasing disgust and repugnance steadily decreased the level of violence tolerated by society, leading to a reduction in violence itself.[4] His arguments also trace the development of the state and how it acquired a monopoly over violence, connecting this to the pacification of society and diminishing levels of individual and group violence.[5] Interdependencies, which derived from nascent globalisation, reinforced the process because economic expansion and growth in international markets engendered a form of "economic interconnectedness", obliging stakeholders to cooperate so that the system could function in its entirety.[6]

[1]Gary L Francione, "The Abolition of Animal Exploitation", in Gary L Francione and Robert Garner (eds), *The Animal Rights Debate: Abolition or Regulation?* 29, 83 Columbia University Press, New York (2010); Discussion, Alexandra Broughton McEwan, *The Concept Of Violence: A Proposed Framework For The Study Of Animal Protection Law And Policy*, Thesis, Australian National University (2016), 32–34, available from https://openresearch-repository.anu.edu.au/handle/1885/112649

[2]Examples include: David N Cassuto, "Bred Meat: The Cultural Foundation of the Factory Farm", (Winter 2007) 70 *Law and Contemporary Problems*, 59, 73, available from https://scholarship.law.duke.edu/lcp/vol70/iss1/3;Gary L Francione, "The Abolition of Animal Exploitation", above 1, 83; generally, Alexandra Broughton McEwan, *The Concept Of Violence: A Proposed Framework For The Study Of Animal Protection Law And Policy*, above 1; Nik Taylor and Heather Fraser, "Slaughterhouses" The Language of Life, the Discourse of Death" in Jennifer Maher, Harriet Pierpoint and Piers Beirne, *The Palgrave International Handbook of Animal Abuse Studies*, 179, 180, 191, Palgrave Macmillan UK (2017). Violence to animals (not necessarily in an agricultural setting) has also been extensively discussed by Jacques Derrida, Jacques Derrida and Elizabeth Roudinesco, in *For What Tomorrow: A Dialogue*, "Violence against Animals", Stanford University Press, Redwood City (2004); generally, Taimie L Bryant, "Sacrificing the Sacrifice of Animals: Legal Personhood for Animals, the Status of Animals as Property, and the Presumed Primacy of Humans", (2008) 39 *Rutgers Law Journal* 247.

[3]Norbert Elias, *The Civilizing Process, Sociogenetic and Psychogenetic Investigations,* Blackwell Publishing, UK (Revised Edition 2000), 157, 161–172, 355.

[4]Stephen Mennel, "Civilizing Processes", (2006) 23 (2–3) *Theory, Culture and Society*, 429.

[5]Stephen Quilley, "The Land Ethic as an Ecological Civilizing Process: Aldo Leopold Norbert Elias, and Environmental Philosophy", (2009) 31 (2) *Environmental Ethics*, 115.

[6]Andrew Linklater and Stephen Mennell, "Norbert Elias, The Civilising Process: Sociogenic and Psychogenetic Investigations – An Overview and Assessment", (2010) 49 (3) *History and Theory, Studies in the Philosophy of History*, 384, doi:https://doi.org/10.1111/j.1468-2303.2010.00550.x

The civilising process is particularly important for farm animals because their reliance on humans links violence against them with society's prevailing stage of civilisation. Consequently, animal harm should decrease in accordance with diminished tolerance for violence, which in turn should lead to kinder treatment for animals, at the very least resulting in society taking into account animal pain and suffering and at best considering the legitimacy of how society uses animals.[7] Moreover, since society's relationship to animals is shaped by the animal welfare paradigm, it also means that the civilising process is important to law and policy because animal welfare is based on state-approved violence.

At the same time, the civilising process does not advance linearly because it is subject to de-civilising forces. These forces shape and sway the process of civilising, and as will be argued, political power wielded by the animal agriculture sector has emerged as a de-civilising force. This includes the introduction of "Ag-Gag" laws, as well as the anomaly of regulatory capture.[8] Ag-Gag laws limit the ability of whistle-blowers to inform regulators and the public of animal cruelty and substandard production practices, exerting de-civilising pressures because repugnance and disgust, critical to Elias's ideas, fail to develop. Regulatory capture exacerbates this weakness, because regulators become sympathetic to industry, acting in its commercial interests.[9]

The chapter commences with an explanation of the civilising process before discussing how the mechanism applies to animals. While civilising processes have led to incrementally decreasing levels of violence towards farm animals, in some jurisdictions these trajectories are in danger of stagnating or even reversing. This situation tends to entrench the commodification of farm animals, creating complex regulatory challenges for the future.

7.2 The Civilising Process

In 1939, Norbert Elias (1897–1990) published his seminal tome *The Civilizing Process* in the German language. Thirty years later, the book was translated into English and re-published, presenting Elias's work to a wider audience.[10] Elias drew together two themes: first, the progression of society from the warrior, "secular

[7] Discussion on kindness, generally, Nik Taylor, and Heather Fraser, "The Cow Project: Analytical and Representational Dilemmas of Dairy Farmers' Conceptions of Cruelty and Kindness", (2019) 8 (2) Animal *Studies Journal*, 133, available from, https://ro.uow.edu.au/asj/vol8/iss2/10

[8] Discussion Sect. 7.4.1 of this chapter.

[9] Jed Goodfellow, *Animal Welfare Regulation in the Australian Agricultural Sector: A Legitimacy Maximising Analysis*, PhD Thesis, Macquarie Law School Macquarie University Research (2015) 8, 162–176, available from https://www.researchonline.mq.edu.au/vital/access/manager/Repository/mq:45113

[10] For a review of the book and a history of its publication, refer to Rod Aya, "Norbert Elias and "The Civilizing Process"", (1978) 5 (2) *Theory and Society*, 219, available from https://www.jstor.org/stable/656697

ruling class[es]", of the Middle Ages to the finesse of the late nineteenth/early twentieth centuries, and, second, the evolution of the state and its authority over violence.[11] These themes broadly focus on forms of "constraint", with the state using its monopoly over violence to pacify society, leading to continual refinement in the way individuals conduct themselves.[12] When analysed over centuries, patterns of behaviour reveal that thresholds of repugnance become progressively lower and manners progressively more genteel.[13] Consequently, violence, public displays of bodily functions and rudimentary eating habits have increasingly become subject to restraint, not only at the level advocated by the state but also through self-discipline by individuals. The totality of these advancements gradually leads to more civilised behaviour.[14]

7.2.1 Elias, Civilisation and the Civilising Process

As a preliminary matter, Elias acknowledged that the concept of civilisation is nuanced and open to interpretation. He observed that the concept looks to:

> the level of technology, to the type of manners, to the development of scientific knowledge, to religious ideas and customs. It can refer to the type of dwelling or the manner in which men and women live together, to the form of judicial punishment or to the way in which food is prepared. Strictly speaking, there is almost nothing which cannot be done in a "civilized" or "uncivilized" way.[15]

Writing in German, Elias also noted that the German word *kultiviert* (cultured) comes closest to the English word civilisation, although the two are not identical. Whereas the former focusses on people's behaviour, their "social quality" including manners, which can vary according to "identity of groups", the latter "emphasizes what is common to all human beings".[16] Moreover, the two words can be used among groups "who share a particular tradition and a particular situation".[17] Consequently, the terms relate to "common experiences" and shared history, a critical

[11]Norbert Elias, *The Civilizing Process, Sociogenetic and Psychogenetic Investigations,* above 3, 162, 109–172, 257–362.

[12]Stanislas Fontaine, "The Civilizing Process Revisited: Interview with Norbert Elias", (1978) 5 (2) *Theory and Society*, 243, available from https://www.jstor.org/stable/656699?seq=1

[13]Stephen Mennel, "Civilizing Processes", above 4, 429.

[14]Stanislas Fontaine, "The Civilizing Process Revisited: Interview with Norbert Elias", above 12, 243.

[15]Norbert Elias, *The Civilizing Process, Sociogenetic and Psychogenetic Investigations*, above 3, 5–6.

[16]Ibid, 6–7.

[17]Ibid, 8.

theme in the civilising process, where across the centuries, internalisation of shame and disgust leads to long-term behavioural changes.[18]

The refining of manners was important to this development, as it restrained how people flaunted their physical functions, including eating, "sexual desire and anger".[19] In the Middle Ages, for example, it was common for people to defecate and urinate in public and eat with their fingers, including from communal plates, where they also spat back unwanted food.[20] However, commencing from the seventeenth century, in Western European society, eating with one's fingers was regarded as crude, while the use of cutlery became a mark of refinement.[21] These changes emanated from the nobility, spreading throughout society so that eventually refinement became a force of habit. It also corresponded to the type of behaviour that would not embarrass or shame people in public.[22] Refinement also covered activities that in the Middle Ages were sport or entertainment but which today would be considered repugnant. Typical of these activities were public executions, burning heretics and witches alive, bear and bull-baiting and other forms of animal cruelty.[23]

Feelings of embarrassment, shame, repugnance and disgust drove the civilising process, prompting society to progress from one standard to another, adopting more polished and sophisticated manners at each stage.[24] One facet of this process led to civility becoming equated with "social rituals", restrained by etiquette.[25] Consequently, in western jurisdictions, the civilising process came to be regarded as a development that disconnected humans from their animalistic side and in the process

---

[18] Ibid 8, 26, 414–435; Siniša Maleševic and Kevin Ryan, "The Disfigured Ontology of Figurational Sociology: Norbert Elias and the Question of Violence", (2012) 39 (2) *Critical Sociology*, 165, 174, doi: https://doi.org/10.1177/0896920511434484

[19] M W Jackson, "Norbert Elias's Theory of the Civilizing Process", (1985) 9 (4) *Bulletin of the Australian Society of Legal Philosophy*, 298, 298, available from http://www.austlii.edu.au/au/journals/AUSocLegPhilB/1985/23.pdf

[20] Stanislas Fontaine, "The Civilizing Process Revisited: Interview with Norbert Elias", above 12, 245.

[21] Marc Trabsky, "Institutionalising the Public Abattoir in Nineteenth Century Colonial Society", (2014) 40 (2) *Australian Feminist Law Journal*, 169, 177, doi: https://doi.org/10.1080/13200968.2014.981357

[22] Kimberley Smith, *Governing Animals Animal Welfare and the Liberal State*, Oxford University Press, New York, (2012), 15.

[23] Stephen Mennel, "Civilizing Processes", above 4, 430; John Pratt, "Civilizing and Decivilizing Characteristics of the Contemporary Penal Field", in *Norbert Elias and Empirical Research,* Tatiana Savoia Landini and François Dépelteau (eds), 63, 66, Palgrave MacMillan, New York (2014).

[24] Norbert Elias, *The Civilizing Process, Sociogenetic and Psychogenetic Investigations,* above 3, 51.

[25] M W Jackson, "Norbert Elias's Theory of the Civilizing Process", above 19, 298; Stephen Hobden and Erika Cudworth, "Civilisation and the Domination of the Animal", (2014) 42 (3) *Millennium: Journal of International Studies*, 746, 747, doi: https://doi.org/10.1177/0305829814540355

separated them from nature.[26] The belief manifests in ideas of domination of nature and animals.[27] It also engendered a form of supremacy among European nations, legitimising colonialism and conquest of "regions occupied by less advanced peoples".[28] Agriculture and raising domesticated animals were crucial to this ideology with farming communities regarded as superior to hunter-gatherer societies. Indeed, as observed in 1871 by George Fleming (1833–1901), pastoralism was revered as mutually beneficial to domesticated animals and humans alike:

> when a nation has passed from a savage condition... of the hunter or fisherman... to the more civilized and civilizing state of a pastoral people, a great change is manifested in its character. The most noteworthy feature... is the high value... [attributed to domesticated animals] which in the former stage of civilization were pursued and destroyed in a wild state, but have now by kindness, and other means founded on motives of economy, become domesticated and soon form the wealth and well-being of their owners.[29]

Similarly, Charles Darwin (1809–1882) considered that savagery was as far removed from civilisation as wild animals were removed from domesticated animals.[30] This reinforced the notion that to be civilised meant being removed from nature, although it did not stop society from giving full vent to the desire to take advantage of all that nature had to offer.[31]

The changes Elias wrote about occurred over centuries, and at first glance, they appear to be linear.[32] However, the civilising process can fluctuate and relapse, leading to de-civilisation, a process discussed in more detail in Sect. 7.4.[33] Progress towards civilising is therefore changeable and uncertain. It is susceptible to how individuals manage their social relationships in an increasingly interdependent society, as well as how the state exercises its growing monopoly on violence.[34] The latter is not only relevant to interpersonal relationships but also to the management of commerce, marketplaces and trade networks. In these cases, clusters of interdependent people converge, necessitating the reduction of disorder and

---

[26] Stephen Hobden and Erika Cudworth, "Civilisation and the Domination of the Animal", above 25, 746, 747.

[27] Ibid, 747.

[28] Ibid, 752.

[29] George Fleming, *Animal Plagues: Their History, Nature, and Prevention*, Chapman and Hall, London (1871), XV. This copy is a reprint.

[30] Tom Ingold, "From Trust to Domination: An Alternative History of Human-Animal Relations" in Manning, Aubrey and Serpell, James, (eds) *Animals and Human Society Changing Perspectives*, 1, 3, Routledge, Abingdon, Oxfordshire (1994) reprinted up to 2011.

[31] Stephen Hobden and Erika Cudworth, "Civilisation and the Domination of the Animal", above 25, 765.

[32] Norbert Elias, *The Civilizing Process, Sociogenetic and Psychogenetic Investigations,* above 3, 157.

[33] Ibid, 157.

[34] Keith Tester, *Animals and Society, the Humanity of Animal Rights*, Routledge, New York (1991), 58.

violence, leading to external restraints in economic and political contexts transforming into individual restraints that allow people to work together.[35]

Elias himself described civilisation, and by implication the civilising process, in terms of restraint, more specifically as the engagement of a "network of limited restrictions which tend toward the attenuation of excesses in pleasure, violence, inequality, and so forth".[36] Consequently, the taming of violence is a key element towards civilisation.

7.2.2 The State, Violence and the Individual

Elias did not define what he meant by violence. He used the German word, *Gewalt*, which "generally indicates force", but which in English describes the "(usually sudden) exercise of physical force so as to inflict injury or damage to humans, animals or things".[37] Violence also carries broader connotations of "actions which infringe physical integrity", such as wounding and torture for humans, as well as arson for objects.[38] Elias used the word in this wider sense and significantly did not appear to limit it to criminal acts.[39]

While Elias recognised that violence was an integral part of society, he also regarded violence as the opposite of civilisation and hence his arguments that the civilising process involves controlling or surmounting violence.[40] Consequently, the way the state manages violence is critical because of the monopoly it holds over violence. This does not mean that the state is the only institution able to use violence, but the state is the only institution that can legitimately control it,[41] so that individuals and associations can legally use violence solely to the extent permitted by the state.[42] In this way, the state's control can potentially pacify humanity, leading society to forge social relations based on influence, control, political power and patterns of behaviour, rather than brute force.[43] At the individual

[35] Brigitte H E Niestroj, "Norbert Elias: A Milestone in Historical The Making of the Social Person", (1989) 2 (2) *Journal of Historical Sociology*, 136, 146.

[36] Stanislas Fontaine, "The Civilizing Process Revisited: Interview with Norbert Elias", above 12, 250.

[37] Jonathan Fletcher, "Towards a Theory of Decivilizing Processes", (1995) 22(2) *Amsterdams Sociologisch Tijdschrift*, 283, 292 fn 1.

[38] Ibid.

[39] Norbert Elias, *The Civilizing Process, Sociogenetic and Psychogenetic Investigations,* above 3, 161–172.

[40] Siniša Maleševic and Kevin Ryan, "The Disfigured Ontology of Figurational Sociology: Norbert Elias and the Question of Violence", above 18, 166.

[41] Karl Dusza, "Max Weber's Conception of the State", (1989) 3 (1) *International Journal of Politics, Culture, and Society*, 71, 88.

[42] Ibid.

[43] Stephen Hobden and Erika Cudworth, "Civilisation and the Domination of the Animal", above 25, 755.

level, the very process of pacification engenders greater reliance on "self-restraint and mutual forbearance", qualities which derive from shame and repugnance, rather than violence.[44]

Elias's views on the state and its powers correlate with Max Weber's (1864–1920) notions of the state as "a human community that successfully claims the monopoly of the legitimate use of physical force within a given territory".[45] The notion of territory and its protection also carries strong commercial underpinnings, important to the civilising process because increased commercial activity supports the nation's economic and military capabilities, cementing the power of the state.[46] This strengthening of power was not only made possible by the exercise of control over the military and taxation but was also interwoven with nascent globalisation.[47] From the eighteenth century, economic integration on an international scale came to be seen as the future, but economists also remarked that a sizeable "territorial monarchy" was needed to secure national prosperity and power.[48] Structural arrangements were crucial to such a realm and they needed to be sufficiently robust to deal with coinage, monetary systems and safety of trade routes, ensuring the security of commercial transactions.[49] Indeed, as commerce and markets expanded, the need for high-level regulation became greater in order to manage the growing and ever more intricate "intertwining [of] human activities".[50]

Addressing the importance of these types of connections was a defining feature of Elias's work because he challenged the belief that individuals and society are distinct, instead focussing on links and networks between the two groups.[51] As the complexities of social interactions increased, so too did the complexities of business transactions, because transacting business with outsiders also required individuals to exercise restraint and develop a sense of concern for others.[52] Over time, this deliberate introspection developed into a habitual exercise of manners that became

[44] Stephen Mennel, "Civilizing Processes", above 4, 429.

[45] Simeon Mitropolitski, "Weber's Definition of the State as an Ethnographic Tool for Understanding the Contemporary Political Science State of the Discipline", (2011) *Canadian Political Science Association*, 1, 1, available from https://papers.ssrn.com/sol3/papers.cfm?abstract_id=1823401

[46] Siniša Maleševic and Kevin Ryan, "The Disfigured Ontology of Figurational Sociology: Norbert Elias and the Question of Violence", above 18, 174.

[47] Norbert Elias, *The Civilizing Process, Sociogenetic and Psychogenetic Investigations,* above 3, 316–317.

[48] Istvan Hont, *Jealousy of Trade, International Competition and the Nation-State in Historical Perspective*, Belknap Press of Harvard University Press, Cambridge MA and London (2005), 4, 188–189.

[49] Norbert Elias, *The Civilizing Process, Sociogenetic and Psychogenetic Investigations,* above 3, 314.

[50] Ibid.

[51] Tânia Quintaneiro, "The Concept of Figuration or Configuration in Norbert Elias' Sociological Theory, Translated by Maya Mtire", (2006) 2 *Teoria and Sociedade*, 0, 2 (pages not numbered within document, but pages numbered according to pdf) http://socialsciences.scielo.org/scielo.php?script=sci_arttext&pid=S1518-44712006000200002&lng=en&nrm=iso

[52] Stephen Mennel, "Civilizing Processes", above 4, 430.

second nature.[53] Elias's work focussed on human-human relationships, but it also has much to say about human-animal relationships.

7.3 The Civilising Process and Animals

The civilising process has been used to analyse humanity's relationship with nature, protection of biodiversity and domesticated animals, being particularly relevant to domesticated animals because of their close association with, and reliance on, humans.[54]

7.3.1 Animals, Repugnance and Violence

Over time, society's interactions with animals have changed in various ways. In some cases, such as rodents, society's increased disgust has potentially led to greater violence against these animals, while in other cases, such as companion animals, society's elevation of their status has reduced violence towards them. For farm animals, violence endures, comprising a complex mixture of state-sanctioned cruelty interspersed with societal concern regarding production and slaughter methods.

In an example of increased societal disgust at rodents, Neil Pemberton describes a mid-nineteenth century prank, where a rat catcher hid ten rats in his shirt, pulling them out while he was in a London pub to demonstrate his prowess at rat catching.[55] Although the practical joke caused a commotion, no one was repulsed by the spectacle of rats being hidden next to a person's body or being handled in a commercial food establishment.[56] In the twenty-first century, the same story would likely induce a different reaction, including a level of disgust or repugnance, extending beyond the health implications of rodents being brought into a business establishment open to the public.[57] Feelings of disgust frequently exist due to the very presence, in the human domain, of an animal regarded as noxious or annoying and are analogous to the fear or repulsion some people feel towards snakes, sharks and spiders, which may result in unnecessary violence towards them.[58] Admittedly,

[53] M W Jackson, "Norbert Elias's Theory of the Civilizing Process", above 19, 300.

[54] Stephen Hobden and Erika Cudworth, "Civilisation and the Domination of the Animal", above 25, 756; Amanda Whitfort, "Justice and the Vulnerable: Extending the Duty to Prevent Serious Crimes Against Children to The Protection of Agricultural and Research Animals", (2018) 39 Adelaide Law Review, 125, 131.

[55] Neil Pemberton, "The Rat-Catcher's Prank: Interspecies Cunningness and Scavenging in Henry Mayhew's London", (2014) 19 (4) *Journal of Victorian Culture*, 520, 520–521, doi: https://doi.org/10.1080/13555502.2014.967548

[56] Ibid, 525.

[57] Ibid, 535.

[58] Generally, Jessica Gall Myrick and Suzannah D Evans, "Do PSAs Take a Bit Out of Shark Week? The Effects of Juxtaposing Environmental Messages with Violent Images of Shark Attacks", (2014)

not all people react this way, and for some, snakes, rats and insects make acceptable companion animals.[59] Nevertheless, the separation from nature discussed above has come to form an influential component of the civilising process, and continuing into the twenty-first century has underscored the idea that society has a right to exploit animals, especially farm animals.[60]

These observations typify the complexity of society's relationship with animals and the fact that the relationship is constantly changing. In the sixteenth century, killing dogs and cats, animal baiting and animal cruelty were acceptable amusements.[61] By the seventeenth and eighteenth centuries, however, the tide was turning. William Hogarth (1697–1764), an artist and engraver, censured animal cruelty in a set of prints, titled the "Four Stages of Cruelty".[62] The engravings depicted the moral degeneration of a young boy from animal cruelty to murder, clearly condemning cruelty to animals and associating it with crimes against society. At a more fundamental level, Hogarth's engravings also warn against the disruption to peace and order that animal violence instigates, an ideological message consistent with Elias's philosophies, particularly the need for strong state control over violence.[63]

Elsewhere, society's perceptions of violence towards animals were influenced by shifts in urban demographics, discussed in Chap. 2 of this book. As urbanisation progressed, urban populations became further removed from farm animals but simultaneously began to enjoy greater rates of companion animal ownership.[64] Prior to the sixteenth century, owning companion animals had not been common, with dogs and cats primarily maintained for their utility rather than their companionship.[65] Indeed, in urban areas dogs and cats were frequently regarded as a nuisance.[66] However, from the sixteenth century, literature and art revealed a nascent tolerance towards keeping companion animals, especially lap dogs, which became a

---

36 (5) *Science Communication*, 544, available from doi: https://doi.org/10.1177/1075547014547159; Vasilios Liordos, Vasileios J Konsiotis, Spyridon Kokoris and Micheala Pienidou, "The Two Faces of Janus, or the Dual Mode of Public Attitudes Towards Snakes", (2018) 621 *Science of the Total Environment*, 670, available from doi: https://doi.org/10.1016/j.scitotenv.2017.11.311

[59]General discussion on the sociology of pet keeping: Jean E Veevers, "The Social Meaning of Pets: Alternative Roles for Companion Animals", (1985) 8 (3–4) *Marriage and Family Review*, 11, doi: https://doi.org/10.1300/J002v08n03_03

[60]Val Plumwood, "Animals and Ecology: Towards a Better Integration", (2003) *ANU Research Publications*, 1, 2, available from https://openresearch-repository.anu.edu.au/handle/1885/41767

[61]R Gordon, "From Pests to Pets: Social and Cultural Perceptions of Animals in Post-medieval Urban Centres in England (AD1500–1900)", (2017) 27(1) Art. 9, *Papers from the Institute of Archaeology*, 1, 3, doi: https://doi.org/10.5334/pia-478

[62]Keith Tester, *Animals & Society the Humanity of Animal Rights*, above 34, 64.

[63]Ibid, 65.

[64]Ibid, 52.

[65]R Gordon, "From Pests to Pets: Social and Cultural Perceptions of Animals in Post-medieval Urban Centres in England (AD1500–1900)", above 61, 1–2, 3.

[66]Ibid.

fashionable accoutrement.[67] This trend continued in the centuries that followed, gathering momentum in the 1970s, by which time dogs, cats and other animals were well-accepted as companions.[68] Although the shift towards keeping companion animals in urban areas prompted feelings of sympathy and kinship towards animals in general,[69] it created a cultural rift between urban and rural populations.[70]

One line of thought accuses urban dwellers of developing "sentimentalised" views of animals,[71] an accusation given full rein in the twentieth century, where companion animals in western urban jurisdictions were regarded as "friends", feted and looked after, sometimes at great cost.[72] For owners of these animals, it would have been inconceivable to deny their companions assistance, such as expensive veterinary treatment. Yet as a generalisation, farmers could ill afford this luxury and would determine the need for veterinary treatment by way of a cost-benefit analysis, multiplied many times over for the large number of animals under their control.[73]

Opposing points of view, however, observe that rather than being sentimental, the physical and cultural distances between urban dwellers and farm animals facilitate a more detached or "scientific" understanding of the issues.[74] Although producers might deal with animals on a day-to-day basis and have detailed knowledge of animal health and production, these matters also have strong commercial underpinnings, which do not automatically equate to repugnance and disgust at cruel treatment.[75] To Elias, the creation of "objective knowledge", chiefly by individuals who are separated from the target of study, is an important part of the

[67] Richard Thomas, "Perceptions Versus Reality: Changing Attitudes Towards Pets in Medieval and Post-medieval England" in A Plukowski (ed) *Just Skin and Bones? New Perspectives on Human-Animal Relations in the Historic Past*, 95, 96–97, 100–101, Archaeopress, Oxford BAR International Series 1410 (2005), available from https://www.academia.edu/151650/Thomas_R_2005b_Perceptions_versus_reality_changing_attitudes_towards_pets_in_medieval_and_post_medieval_England_pp_95_105_in_Plukowski_A_ed_Just_Skin_and_Bones_New_Perspectives_on_Human_Animal_Relations_in_the_Historic_Past_BAR_International_Series_1410_Archaeopress_Oxford (subscription required).

[68] Jean E Veevers, "The Social Meaning of Pets: Alternative Roles for Companion Animals", above 59, 11–12.

[69] Keith Tester, *Animals & Society the Humanity of Animal Rights*, above 34, 52.

[70] R Gordon, "From Pests to Pets: Social and Cultural Perceptions of Animals in Post-medieval Urban Centres in England (AD1500–1900)", above 61, 1–2.

[71] Stephen Hobden and Erika Cudworth, "Civilisation and the Domination of the Animal", above 25, 756.

[72] R G Beilharz, "Science and the Politics of Animal Use in Food Production: The Situation in Australia", (1988) 20 *Applied Animal Behaviour Science*, 143, 145.

[73] Ibid.

[74] Keith Tester, *Animals & Society the Humanity of Animal Rights*, above 34, 57.

[75] F Bailey Norwood and Jayson L Lusk, *Compassion, by the Pound: The Economics of Farm Animal Welfare* Oxford University Press, New York (2011) 33; Chap. 3, part 3.5 of this book, discussion of dishorning.

civilising process.[76] It provides a means of determining how individual human needs and wants should be satisfied in the context of interconnected, though geographically and culturally divided, communities where not all expectations can be met.[77] For farm animals, a focal point derives from the normalisation of cruelty, which brings the discussion back to the state's control over violence.

As already noted, the catalyst for legitimising state control over violence starts with repugnance or disgust at unrestrained violence. Before Elias, Henry Salt in the nineteenth century had reached an analogous conclusion by referring to the "instinctive shock, or natural feelings of disgust" that humans feel when they see pain and suffering, whether it targets fellow humans or animals.[78] Salt was also confident that the "humanitarian tendency" which led to the end of slavery would benefit animals.[79] This tendency derived from social progress, a movement consistent with the civilising process because both deal with improved standards of behaviour.

However, protection for animals and repugnance at their treatment has developed in a piecemeal way, questioning the meaning of violence, and therefore the civilising process in the context of farm animals. Violence would include criminal acts, such as infliction of deliberate and illegal cruelty. However, given that the state controls violence, it is not clear whether cruel, but legal acts would fall within the concept of violence posited by Elias. This point is particularly important for farm animals because production practices involve painful and cruel procedures, such as beak trimming and dishorning, accepted as necessary by producers and the state alike. In Australia, beak trimming is allowable, and although it is to be carried out by experienced operators, there is no requirement that birds be given pain relief.[80] In a similar way, beak trimming is also permitted in the European Union (EU)[81]

---

[76] Norbert Elias, Stephen Quilley (ed), *Involvement and Detachment*, University Dublin Press, Dublin (2007), 228, as cited in Keith Tester, *Animals & Society the Humanity of Animal Rights*, above 34, 56.

[77] Robert van Krieken, "Law and Civilization: Norbert Elias as a Regulation Theorist", (2019) 15 *Annual Review of Law and Social Science*, 267, 274.

[78] Henry Salt, *Animal Rights Considered in Relation to Social Progress*, G Bell and Sons Ltd., London, Revised Edition (1922), first printed 1892, 15–16, available from available from https://archive.org/details/cu31924030305332

[79] Ibid.

[80] Commonwealth of Australia, Primary Industries Standing Committee Model Code of Practice for the Welfare of Animals, Domestic Poultry, fourth Edition. SCARM Report 83, Commonwealth of Australia (2002), part 13.2, available from https://www.publish.csiro.au/ebook/download/pdf/3451. This code is under review but was current at the time of writing.

[81] *Treaty on European Union,* opened for signature 1 February 1992, [2002] *Official Journal of the European Communities* C 325/5, 24 December 2002, (entered into force 1 November 1993). The European Union has 27 members, available from https://eur-lex.europa.eu/legal-content/EN/TXT/PDF/?uri=OJ:C:2002:325:FULL&from=EN; Council Directive 1999/74/EC of 19 July 1999, Laying down Minimum Standards for the Protection of Laying Hens, [1999] Official Journal, L 203/53, 3/8/1999, available from https://eur-lex.europa.eu/LexUriServ/LexUriServ.do?uri=OJ:L:1999:203:0053:0057:EN:PDF; Council Directive 2007/43/EC of 28 June 2007, laying down Minimum Rules for the Protection of Chickens Kept for Meat Production (Text with EEA

although some jurisdictions within the EU prohibit it.[82] Francione cogently argues that violence against animals includes these types of practices, which are inherently cruel, even if authorised by the state.[83] Moreover, some commentators expand the notion of violence even further, concluding that it can be symbolic in the sense that it extends to acts of domination over animals.[84] These developments are complicated by the fact that animal welfare can be viewed through differing lenses. While those involved in animal protection might see practices like beak trimming as cruel,[85] those involved in the industry would see the phasing out of beak trimming as involving a "nasty transition period where bird welfare is terrible" because of increased feather pecking and hens pecking at each other (cannibilisation).[86]

Nevertheless, it is a key tenet of the civilising process that individuals and society push against the boundaries of accepted, even legal, traditions which are repugnant or violent because in this way society advances towards higher levels of civilisation. Francione's arguments, for example, extend to eating and using animal products because they exist as a result of violence.[87] These assertions, while potentially controversial, are still consistent with the tenor of Elias's arguments, if one keeps in mind how pastimes involving animal cruelty have fallen out of favour. In addition, while Elias claimed that violence should occur within parameters authorised by the state, he did not argue that the state monopoly on violence is static.[88] Certainly, as discussed in Chap. 3 of this book, regulation of farm animals, and therefore the

relevance) [2007] Official Journal, L 182/19, 12/7/2007, available from https://eur-lex.europa.eu/LexUriServ/LexUriServ.do?uri=OJ:L:2007:182:0019:0028:EN:PDF

[82] Anaëlle Laravoire, "Status of Beak Trimming", Danish poultry Congress 2016, Chamber of AGriculure Pays de la Loire, power point slides: the countries which prohibit debeaking are Norway (1974), Finland (1986), Sweden (1988), Austria (2000), Denmark (2013), available from, http://www.barnhealth.com/wp-content/uploads/2016/04/BeakTrimming_status.pdf

[83] Gary L Francione, "The Abolition of Animal Exploitation" above 1, 83; Discussion, Alexandra Broughton McEwan, *The Concept Of Violence: A Proposed Framework For The Study Of Animal Protection Law And Policy*, above 1, 119–122, 123.

[84] Alexandra Broughton McEwan, *The Concept Of Violence: A Proposed Framework For The Study Of Animal Protection Law And Policy*, above 1, 32–34.

[85] Humane Society International, *An HSI Report: The Welfare of Animals in the Egg Industry*, (2011), available from https://www.hsi.org/wp-content/uploads/assets/pdfs/welfare_of_animals_in_the_egg.pdf;

Voiceless, *The Life of the Dairy Cow*, Voiceless (2015), 33, 39, 42, available from https://voiceless.org.au/reading-list-item/the-life-of-the-dairy-cow/; Voiceless, *Unscrambled: The Hidden Truth of Hen Welfare in the Australian Egg Industry,* Voiceless (2017), 3, available from https://voiceless.org.au/wp-content/uploads/2017/06/VoicelessReport_Unscrambled_June2017.pdf

[86] Melanie Epp, "Beak Trimming Trends", (2019) April, *Canadian Poultry*, Quoting Christine Nicol, a professor animal welfare at the University of London, available from http://magazine.canadianpoultrymag.com/publication/?m=1191&i=574088&view=articleBrowser&article_id=3331179&ver=html5

[87] Gary L Francione, "The Abolition of Animal Exploitation" above 1, 83; Discussion, Alexandra Broughton McEwan, *The Concept Of Violence: A Proposed Framework For The Study Of Animal Protection Law And Policy*, above 1, 32–34.

[88] Keith Tester, *Animals and Society, the Humanity of Animal Rights*, above 34, 63.

violence against them, has changed throughout the nineteenth and twentieth centuries. Yet, at its core, regulation still treats farm animals as food and commodities in trade, leading to the next part of the discussion, which focusses on how the civilising process influences the raising of animals for food.

7.3.2 Animals as Food

Musician, Sir Paul McCartney, once quipped that "If slaughterhouses had glass walls, everyone would be vegetarian".[89] This off-the-cuff statement embodies two themes meaningful to the civilising process: first, that people have become, or are likely to become, squeamish about the consumption of animal products, leading to repugnance, and, second, that violence to farm animals largely occurs in secret.

Concerning the first point, in *The Civilizing Process*, Elias discussed how meat eating has transformed over the centuries, especially the way serving either an entire animal or large parts of it at the table has fallen out of favour.[90] This trend is partly explainable by the fact that as family units became smaller, they did not require a large amount of meat to be served in one sitting.[91] However, another explanation stems from the discomfort many people feel at eating animal products that provide clear reminders of their origins. The transition in table manners began as an aversion to pulling apart the cooked animal by hand, which gave way to the more refined practice of carving, where the animal was presented in slices and morsels.[92] However, even in this case, it frequently involved "carving half a calf or pig" or removing "meat from a pheasant still adorned with its feathers".[93]

Such images recalled that the dish was once a living animal that had to be killed and butchered for a diner's enjoyment, so consumers found ways of sidestepping these cues.[94] One approach lay in swearing off animal products, although in 1939 Elias described vegetarianism as a "threshold of repugnance that go[es] beyond the standard of civilised society in the twentieth century" and which consequently may be regarded as "abnormal".[95] Elias did, however, concede that in the past, it was this very type of change that had led to improvement and refinement of standards and in his opinion; the trajectory towards vegetarianism was following this path.[96] Elias's

---

[89] Paul McCartney, *Glass Walls* (PETA 2009), available at http://www.peta.org/tv/videos/celebritiesvegetarianism/86975251001.aspx

[90] Norbert Elias, *The Civilizing Process, Sociogenetic and Psychogenetic Investigations,* above 3, 102–103.

[91] Ibid, 102.

[92] M W Jackson, "Norbert Elias's Theory of the Civilizing Process", above 19, 298.

[93] Norbert Elias, *The Civilizing Process, Sociogenetic and Psychogenetic Investigations,* above 3, 102.

[94] Ibid.

[95] Ibid.

[96] Ibid.

point was very valid, because, at the beginning of the twenty-first century, rates of vegetarianism may have been increasing, but they were not high, varying from 8% of the population in Canada to 3% in the UK, 6% in Ireland, 9% in Germany and 40% in India.[97]

These statistics reveal that the world comprises a large number of meat eaters, who would have been in an even greater majority during the times discussed by Elias. For these people, the easier approach lay in removing as many reminders as possible, of the origin of their meal, and the violence that preceded its appearance at the dinner table.[98] Consequently, in addition to the decline of serving animals in their entirety, smaller animals such as rabbits, pigs and fish were increasingly served without their heads, and sauces were used to hide the gamey taste of wild animals.[99] By the twenty-first century, empirical research confirmed that consumers concerned about how their meat was produced tended to use it as an ingredient, or smaller part of a meal, rather than the main part.[100] The tendency towards obscuring the origins of meat also accelerated from the second half of the twentieth century, as supermarkets marketed animal products in the form of patties, nuggets, devon, pâté and other ready-to-eat goods, far removed in appearance from their animal sources.[101]

Another aspect of this trend, the slaughter of animals for food, has been discussed in Chap. 6 of this book. Notwithstanding the increasing geographical separation of rural and urban dwellers, cruelty at slaughter was highly visible, due to the growth of markets in animal products and the increase in popularity of meat eating.[102] Throughout the nineteenth century in the UK, slaughterhouses were still located in urban areas, resulting in animal slaughter emerging as a social issue.[103] One facet of the calls for reform was consistent with a generalised trend against animal cruelty[104]; however, in practice, slaughterhouse reforms tended to focus on safeguarding

[97] Matthew B Ruby, "Vegetarianism. A Blossoming Field of Study", (2012) 58 *Appetite*, 141, 142 (references omitted), available from https://doi.org/10.1016/j.appet.2011.09.019

[98] Norbert Elias, *The Civilizing Process, Sociogenetic and Psychogenetic Investigations,* above 3, 102; Christopher Otter "Cleansing and Clarifying: Technology and Perception in Nineteenth-Century London'", (2004) 43 (1) *Journal of British Studies* 40, 46.

[99] M W Jackson, "Norbert Elias's Theory of the Civilizing Process", above 19, 298; Piers Beirne, "Theriocide: Naming Animal Killing", (2014) 3 (2) *International Journal for Crime, Justice and Social Democracy*, 49, 53.

[100] Maria Font-i-Furnols and Luis Guerrero, "Consumer Preference, Behavior and Perception about Meat and Meat Products: An Overview", (2014) 98 (3) *Meat Science*, 361, 363, available from https://www.sciencedirect.com/science/article/abs/pii/S0309174014001934

[101] Ramona Ilea, "Nussbaum's Capabilities Approach and Nonhuman Animals: Theory and Public Policy", (2008) 39 (4) *Journal of Social Philosophy*, 547, 553.

[102] Ian MacLachlan, "Humanitarian Reform Slaughter Technology, and Butcher Resistance in Nineteenth-Century Britain", in *Meat, Modernity and the Rise of the Slaughterhouse* (Paula Young Lee ed), University of New Hampshire Press, Hanover (2008) 107, 107.

[103] Ibid.

[104] Ibid.

sanitary conditions while ignoring matters such as animal cruelty, which did not affect the public as consumers.[105]

Nevertheless, many people were still uncomfortable about killing animals for food, leading to slaughterhouses and the death they signified, being hidden from the public.[106] Yet, animal death had not always been viewed in this way. Noëlie Vialles points out that it was once celebrated by way of sacrifices, and butchers had freely and publicly operated in urban areas.[107] In one sense, the death of animals mirrors how society came to feel about death in general, a point underscored by Elias when he said that death is an "embarrassment felt by the living in the presence of [the dying]".[108] Yet, for some, it was not just death itself that was at issue but the fact that when humans eat meat, they must forget the "bloody, pitiless process through which [the] animal became meat".[109] In other words, although society knows that the slaughter of animals occurs, it is hidden, it happens in secret and there is a psychological disconnect between the living animal and food products.[110] Salt acquiesced with this perspective, pointing out that, notwithstanding improvements in regulation, at the end of the nineteenth century, animal cruelty had increased, referring in particular to substandard transportation and slaughter methods.[111] He pointedly noted that "people of sensibility and refinement" feel repugnance at the slaughter and butchering of animals for meat and consequently want "details of the revolting process… as far as possible, kept carefully out of sight and out of mind".[112]

This feeling of repugnance correlates precisely with Elias's analysis of disgust and violence. Animal cruelty, even if legal, is a "public bad", which can stimulate activism in an effort to persuade governments to exercise their monopoly on violence by enacting, and enforcing, ameliorating regulation.[113] The alternative is the type of institutionalised cruelty depicted by Ruth Harrison in *Animal Machines*, resulting from market-driven production methods.[114] The Brambell Report, which

---

[105] Richard Perren, "Filth and Profit, Disease and Health: Public and Private Impediments to Slaughterhouse Reform in Victorian Britain", in *Meat, Modernity and the Rise of the Slaughterhouse* (Paula Young Lee ed), University of New Hampshire Press, Hanover (2008) 127, 128.

[106] Keith Tester, *Animals and Society, the Humanity of Animal Rights*, above 34, 56; Stephen Hobden and Erika Cudworth, "Civilisation and the Domination of the Animal", above 25, 756.

[107] Noëlie Vialles, *Animal to Edible*, Cambridge University Press, Cambridge (1994). 5.

[108] Norbert Elias, *The Loneliness of the Dying*, Translated by Edmund Jephcott, Continuum, New York (2001), 23, available from https://monoskop.org/images/f/fc/Elias_Norbert_Loneliness_of_the_Dying_2001.pdf

[109] Christopher Otter "Cleansing and Clarifying: Technology and Perception in Nineteenth-Century London", above 97, 46.

[110] Noëlie Vialles, *Animal to Edible*, above 107, 5.

[111] Henry Salt, *Animals' Rights: Considered in Relation to Social Progress*, above 78, 45.

[112] Ibid, 45–46.

[113] David Harvey and Carmen Hubbard, "Reconsidering the Political Economy of Farm Animal Welfare: An Anatomy of Market Failure", (2013) 38 *Food Policy*, 105, 108.

[114] Henry Buller and Emma Roe, "Modifying and Commodifying Farm Animal Welfare: The Economisation of Layer Chickens", (2014) 33 *Journal of Rural Studies*, 141, 144; Ruth Harrison,

followed the publication of *Animal Machines*, also examined the notion of repugnance, referring to it three times, once in conjunction with the use of battery cages for egg production, and twice with respect to pig production.[115]

Members of the public felt repugnant towards intensive methods used to produce eggs, as well as the use of "sweat-houses" to accommodate pigs. Sweat-houses were structures that housed a large number of pigs in close proximity, using the pigs' body heat to maintain a high ambient temperature in the hope of diminishing the incidence of disease.[116] The structures were abhorred by producers and the public alike, with the Brambell Report concluding:

> There can be no doubt that sweat-houses are aesthetically repugnant to many people. The conditions often appear to be unpleasantly dirty... [and t]his repugnance is... shared by many pigfarmers and others connected with the pig industry.[117]

Events after the release of the report were consistent with the progress of the civilising process, because, as discussed in Chap. 6 of this book, the Brambell report introduced the concept of animal welfare into the discourse.[118] The concept had the potential to lead to increasingly better treatment for animals and to some extent this did occur.[119] However, the civilising process is subject to de-civilising forces, which for farm animals derive from the way the state exercises its perogatives regarding violence.

7.4 De-civilising, Violence and the State

One of the themes that weaves throughout this book is that farm animals were, and are, regulated as food products and as commodities in the marketplace, which means that violence towards them is influenced by the demands of production and the market. This part argues that such forces result in de-civilising pressures, more specifically circumstances where the state attenuates its power relative to the

Animal Machines, first published 1964 by Vincent Stuart Publishers, 2013 Edition J Harrison and J Wilson, CABI Oxfordshire (2013).

[115] F W Rogers Brambell, *Report of the Technical Committee to Inquire into the Welfare of Animals Kept under Intensive Livestock Husbandry Systems*, Her Majesty's Stationery Office, (1965, Reprinted 1967), available from https://edepot.wur.nl/134379. Professor F W Rogers Brambell, the *Report of the Technical Committee to Inquire into the Welfare of Animals Kept under Intensive Livestock Husbandry Systems* (the Brambell Report), 18, 31.

[116] Ibid, 30–31.

[117] Ibid, 31.

[118] Ibid, 25.

[119] Steven P McCulloch, "A Critique of FAWC's Five Freedoms as a Framework for the Analysis of Animal Welfare", (2013) 26 (5) *Journal of Agricultural and Environmental Ethics*, 959, 959, 974, DOI https://doi.org/10.1007/s10806-012-9434-7

"coercive capacity" and power of stakeholders, especially those with commercial interests who exercise economic and political influence.[120]

7.4.1 De-civilising and Farm Animals

Although Elias's work did not provide a hypothesis around de-civilising, he acknowledged that the "civilization of which I speak is never completed and [is] constantly endangered".[121] This speaks to the fragile nature of the process, which can advance and regress to norms of earlier ages.[122] Consequently, de-civilisation also leads to the breakdown of community benchmarks and can occur at the individual, group and global levels.[123]

While this deterioration can be seen in terms of a reversal, it is more apt to think of it as a struggle between dominant and secondary forces, which moulds the debate and also leads the state to adjust its stance on violence.[124] Hence, the prevailing order stems from historical conflicts where stakeholders endeavoured to maintain their standing on issues important to them.[125] In this way, civilising and de-civilising forces are ever present, so that the construct of regimes depends on which succeeds. If civilising forces predominate, the civilising process continues at a faster rate. Where, however, de-civilising forces predominate, it can lead to diminished self-restraint and a failure of government control.[126] Additionally, de-civilising is frequently accompanied by the collapse of "interdependency chains",[127] correlating with dwindling societal trust, not only among individuals but also with respect to the exercise of state authority.[128] The realignment of regulation that follows leads to governance frameworks that engage more with non-government institutions, rather than relying on centralised authority.[129]

In the 1990s, an analogous form of realignment, "smart" regulation, was proposed as an alternative to command and control systems, which apply rules and

[120] Tânia Quintaneiro, "The Concept of Figuration or Configuration in Norbert Elias' Sociological Theory, Translated by Maya Mtire", above 51, 5.

[121] Jonathan Fletcher, "Towards a Theory of Decivilizing Processes", above 37, 285, 287.

[122] John Pratt, "Civilizing and Decivilizing Characteristics of the Contemporary Penal Field", above 22, 64.

[123] Jonathan Fletcher, "Towards a Theory of Decivilizing Processes", above 37, 290.

[124] Jonathan Fletcher, "Towards a Theory of Decivilizing Processes", above 37, 288–289; John Pratt, "Civilizing and Decivilizing Characteristics of the Contemporary Penal Field", above 22, 64, 78.

[125] Karl Dusza, "Max Weber's Conception of the State", above 41, 77.

[126] John Pratt, "Civilizing and Decivilizing Characteristics of the Contemporary Penal Field", above 22, 64, 76.

[127] Jonathan Fletcher, "Towards a Theory of Decivilizing Processes", above 37, 290.

[128] John Pratt, "Civilizing and Decivilizing Characteristics of the Contemporary Penal Field", above 22, 70.

[129] Ibid, 70.

standards followed by sanctions and penalties for breach.[130] Although smart regulation encourages government to share regulatory authority, the role of the state is still critical because it occupies a dominant position as the architect and enabler of regimes.[131] Smart regulation also involves a type of social control, which builds on stakeholders' common value systems.[132] By way of contrast, de-civilising forces undermine society's interdependencies leading to less trust and more violence.

The previous chapters of this book have examined the commodification of farm animals, as well as the evolution of animal welfare regulation and in the process, have revealed the ebb and flow of the civilising process. Briefly, the earliest laws targeted cruelty to cattle, acting as the impetus throughout the nineteenth century for further regulation and the establishment of animal protection organisations.[133] However, for farm animals, this trend was always at odds with economic imperatives, which attenuated anti-cruelty regulation, culminating in the mid-twentieth century in the type of institutionalised cruelty critiqued by Ruth Harrison in *Animal Machines*.[134] Still, it cannot be said that civilising forces were absent during this time. The 1935 *International Convention Concerning the Transit of Animals, Meat and Other Products of Animal Origin* and a raft of anti-cruelty laws evinced the desire to rein in animal cruelty.[135] However, commercial imperatives had gained a secure foothold, which led to stagnated reform.

Nevertheless, from the 1960s, society began engaging with animal wellbeing, with the 1970s and 1980s seeing the philosophies of Singer's animal welfare and Regan's animal rights join the debate.[136] In jurisdictions that form part of the EU, which at the time included the United Kingdom, compliance with EU directives

[130] N Gunningham, D Sinclair, P Grabosky and N Peter, *Smart Regulation: Designing Environmental Policy* (Oxford University Press) Oxford: Clarendon Press; New York: Oxford University Press 1998; Neil Gunningham and Cameron Holley, "Next-Generation Environmental Regulation: Law, Regulation, and Governance", (2016) *The Annual Review of Law and Social Science*, 273, DOI 10.1146/annurev-lawsocsci-110615-084651

[131] Cameron Holley and Clifford Shearing, "A Nodal Perspective of Governance: Advances in Nodal Governance Thinking", in *Regulatory Theory: Foundations and Applications* (Peter Drahos ed) ANU Press. ACT (2017), 163, 165, 171.

[132] Ildiko Erdei, "Television, Rituals, Struggle for Public Memory", (2008) 3 (3) *Issues in Ethnology and Anthropology*, 145, 158.

[133] Katharine Gelber and Siobhan O'Sullivan, "Cat got Your Tongue? Free speech, Democracy and Australia's 'Ag-gag' Laws", (2020) *Australian Journal of Political Science*, 1, 2, 3, https://doi.org/10.1080/10361146.2020.1799938

[134] Servière Jacques, "Science and Animal Welfare in France and European Union: Rules, Constraints, Achievements", (2014) 98 (3) *Meat Science*, 486, available from https://www.sciencedirect.com/science/article/abs/pii/S0309174014002216

[135] 1935, *International Convention Concerning the Transit of Animals, Meat and Other Products of Animal Origin*, opened for signature, 20 February 1935, 193 *LNTS* 37, entered into force 6 Dec 1938, signed by nine parties and entered into force with five ratifications. Available from https://treaties.un.org/doc/Publication/UNTS/LON/Volume%20193/v193.pdf; Discussion Chap. 3, Sect. 3.4 and Chap. 5, Sect. 5.5 of this book.

[136] Peter Singer, *Animal Liberation*, fourth Edition Harper Perennial New York (2009), 8. The first edition was published as Peter Singer, *Animal Liberation*, Harper Collins, first Edition, New York

concerning farm animals became mandatory. This includes Council Directive 98/58/EC concerning the protection of animals kept for farming purposes, a directive that reflects the Five Freedoms, discussed in Chap. 6 of this book.[137] Moreover, the Treaty of Amsterdam, which amended the Treaty on European Union, incorporated a protocol on animal welfare, emphasising that animals are sentient beings.[138] By the twenty-first century, the World Organisation for Animal Health (OIE)[139] had expanded its remit to include animal welfare.[140]

At the same time, in some jurisdictions, de-civilising pressures exerted by "Ag-Gag" laws and regulatory capture counterbalanced this progress. Mark Bittman first used the term Ag-Gag in his 2011 opinion piece "Who Protects the Animals?" to describe criminal offences created by legislatures for the surveillance and recording of animal abuse in production systems.[141] Ag-Gag laws have been enacted in several jurisdictions, including the United States of America (USA) and Australia.[142]

(1975); Tom Regan, *The Case for Animal Rights*", University of California Press, Los Angeles, (2004), first published in 1983.

[137] Council Directive 98/58/EC of 20 July 1998, Concerning the Protection of Animals Kept for Farming Purposes, [1998] Official Journal, L 221, 08/08/1998 P 0023–0027, available from https://eur-lex.europa.eu/legal-content/EN/ALL/?uri=CELEX%3A31998L0058

[138] Treaty of Amsterdam amending the Treaty on European Union, the Treaties Establishing the European Communities and Certain Related Acts, opened for signature 22 October 1997, *Official Journal of the European Communities* C 340, P. 0001–0144, 10 November 1997 (entered into force 1 May 1999 having been ratified by all 15 members). Available from https://eur-lex.europa.eu/legal-content/EN/TXT/?uri=CELEX:11997D/TXT

[139] International Agreement for the Creation at Paris of an International Office for Dealing with Contagious Diseases of Animals and Annex, opened for signature 25 January 1924, [1925] *ATS* 15, (entered into force 12 January 1925). As at May 2020 the organisation had 182 members.

[140] OIE, *OIE Global Animal Welfare Strategy*, (2017), available from https://www.oie.int/fileadmin/Home/eng/Animal_Welfare/docs/pdf/Others/EN_OIE_AW_Strategy.pdf; OIE, *Terrestrial Animal Health Code*, Article 7.1.1, 28th edition (2019) available from https://www.oie.int/index.php?id=169&L=0&htmfile=chapitre_aw_introduction.htm

[141] Mark Bittman, "Who Protects the Animals?", *The New York Times*, 26 April, 2011, available from https://opinionator.blogs.nytimes.com/2011/04/26/who-protects-the-animals/ (subscription needed).

[142] Amanda Whitfort, "Justice and the Vulnerable: Extending the Duty to Prevent Serious Crimes Against Children to The Protection of Agricultural and Research Animals", above 54, 127; Examples include: Kansas General Assembly, Kansas Statutes, Chap. 47, Livestock and Domestic Animals, § 47–1827, available from https://www.ksrevisor.org/statutes/chapters/ch47/047_018_0027.html; Montana General Assembly, Montana Code Annotated 2019, Title 81 Livestock, Chap. 30 Protection of Farm Animals and Research Facilities, § 81–30-103(2)(e) (West 2015) https://leg.mt.gov/bills/mca/title_0810/chapter_0300/part_0010/sections_index.html; *Inclosed Lands Protection Act 1901* (NSW), ss 4AA, 4A-4C, available from http://www.austlii.edu.au/cgi-bin/viewdb/au/legis/nsw/consol_act/ilpa1901264/; *Criminal Code Amendment (Agricultural Protection) Act 2019* (Cth) available from https://parlinfo.aph.gov.au/parlInfo/download/legislation/bills/r6351_aspassed/toc_pdf/19117b01.pdf;fileType=application%2Fpdf; *Criminal Code Act 1995* (Cth), Division J, ss 474.6 and 474.7, available from http://www.austlii.edu.au/cgi-bin/viewdb/au/legis/cth/consol_act/cca1995115/. A table of USA laws is found in Caitlin A Ceryes and Christopher D. Heaney, "Ag-Gag" Laws: Evolution, Resurgence, and Public Health Implications",

Although these laws vary in detail, their general effect is to restrain discourse that censures production systems in animal agriculture.[143]

The Montana legislature, for example, has created offences for trespassing and/or remaining concealed in an animal facility with the intent of damaging the facility or its business.[144] In Missouri, video or other surveillance depicting animal abuse or neglect needs to be submitted to law enforcement agencies within 24 hours.[145] These laws seem to target activities that would otherwise verify "a pattern of behavior" demonstrating animal abuse.[146] Australian laws follow a similar trajectory. Amendments to the *Inclosed Lands Protection Act 1901* (NSW) create offences for unlawful entry and interference with or attempts to interfere with the conduct of a business being conducted on inclosed lands.[147] In the state of Queensland, Section 13 of the *Summary Offences Act 2005* (Qld) 2005 criminalises the unlawful entry onto agricultural land,[148] while in South Australia, the *Surveillance Devices Act 2016* (SA) creates offences for surveillance without a warrant, which extends to individuals who take videos and procure other evidence of agricultural practices.[149] At the Commonwealth level, parliament amended the *Criminal Code Act 1995* (Cth) in 2019 to prohibit the use of communication services to incite trespass on agricultural land, a move aimed at agricultural whistle-blowers.[150]

(2018) 28(4) *New Solutions: A Journal of Environmental and Occupational Health Policy*, 664, 665, doi: https://doi.org/10.1177/1048291118808788

[143] Justin F Marceau, "Ag Gag Past, Present, and Future", (2015) 38 *Seattle University Law Review*, 1317, 1332.

[144] Montana General Assembly, Montana Code Annotated 2019, Title 81 Livestock, Chap. 30 Protection of Farm Animals and Research Facilities, § 81-30-103(2)(e) (West 2015) https://leg.mt.gov/bills/mca/title_0810/chapter_0300/part_0010/sections_index.html

[145] Montana General Assembly, Title XXXVIII Crimes and Punishment; Peace Officers and Public Defenders, Chap. 578, § 578.013, available from https://revisor.mo.gov/main/OneSection.aspx?section=578.013, discussion, Justin F Marceau, "Ag Gag Past, Present, and Future", above 143, 1340.

[146] Caitlin A Ceryes and Christopher D. Heaney, "Ag-Gag" Laws: Evolution, Resurgence, and Public Health Implications", above 142, 667.

[147] *Inclosed Lands Protection Act 1901* (NSW), ss 4AA, 4A-4C, available from http://www.austlii.edu.au/cgi-bin/viewdb/au/legis/nsw/consol_act/ilpa1901264/; discussion, Katharine Gelber and Siobhan O'Sullivan, "Cat got Your Tongue? Free speech, Democracy and Australia's 'Ag-gag' Laws", above 133, 2.

[148] *Summary Offences Act 2005* (Qld), s 13, available from http://www.austlii.edu.au/cgi-bin/viewdb/au/legis/qld/consol_act/soa2005189/

[149] *Surveillance Devices Act 2016* (SA), ss 16–19, available from http://www.austlii.edu.au/cgi-bin/viewdb/au/legis/sa/consol_act/sda2016210/

[150] *Criminal Code Amendment (Agricultural Protection) Act 2019* (Cth) available from https://parlinfo.aph.gov.au/parlInfo/download/legislation/bills/r6351_aspassed/toc_pdf/19117b01.pdf;fileType=application%2Fpdf; *Criminal Code Act 1995* (Cth), Division J, ss 474.6 and 474.7, available from http://www.austlii.edu.au/cgi-bin/viewdb/au/legis/cth/consol_act/cca1995115/; discussion Katharine Gelber and Siobhan O'Sullivan, "Cat got Your Tongue? Free speech, Democracy and Australia's 'Ag-gag' Laws", above133, 2.

Those in favour of Ag-Gag laws point to the fact that information and evidence are collected in an illegal and misleading manner, which also carries the possibility of introducing disease into the production chain.[151] However, not all industry producers are in favour of Ag-Gag laws, including small operators and those sympathetic to sustainable practices, because they regard Ag-Gag laws as promoting the deterioration of standards.[152] What is more, the tenor of Ag-Gag laws has a "chilling effect", because the laws proscribe the exposure of unlawful activities, including animal abuse.[153] Given that such exposure is normally in the public interest,[154] it is not surprising that Ag-Gag laws tend to diminish public confidence in the sector – a classic sign of de-civilisation.[155]

Another warning sign stems from regulatory capture, a phenomenon where "a regulatory agency acts in the interests of the industry it is responsible for regulating in a way that deviates from the public interest."[156] This situation predominantly derives from the fact that in typical governance systems, animal welfare regulation operates within a department of agriculture.[157] Consequently, the same agency is charged with enhancing industry production while simultaneously overseeing animal welfare. The resulting lack of public confidence stems not so much from the possibility of conflict, but from the way that agencies prioritise their obligations.[158] Decisions are inevitably made in the shadow of the bureau's core objectives where determinations signal the relative worth or value of each obligation.[159] The practical outcome all but reinforces the lesser importance of animal welfare because it clashes

[151] Sonci Kingery, "The Agricultural Iron Curtain: Ag Gag Legislation and the Threat To Free Speech, Food Safety, and Animal Welfare", (2013) 17 (3) *Drake Journal of Agricultural Law*, 645, 657.

[152] Garrett M Broad, "Animal Production, Ag-gag Laws, and the Social Production of Ignorance: Exploring the Role of Storytelling", (2016) 10 (1) *Environmental Communication*, 43, 53, https://doi.org/10.1080/17524032.2014.968178

[153] Caitlin A Ceryes and Christopher D. Heaney, "Ag-Gag" Laws: Evolution, Resurgence, and Public Health Implications", above 142, 664; Amanda Whitfort, "Justice and the Vulnerable: Extending the Duty to Prevent Serious Crimes Against Children to The Protection of Agricultural and Research Animals", above 54, 127.

[154] Katharine Gelber and Siobhan O'Sullivan, "Cat got Your Tongue? Free speech, Democracy and Australia's 'Ag-gag' Laws", above 133, 2.

[155] J A Robbins, B Franks, D M Weary and M A G von Keyserlingk, "Awareness of Ag-gag Laws Erodes Trust in Farmers and Increases Support for Animal Welfare Regulations", (2016) 61 *Food Policy*, 121, 124; John Pratt, "Civilizing and Decivilizing Characteristics of the Contemporary Penal Field", above 22, 70.

[156] Jed Goodfellow, *Animal Welfare Regulation in the Australian Agricultural Sector: A Legitimacy Maximising Analysis*, above 9, 8, 162–176.

[157] For example, In the United Kingdom, animal welfare is governed by the Department for Environment, Food and Rural Affairs, https://www.gov.uk/government/organisations/department-for-environment-food-rural-affairs; in Australia, animal welfare is governed by the Department of Agriculture, Water and the Environment, https://www.agriculture.gov.au/animal/welfare

[158] Jed Goodfellow, *Animal Welfare Regulation in the Australian Agricultural Sector: A Legitimacy Maximising Analysis*, above 9, 169–170.

[159] Ibid.

with the promotion of the sector.[160] Indeed, this last point forms the crux of regulatory capture, because animal welfare and public interest in its enforcement are downgraded relative to commercial priorities.[161] As with Ag-Gag laws, this contributes to decreasing community trust in regulatory and governance mechanisms initiated by the state.[162]

A corollary to the breakdown of public confidence is the way the state manages violence. This is significant because one of the key tenets of the civilising process is predicated on the state exercising strong controls over violence, which may be delegated, but not relinquished, to others.

7.4.2 De-civilising and Violence Against Animals

As discussed, violence against animals includes abject cruelty as well as practices common in production and slaughter systems; additionally, commentators such as Francione extend the concept to the consumption and use of animal products. Where government does not holistically manage violence, it means that bias or inertia can unleash de-civilising forces. The state's role is particularly critical because intensive agricultural systems operate behind closed doors, engendering an air of secrecy that has increasingly led society to question the probity of the sector.[163] This is matched by concern at the effectiveness of government regulation and enforcement systems, especially following the release of surveillance media that reveal animal abuse.[164] Overall, the way the sector operates exerts de-civilising pressures in several ways.

To start with, the sector wields considerable political power, being instrumental in setting its own codes of conduct and otherwise influencing the design and operation of regulation.[165] In countries such as the USA and Australia, Ag-Gag laws are invariably the result of lobbying by producers, which not only protects the industry against evidence obtained by whistle-blowers but also shields the industry from lawsuits and criticism in general.[166] The latter also means that government and the sector control whatever information is available to consumers, leading to what has been described as "the cultural production of ignorance".[167] In other words, by controlling the narrative, government and industry may disseminate misleading

160 Ibid.

161 Ibid.

162 Ibid, 8.

163 Sonci Kingery, "The Agricultural Iron Curtain: Ag Gag Legislation and the Threat To Free Speech, Food Safety, and Animal Welfare", above 151, 680.

164 Ibid,

165 Ibid, 677; Kyle H Landis-Marinello, "The Environmental Effects of Cruelly to Agricultural Animals", (2008) 106 *Michigan Law Review First Impressions*, 147, 147.

166 Justin F Marceau, "Ag Gag Past, Present, and Future", above 143, 1318, 1332.

167 Garrett M Broad, "Animal Production, Ag-gag Laws, and the Social Production of Ignorance: Exploring the Role of Storytelling", above 152, 44–45.

information. Accounts of farm animals invariably ignore institutionalised cruelty in favour of creating fictions of "happy animals" contently grazing in lush meadows, thereby establishing a "cognitive barrier" between living animals and their products.[168] This barrier is not new, for as discussed, consumers are progressively being distanced from the realities of animal death.[169]

Yet, the market proceeds on the basis that people make meaningful choices.[170] The problem, however, is that such choices are at least partly based on "information asymmetries" because producers know how their products are manufactured and in some cases work hard to keep this information from consumers.[171] This situation also tends to normalise cruelty because the public is not aware of practices that are considered routine or standard or indeed simply incidents of animal abuse that occur regularly.[172] Surveillance videos in the USA, for example, revealed "employees on... farms or packing plants abusing animals by beating, throwing, or kicking them".[173] The regulatory challenges presented by these difficulties are aggravated by Ag-Gag laws because they prevent the use of pictorial evidence, which would otherwise link violence perpetrated against animals with consumer's dietary choices.[174] In this way, Ag-Gag laws also impede improvement to animal welfare regulation because they suppress the very evidence that supports the need for reform.[175] In the parlance of the civilising process, the hidden nature of production does not allow disgust or repugnance to develop in the way envisaged by Elias.

These failings also solidify the commodification of farm animals because the violence against them is largely appraised in the context of production and trade. This is not to say that societal input is lacking. However, violence provides the foundation for animal production, so from that perspective violence cannot be reduced to a commercially unviable level, in case it leaves the sector unprofitable. This mindset reinforces the discussion in Chap. 6 of this book that animals are the raw materials of production, making the future direction of the civilising process uncertain.

---

[168] Ibid, 47.

[169] Piers Beirne, "Theriocide: Naming Animal Killing", above 99, 51.

[170] M W Jackson, "Norbert Elias's Theory of the Civilizing Process", above 19, 307.

[171] F Bailey Norwood and Jayson L Lusk, *Compassion, by the Pound: The Economics of Farm Animal Welfare*, above 75, 328.

[172] Garrett M Broad, "Animal Production, Ag-gag Laws, and the Social Production of Ignorance: Exploring the Role of Storytelling", above 152, 52; Temple Grandin, "Animal Welfare and Society Concerns Finding the Missing Link", (2014) 98 (3) *Meat Science*, 461, available from https://www.sciencedirect.com/science/article/pii/S0309174014001314

[173] Temple Grandin, "Animal Welfare and Society Concerns Finding the Missing Link", above 172, 461.

[174] Pamela Fiber-Ostrow and Jarret S Lovell, "Behind a Veil of Secrecy: Animal Abuse, Factory Farms, and Ag-Gag Legislation", (2016) 19 (2) *Contemporary Justice Review*, 230, 234, doi: https://doi.org/10.1080/10282580.2016.1168257

[175] Larissa Wilson, "Ag-Gag Laws: A Shift in the Wrong Direction for Animal Welfare on Farms", (2014) 44 (3) *Golden Gate University Law Review*, 311, 312.

7.5 The Future

On one view, it is logical to conclude that the prospect of improved wellbeing for farm animals lies in incremental advancements to regulation implemented by the state and adopted by industry.[176] This would be consistent with the civilising process and its progressive diminishing of violence. However, civilising and de-civilising forces exert conflicting regulatory pressures, which largely depend on how far government is prepared to manage violence in a commercial context which also includes many stakeholders.

The discussion in earlier chapters of this book has identified a range of these stakeholders, including government, consumers and animal protection organisations. Their understanding of the civilising process and the legitimate level of a state's authority will likely vary. Those directly involved in animal production will probably view their industry as consistent with the civilising process because they aim to conform to law and policy initiated by the state. From the consumer's perspective, civilising has led to greater awareness of ethical considerations in animal production.[177] For those interested in animal welfare, the civilising process is consistent with regimes based on animal sentience, because such systems attenuate pain and suffering inflicted on animals.[178] However, for those interested in animal rights, welfare regulation will not assuage concerns, because welfare principles still envisage the use of animals, notwithstanding the fact that this occurs within allowable parameters.[179]

Aside from these variances, as society becomes more affluent and better educated, calls for improved animal welfare and tighter regulation of production methods increase.[180] This was apparent from at least the nineteenth century, where social progress benefitted both humans and animals because it not only improved the lives of humans but also elevated concern for animals.[181] Nevertheless, farm animals continue to lead lives that have increasingly become "shorter and more impoverished", notwithstanding advances in veterinary care.[182] Yet, this very point also underscores that society, especially consumers, can wield political pressure, prompting government to respond.[183] Whether this translates to better treatment for farm animals depends on several things.

---

[176] David Harvey and Carmen Hubbard, "Reconsidering the Political Economy of Farm Animal Welfare: An Anatomy of Market Failure", above 113, 107.

[177] Ibid, 108.

[178] Ibid.

[179] Gaverick Matheny, "Utilitarianism and Animals", in Peter Singer (ed) *In Defense of Animals the Second Wave*, 13, 15, Blackwell Publishing, New Jersey (2006).

[180] David Harvey and Carmen Hubbard, "Reconsidering the Political Economy of Farm Animal Welfare: An Anatomy of Market Failure", above 113, 108.

[181] Henry Salt, *Animals' Rights: Considered in Relation to Social Progress*, above 78, 13, 115.

[182] Steven McMullen, *Animals and the Economy*, Palgrave Macmillan, Basingstoke, UK (2016), 1.

[183] David Harvey and Carmen Hubbard, "Reconsidering the Political Economy of Farm Animal Welfare: An Anatomy of Market Failure", above 113, 111.

First, information asymmetries need to be corrected. As discussed, the narrative around farm animals tends to be heavily influenced by producers, so government intervention to stop misinformation is critical. In some North American jurisdictions, independent auditors monitor surveillance video of production systems, so that the potential for distortion is substantially reduced.[184] Second, although the state has the power to use its monopoly on violence and enact higher standards, it may lack the political will to do so.[185] Even if it does act, it may find itself enmeshed in arguments concerning what those standards should be.[186] Therefore, the development of independently verifiable standards is an important objective. Third, economic arguments, which abound in the sector, focus on whether consumers are prepared to pay for higher welfare measures.[187] However, economic arguments are at least partly bound with consumers' understanding of the meat industry, which as previously observed, is gleaned from the material disseminated by industry. Consequently, whether consumers are prepared to pay for high-level animal welfare hinges on their outlooks, which are shaped by the sector.[188] More to the point, arguments about the preparedness of consumers to pay for animal welfare misconstrue the issues, because if society sanctions the production and consumption of animals, their welfare should not be negotiable.[189] Otherwise, animals are treated as little more than commodities.[190]

This then leads to the question of how far the state should intervene. Philosophies that underpin western liberal democracies ascribe to policies that involve the least interference by the state and instead rely on individuals and groups, such as producers, consumers and animal protection societies to achieve stable societies.[191] This approach is not necessarily inconsistent with Elias's theories, because he acknowledged such interdependencies. However, stakeholder inputs still need to occur within the boundaries determined by the state; hence, the way the state intervenes and deals with power plays is important.[192] Of course, and as mentioned several times throughout this book, none of these issues deals with more fundamental questions relating to society's right to use animals, no matter how gentle, or harsh that use is. This type of threshold question is important to the animal rights movement and is an issue that Ruth Harrison identified as significant to broader

---

[184] Temple Grandin, "Animal Welfare and Society Concerns Finding the Missing Link", above 172, 467.

[185] David Harvey and Carmen Hubbard, "Reconsidering the Political Economy of Farm Animal Welfare: An Anatomy of Market Failure", above 113, 107.

[186] Ibid, 108.

[187] Kimberley Smith, *Governing Animals Animal Welfare and the Liberal State*, above 22, 94.

[188] Maria Font-i-Furnols and Luis Guerrero, "Consumer Preference, Behavior and Perception about Meat and Meat Products: An Overview", above 100, 365, 367.

[189] Kimberley Smith, *Governing Animals Animal Welfare and the Liberal State*, above 22, 94.

[190] Steven McMullen, *Animals and the Economy*, above 182, 55.

[191] Kimberley Smith, *Governing Animals Animal Welfare and the Liberal State*, above 22, 94.

[192] Robert van Krieken, "Law and Civilization: Norbert Elias as a Regulation Theorist", above 77, 269.

questions of humanity's relationship with animals.[193] The fact that these matters are part of the debate signifies that the civilising process is still pertinent. However, its practical import and the effectiveness of regimes partly depend on the role each stakeholder plays and partly on government exercising its control over violence in ways that do not destabilise progress of the civilising process.[194]

7.6 Conclusion

The civilising process had the potential to reduce violence towards animals in a more concrete way than has occurred. By the beginning of the nineteenth century, the formal steppingstones of this process were established with the introduction of anti-cruelty laws and the advent of animal protection organisations. However, civilising forces do not operate in a vacuum and are subject to de-civilising pressures. For farm animals, the latter primarily derive from the economic needs of animal agriculture as it responds to increased demand for its products. Regimes in western jurisdictions are the outcome of inputs from a range of stakeholders, including consumers, animal protection organisations, government and producers. Regulation, however, also reflects political struggles among these groups, and depending on which group gains the ascendancy, regulation can either advance towards the civilised, kinder treatment of animals or entrench their commodification.[195]

Currently, de-civilising forces in some jurisdictions are strengthening. The introduction of Ag-Gag laws, developments in regulatory capture and the political power of the agricultural product sector aggravate power imbalances.[196] Although government has the power to correct these imbalances, it may lack the political will to do so, leading to greater de-civilising pressures, not only among competing human groups but also with respect to the human-animal relationship itself. At the end of the day, this reinforces Robert Garner's remarks that the state is unlikely to create and enforce "moral obligations to protect animals. . . when they clash" with human demands.[197] Yet, this is precisely what needs to happen if animals are to be treated less like commodities and more like living beings.

[193] Abigail Woods, "From Cruelty to Welfare: The Emergence of Farm Animal Welfare in Britain, 1964–71", (2011) 36 (1) *Endeavour*, 14, 19.

[194] Kimberley Smith, *Governing Animals Animal Welfare and the Liberal State*, above 22, 34.

[195] S E Curtis, "Animals in Food Production – American Issues", (1988) 20 *Applied Animal Behaviour Science*, 151, 153.

[196] Pamela Fiber-Ostrow and Jarret S Lovell, "Behind a Veil of Secrecy: Animal Abuse, Factory Farms, and Ag-Gag Legislation", above 172, 232.

[197] Robert Garner, *A Theory of Justice for Animals: Animal Rights in a Nonideal World*, Oxford University Press, New York (2013) 61.

Bibliography

Aya R (1978) Norbert Elias and "the civilizing process". Theory Soc 5(2):219. https://www.jstor.org/stable/656697

Beilharz RG (1988) Science and the politics of animal use in food production: the situation in Australia. Appl Anim Behav Sci 20:143

Beirne P (2014) Theriocide: naming animal killing. Int J Crime, Justice Soc Democracy 3(2):49

Bittman M (2011) Who protects the animals? The New York Times, 26 April. https://opinionator.blogs.nytimes.com/2011/04/26/who-protects-the-animals/ (subscription needed)

Brambell FWR (1965) Report of the technical committee to inquire into the welfare of animals kept under intensive livestock husbandry systems. Her Majesty's Stationery Office (Reprinted 1967). https://edepot.wur.nl/134379

Broad GM (2016) Animal production, ag-gag Laws, and the social production of ignorance: exploring the role of storytelling. Environ Commun 10(1):43. https://doi.org/10.1080/17524032.2014.968178

Bryant TL (2008) Sacrificing the sacrifice of animals: legal personhood for animals, the status of animals as property, and the presumed primacy of humans. Rutgers Law J 39:247

Buller H, Roe E (2014) Modifying and commodifying farm animal welfare: the economisation of layer chickens. J Rural Stud 33:141

Cassuto DN (2007) Bred meat: the cultural Foundation of the factory farm. Law Contemp Probl 70: 59. https://scholarship.law.duke.edu/lcp/vol70/iss1/3

Ceryes CA, Heaney CD (2018) "Ag-gag" laws: evolution, resurgence, and public health implications. New Solut 28(4):664. https://doi.org/10.1177/1048291118808788

Commonwealth of Australia, Primary Industries Standing Committee Model Code of Practice for the Welfare of Animals, Domestic Poultry, 4th edn. SCARM Report 83, Commonwealth of Australia (2002), part 13.2. https://www.publish.csiro.au/ebook/download/pdf/3451

Curtis SE (1988) Animals in food production—American issues. Appl Anim Behav Sci 20:151

Derrida J, Roudinesco E (2004) For what tomorrow: a dialogue, 'violence against animals'. Stanford University Press, Redwood City

Dusza K (1989) Max Weber's conception of the state. Int J Pol Culture Soc 3(1):71

Eccles T, Pointing J (2013) A smart regulation, shifting architectures and changes in governance. Int J Build Environ 5(1):71

Elias N (2000) The civilizing process, sociogenetic and psychogenetic investigations, revised edn. Blackwell Publishing, UK

Elias N (2001) The loneliness of the dying. Translated by Edmund Jephcott, Continuum, New York. https://monoskop.org/images/f/fc/Elias_Norbert_Loneliness_of_the_Dying_2001.pdf

Epp M (2019) Beak trimming trends, Canadian Poultry. http://magazine.canadianpoultrymag.com/publication/?m=1191&i=574088&view=articleBrowser&article_id=3331179&ver=html5

Erdei I (2008) Television, rituals, struggle for public memory. Issues Ethnol Anthropol 3(3):145

Fiber-Ostrow P, Lovell JS (2016) Behind a veil of secrecy: animal abuse, factory farms, and ag-gag legislation. Contemp Justice Rev 19(2):230. https://doi.org/10.1080/10282580.2016.1168257

Fleming G (1871) Animal plagues: their history, nature, and prevention. Chapman and Hall, London. This copy is a reprint

Fletcher J (1995) Towards a theory of decivilizing processes. Amsterdams Sociol Tijdschr 22(2): 283

Fontaine S (1978) The civilizing process revisited: interview with Norbert Elias. Theory Soc 5(2): 243. https://www.jstor.org/stable/656699?seq=1

Francione GL (2010) The abolition of animal exploitation. In: Francione GL, Garner R (eds) The animal rights debate: abolition or regulation? Columbia University Press, New York, p 29

Furnols MF, Guerrero L (2014) Consumer preference, behavior and perception about meat and meat products: an overview. Meat Sci 98(3):361. https://www.sciencedirect.com/science/article/abs/pii/S0309174014001934

Garner R (2013) A theory of justice for animals: animal rights in a nonideal world. Oxford University Press, New York

Gelber K, O'Sullivan S (2020) Cat got your tongue? Free speech, democracy and Australia's 'ag-gag' laws. Aust J Polit Sci. https://doi.org/10.1080/10361146.2020.1799938

Goodfellow, Jed (2015) Animal welfare regulation in the australian agricultural sector: a legitimacy maximising analysis. PhD Thesis, Macquarie Law School Macquarie University Research. https://www.researchonline.mq.edu.au/vital/access/manager/Repository/mq:45113

Gordon R (2017) From pests to pets: social and cultural perceptions of animals in post-medieval urban centres in England (AD1500–1900). Papers Inst Archaeol 27(1):1. https://doi.org/10.5334/pia-478

Grandin T (2014) Animal welfare and society concerns finding the missing link. Meat Sci 98(3): 461. https://www.sciencedirect.com/science/article/pii/S0309174014001314

Gunningham N, Sinclair D, Grabosky P, Peter N (1998) Smart regulation: designing environmental policy. Clarendon Press/Oxford University Press, Oxford/New York

Gunningham N, Holley C (2016) Next-generation environmental regulation: law, regulation, and governance. Annu Rev Law Social Sci:273. https://doi.org/10.1146/annurev-lawsocsci-110615-084651

Harrison R (2013) Animal machines, first published 1964 by Vincent Stuart Publishers. Edition J Harrison and J Wilson, CABI Oxfordshire

Harvey D, Hubbard C (2013) Reconsidering the political economy of farm animal welfare: an anatomy of market failure. Food Policy 38:105

Hobden S, Cudworth E (2014) Civilisation and the domination of the animal. Millenn J Int Stud 42(3):746. https://doi.org/10.1177/0305829814540355

Holley C, Shearing C (2017) A nodal perspective of governance: advances in nodal governance thinking. In: Drahos P (ed) Regulatory theory: foundations and applications. ANU Press, ACT

Hont I (2005) Jealousy of trade, international competition and the nation-state in historical perspective. Belknap Press of Harvard University Press, Cambridge, MA/London

Humane Society International (2011) An HSI report: the welfare of animals in the egg industry. https://www.hsi.org/wp-content/uploads/assets/pdfs/welfare_of_animals_in_the_egg.pdf

Ilea R (2008) Nussbaum's capabilities approach and nonhuman animals: theory and public policy. J Soc Philos 39(4):547

Ingold T (1994) From trust to domination: an alternative history of human-animal relations. In: Manning A, Serpell J (eds) Animals and human society changing perspectives. Routledge, Abingdon, p 1, reprinted up to 2011

Jackson MW (1985) Norbert Elias's theory of the civilizing process. Bull Austral Soc Legal Philos 9(4):298. http://www.austlii.edu.au/au/journals/AUSocLegPhilB/1985/23.pdf

Jacques S (2014) Science and animal welfare in France and European Union: rules, constraints, achievements. Meat Sci 98(3):484. https://www.sciencedirect.com/science/article/abs/pii/S0309174014002216

Kingery S (2013) The agricultural Iron curtain: ag gag legislation and the threat to free speech, food safety, and animal welfare. Drake J Agric Law 17(3):645

van Krieken R (2019) Law and civilization: Norbert Elias as a regulation theorist. Ann Rev Law Soc Sci 15:267

Landis-Marinello KH (2008) The environmental effects of cruelty to agricultural animals. Michigan Law Rev First Impress 106:147

Linklater A, Mennell S (2010) Norbert Elias, the civilising process: sociogenic and psychogenetic investigations—an overview and assessment. Hist Theory, Stud Philos Hist 49(3):384. https://doi.org/10.1111/j.1468-2303.2010.00550

Liordos V, Konsiotis VJ, Kokoris S, Pienidou M (2018) The two faces of Janus, or the dual mode of public attitudes towards snakes. Sci Total Environ 621:670. https://doi.org/10.1016/j.scitotenv.2017.11.311

MacLachlan I (2008) Humanitarian reform slaughter technology, and butcher resistance in nineteenth-century Britain. In: Lee PY (ed) Meat, modernity and the rise of the slaughterhouse. University of New Hampshire Press, Hanover, p 107
Maleševic S, Ryan K (2012) The disfigured ontology of figurational sociology: Norbert Elias and the question of violence. Crit Sociol 39(2):165. https://doi.org/10.1177/0896920511434484
Marceau JF (2015) Ag gag past, present, and future. Seattle Univ Law Rev 38:1317
Matheny G (2006) Utilitarianism and animals. In: Singer P (ed) In defense of animals the second wave. Blackwell Publishing Ltd, New Jersey, p 13
McCulloch SP (2013) A critique of FAWC's five freedoms as a framework for the analysis of animal welfare. J Agric Environ Ethics 26(5):959. https://doi.org/10.1007/s10806-012-9434-7
McEwan AB (2016) The concept of violence: a proposed framework for the study of animal protection law and policy. Thesis, Australian National University. https://openresearch-repository.anu.edu.au/handle/1885/112649
McMullen S (2016) Animals and the economy. Palgrave Macmillan, Basingstoke
Mennel S (2006) Civilizing Processes. Theory, Culture Soc 23(2–3):429
Mitropolitski S (2011) Weber's definition of the state as an ethnographic tool for understanding the contemporary political science state of the discipline. Canad Polit Sci Assoc 1. https://papers.ssrn.com/sol3/papers.cfm?abstract_id=1823401
Myrick JG, Evans SD (2014) Do PSAs take a bit out of shark week? The effects of juxtaposing environmental messages with violent images of shark attacks. Sci Commun 36(5):544. https://doi.org/10.1177/1075547014547159
Niestroj BHE (1989) Norbert Elias: a milestone in historical the making of the social person. J Hist Sociol 2(2):136
Norwood FB, Lusk JL (2011) Compassion, by the pound: the economics of farm animal welfare. Oxford University Press, New York
OIE (2017) OIE global animal welfare strategy. https://www.oie.int/fileadmin/Home/eng/Animal_Welfare/docs/pdf/Others/EN_OIE_AW_Strategy.pdf
OIE (2019) Terrestrial animal health code, article, 28th edn. https://www.oie.int/index.php?id=169&L=0&htmfile=chapitre_aw_introduction.htm
Otter C (2004) Cleansing and clarifying: technology and perception in nineteenth-century London. J Brit Stud 43(1):40
Pemberton N (2014) The rat-Catcher's prank: interspecies cunningness and scavenging in Henry Mayhew's London. J Vict Culture 19(4):520. https://doi.org/10.1080/13555502.2014.967548
Perren R (2008) Filth and profit, disease and health: public and private impediments to slaughterhouse reform in Victorian Britain. In: Lee PY (ed) Meat, modernity and the rise of the slaughterhouse. University of New Hampshire Press, Hanover, p 127
Plumwood V (2003) Animals and ecology: towards a better integration, ANU Res Publ. 1. https://openresearch-repository.anu.edu.au/handle/1885/41767
Pratt J (2014) Civilizing and Decivilizing characteristics of the contemporary penal field. In: Landini TS, Dépelteau F (eds) Norbert Elias and Empirical research. Palgrave Macmillan, New York, p 63
Quilley S (2009) The land ethic as an ecological civilizing process: Aldo Leopold, Norbert Elias, and environmental philosophy. Environ Ethics 31(2):115
Quintaneiro T (2006) The concept of figuration or configuration in norbert elias' sociological theory, translated by Maya Mtire. Teoria Soc 2. http://socialsciences.scielo.org/scielo.php?script=sci_arttext&pid=S1518-44712006000200002&lng=en&nrm=iso
Regan T (2004) The case for animal rights. University of California Press, Los Angeles (first published in 1983)
Robbins JA, Franks B, Weary DM, von Keyserlingk MAG (2016) Awareness of ag-gag Laws erodes Trust in Farmers and Increases Support for animal welfare regulations. Food Policy 61:121
Ruby MB (2012) Vegetarianism. A blossoming field of study. Appetite 58:141. https://doi.org/10.1016/j.appet.2011.09.019

Salt H (1922) Animal rights considered in relation to social progress, revised edn. G Bell and Sons Ltd, London (first printed 1892). https://archive.org/details/cu31924030305332

Singer P (2009) Animal liberation, 4th edn. Harper Perennial, New York (The first edition was published as Peter Singer, Animal Liberation, Harper Collins, 1st Edition, New York (1975))

Smith K (2012) Governing animals animal welfare and the Liberal state. Oxford University Press, New York

Staaveren NE, Jennifer B, Christine F, Harlander-Matauschek A (2021) A meta-analysis on the effect of environmental enrichment on feather pecking and feather damage in laying hens. Poult Sci 100(2):407–409

Taylor N, Fraser H (2017) "Slaughterhouses" the language of life, the discourse of death. In: Maher J, Pierpoint H, Beirne P, The Palgrave international handbook of animal abuse studies. Palgrave Macmillan UK, pp 179, 180, 191

Taylor N, Fraser H (2019) The cow project: analytical and representational dilemmas of dairy farmers' conceptions of cruelty and kindness. Anim Stud J 8(2):133. https://ro.uow.edu.au/asj/vol8/iss2/10

Tester K (1991) Animals and society, the humanity of animal rights. Routledge, New York

Thomas R (2005) Perceptions versus reality: changing attitudes towards pets in medieval and post-medieval England. In: Plukowski A (ed) Just Skin and Bones? New perspectives on human-animal relations in the historic past. Archaeopress, Oxford BAR International Series 1410, p 95. https://www.academia.edu/151650/Thomas_R_2005b_Perceptions_versus_reality_changing_attitudes_towards_pets_in_medieval_and_post_medieval_England_pp_95_105_in_Plukowski_A_ed_Just_Skin_and_Bones_New_Perspectives_on_Human_Animal_Relations_in_the_Historic_Past_BAR_International_Series_1410_Archaeopress_Oxford (subscription required)

Trabsky M (2014) Institutionalising the public abattoir in nineteenth century colonial society. Aust Femin Law J 40(2):169. https://doi.org/10.1080/13200968.2014.981357

Veevers JE (1985) The social meaning of pets: alternative roles for companion animals. Marr Family Rev 8(3–4):11. https://doi.org/10.1300/J002v08n03_03

Vialles N (1994) Animal to edible. Cambridge University Press, Cambridge

Voiceless (2015) The life of the dairy cow. Voiceless. https://voiceless.org.au/reading-list-item/the-life-of-the-dairy-cow/

Voiceless (2017) Unscrambled: the hidden truth of hen welfare in the Australian egg industry. Voiceless. https://voiceless.org.au/wp-content/uploads/2017/06/VoicelessReport_Unscrambled_June2017.pdf

Whitfort A (2018) Justice and the vulnerable: extending the duty to prevent serious crimes against children to the protection of agricultural and research animals. Adelaide Law Rev 39:125

Wilson L (2014) Ag-gag laws: a shift in the wrong direction for animal welfare on farms. Golden Gate Univ Law Rev 44(3):311

Woods (2011) Abigail "from cruelty to welfare: the emergence of farm animal welfare in Britain, 1964–71". Endeavour 36(1):14

Websites, Presentations and Quotes

Australia, Department of Agriculture, Water and the Environment, https://www.agriculture.gov.au/animal/welfare

Laravoire A (2016) Status of beak trimming. Danish poultry Congress, Chamber of Agriculture Pays de la Loire, power point slides. http://www.barnhealth.com/wp-content/uploads/2016/04/BeakTrimming_status.pdf

McCartney P (2009) Glass Walls (PETA). http://www.peta.org/tv/videos/celebritiesvegetarianism/86975251001.aspx

United Kingdom, the Department for Environment, Food and Rural Affairs, https://www.gov.uk/government/organisations/department-for-environment-food-rural-affairs

Legislation and Regulations

1901, Inclosed Lands Protection Act 1901 (NSW). http://www.austlii.edu.au/cgi-bin/viewdb/au/legis/nsw/consol_act/ilpa1901264/

1990, Kansas General Assembly, Kansas Statutes, Chapter 47, Livestock and Domestic Animals, § 47-1827. https://www.ksrevisor.org/statutes/chapters/ch47/047_018_0027.html

1995, Criminal Code Act 1995 (Cth). http://www.austlii.edu.au/cgi-bin/viewdb/au/legis/cth/consol_act/cca1995115/

2005, Summary Offences Act 2005 (Qld). http://www.austlii.edu.au/cgi-bin/viewdb/au/legis/qld/consol_act/soa2005189/

2012, Montana General Assembly, Title XXXVIII Crimes and Punishment; Peace Officers and Public Defenders, Chapter 578, § 578.013. https://revisor.mo.gov/main/OneSection.aspx?section=578.013

2016, Surveillance Devices Act 2016 (SA). http://www.austlii.edu.au/cgi-bin/viewdb/au/legis/sa/consol_act/sda2016210/

2019, Criminal Code Amendment (Agricultural Protection) Act 2019 (Cth). https://parlinfo.aph.gov.au/parlInfo/download/legislation/bills/r6351_aspassed/toc_pdf/19117b01.pdf;fileType=application%2Fpdf

2019, Montana General Assembly, Montana Code Annotated 2019, Title 81 Livestock, Chapter 30 Protection of Farm Animals and Research Facilities, § 81-30-103(2)(e) (West 2015) https://leg.mt.gov/bills/mca/title_0810/chapter_0300/part_0010/sections_index.html

European Union Council Directive 98/58/EC of 20 July 1998, Concerning the Protection of Animals Kept for Farming Purposes, [1998] Official Journal, L 221, 08/08/1998 P 0023–0027. https://eur-lex.europa.eu/legal-content/EN/ALL/?uri=CELEX%3A31998L0058

European Union Council Directive 1999/74/EC of 19 July 1999, Laying down Minimum Standards for the Protection of Laying Hens, [1999] Official Journal, L 203/53, 3/8/1999. https://eur-lex.europa.eu/LexUriServ/LexUriServ.do?uri=OJ:L:1999:203:0053:0057:EN:PDF

European Union Council Directive 2007/43/EC of 28 June 2007, laying down Minimum Rules for the Protection of Chickens Kept for Meat Production (Text with EEA relevance) [2007] Official Journal, L 182/19, 12/7/2007. https://eur-lex.europa.eu/LexUriServ/LexUriServ.do?uri=OJ:L:2007:182:0019:0028:EN:PDF

International Instruments

1924, International Agreement for the Creation at Paris of an International Office for Dealing with Contagious Diseases of Animals and Annex, opened for signature 25 January 1924, [1925] ATS 15, (entered into force 12 January 1925)

1935, International Convention Concerning the Transit of Animals, Meat and Other Products of Animal Origin, opened for signature, 20 February, 1935, 193 LNTS 37, entered into force 6 Dec 1938, signed by 9 parties and entered into force with 5 ratifications. https://treaties.un.org/doc/Publication/UNTS/LON/Volume%20193/v193.pdf

1992, Treaty on European Union, opened for signature 1 February 1992, [2002] Official Journal of the European Communities C 325/5, 24 December 2002, (entered into force 1 November 1993) available from https://eur-lex.europa.eu/legal-content/EN/TXT/PDF/?uri=OJ:C:2002:325:FULL&from=EN

1997, Treaty of Amsterdam amending the Treaty on European Union, the Treaties Establishing the European Communities and Certain Related Acts, opened for signature 22 October 1997, Official Journal of the European Communities C 340, P. 0001—0144, 10 November 1997, available from https://eur-lex.europa.eu/legal-content/EN/TXT/?uri=CELEX:11997D/TXT

Conclusion: Decommodifying Farm Animals 8

Abstract

The primary issue addressed by this book has been the commodification of farm animals, including how it occurred, its consequences and the need for change. From the nineteenth century, the demand for animal products encouraged economies of scale and maximisation of profits, so that by the time of World War II, livestock production had transformed from a system based on animal stewardship to one concentrated in continuous production systems. Both anti-cruelty regulation and animal welfare, which limits how humans use animals, are based on utilitarianism, a philosophy that balances animal interests against human interests. The practical outcome has been an inequitable bias in favour of human interests, where a pragmatic approach has led to farm animals being treated as commodities. This has led to the entrenchment of commercial biases because society has long accepted the fact that animals and their products are goods to be traded. Moreover, these assumptions accept the validity of animals' classification as property and the legitimacy of using animals in an increasingly intensified way. This situation is unlikely to change without government exercising its political will and power to correct imbalances in the sector. Only then will society evince a view of farm animals as living beings worthy of protection, rather than seeing them as commodities to be exploited.

Keywords

Animal commodification · Commercial bias

This book has taken the reader on a journey that started in the meat markets of the nineteenth century and ended with the philosophical optimism of Norbert Elias

S. Riley, *The Commodification of Farm Animals*, Animal Welfare 21,
https://doi.org/10.1007/978-3-030-85870-4_8

(1897–1990).[1] The primary issue addressed has been the commodification of farm animals, including how this occurred, its consequences and the need for change.

The demand for animal products in the nineteenth century led to widespread commercialisation of farm animals, which encouraged economies of scale and maximisation of profits; by the time of World War II, livestock production had altered from a system based on animal stewardship to one concentrated in "continuous production systems".[2] The emphasis lay on ensuring that supply met demand, engendering a commercial bias, which ultimately disconnected farm animals from emerging ideologies proscribing animal cruelty. These ideologies initially derived from the Enlightenment, an era that heralded intellectual freedom and debate, and which by the end of the eighteenth century saw beliefs in animal sentience gain traction. In response, government initiated anti-cruelty regulation that limited how society used animals.

However, this legislation also intersected with increasing urbanisation, industrialisation and the growing appetite for meat and other animal products. Consequently, farmers turned to production methods that would make animals ready for market as quickly as possible. Where this involved animal cruelty, the legality of industry practices depended on whether they were deemed necessary, resulting in industry standards assuming a high degree of significance. This led to notions of anti-cruelty legislation becoming tailored to market conditions, obscuring deeper analyses of humanity's relationship to animals.[3] Importantly, society accepted that animals and their products were commodities to be traded and as a result assumed the validity of foundational principles, including animals' classification as property and the legitimacy of using animals in an increasingly intensified way.

By 1894, when Henry Salt (1851–1939) proposed rights for animals, he faced a society that had come to expect easy access to animal products, supported by a regulatory system that interpreted cruelty in a way that facilitated production.[4] This was reflected in the fact that anti-cruelty laws created offences for animal abuse that had already occurred but largely did not stipulate obligations of care. It was an omission that became critical as increasing trade led to the introduction and spread of animal disease, where the health of animals in trade became conflated with the philosophical and moral footings of anti-cruelty.[5]

[1]Norbert Elias, *The Civilizing Process, Sociogenetic and Psychogenetic Investigations,* Blackwell Publishing, UK, (Revised Edition 2000).

[2]Harriet Friedman and Philip McMichael, "Agriculture and the State System, the Rise and Decline of National Agricultures, 1870 to the Present", (1989) XXIX-2 *Sociologia Ruralis*, 93, 106.

[3]Cassandra D Williams, "Liberating the Enlightenment: how a Transformed Relationship with Animals can help us Transcend Modernity", (2003) 29, 1 *Religious Education*, 95, 98–99.

[4]Henry Salt, *Animals' Rights: Considered in Relation to Social Progress*, (1894), revised edition G Bell and Sons Ltd. (1922), available from https://ia600901.us.archive.org/32/items/cu31924030305332/cu31924030305332.pdf

[5]John McEldowney, Wyn Grant, Graham Medley et al, *The Regulation of animal Health and Welfare*, Routledge, Taylor & Francis Group (2013), 20.

The cattle plagues of the 1860s were a turning point because animal diseases were only brought under control once regulators implemented strict orders that had been proposed by John Gamgee (1830–1894). Measures were based on stamping out disease and involved stringent border controls, restrictions on the movement of animals and slaughter of infected animals. Although these measures succeeded in controlling the cattle plague, they were also reactive, treating animals as expendable. Moreover, stamping out disease had consequences for veterinary medicine, because it became shaped by the needs of the sector.

In consequence, disease control and veterinary medicine came to be viewed through the lens of food security and trade protection. Moreover, because strict interventions had triumphed, this reinforced the need for government involvement.[6] In practical terms, it meant that owners exchanged their property rights in animals for benefits conferred by the state, such as protection of the sector from disease, and the right to receive compensation for animals that were destroyed.[7] Farm animals thus became ever more fungible, a commodity interchangeable with money, either through market transactions or by way of compensation.

As trade and commerce in farm animals grew, so too did the likelihood of disease being transmitted, a situation that was exacerbated where trade crossed international boundaries. This made it increasingly necessary to find an international solution to the problem of animal disease.[8] At the same time, by at least the eighteenth century, international trade had acquired a political dimension that merged private contractual arrangements with the national interest, which favoured expanding volumes of trade.[9] However, because international trade increased the likelihood of introducing and spreading animal disease, the national interest was overlaid with conceptions of the global good, represented by international cooperation and agreement. In this way, the legitimacy of unilateral national measures, particularly surrounding animal quarantine, slowly gave way to calls for more harmonised and standardised regimes based on veterinary science and international oversight.

While veterinary science was critical to the health of animals, the fact that animal disease, and by implication animal wellbeing, was categorised as an issue of international trade meant that the regime contributed greatly to the commodification of farm animals. Animal health came to be regarded less as a matter of animal wellbeing and more as an issue of protecting national territory against diseased

---

[6] Abigail Woods, "The Construction of and Animal Plague, Foot and Mouth Disease in Nineteenth-Century Britain" (2004) 17 (1) *Social History of Medicine*, 23, 30–32; Joanna Swabe, *Animals, Disease and Human Society, Human-animal Relations and the Rise of Veterinary Medicine*, Routledge (1999), 104, 108.

[7] Abigail Woods, "The Construction of and Animal Plague, Foot and Mouth Disease in Nineteenth-Century Britain" above 6, 30–32; Joanna Swabe, *Animals, Disease and Human Society, Human-animal Relations and the Rise of Veterinary Medicine*, above 6, 108.

[8] Joanna Swabe, *Animals, Disease and Human Society, Human-animal Relations and the Rise of Veterinary Medicine*, above 6, 105.

[9] Istvan Hont, *Jealousy of Trade, International Competition and the Nation-State in Historical Perspective*, Belknap Press of Harvard University Press, Harvard (2010), 4–6.

shipments. Only a handful of international instruments from the 1930s considered that animal suffering was part of their remit, notwithstanding the fact that at the national level, jurisdictions in Europe and the Americas had already passed a raft of anti-cruelty laws.[10] This meant that while animal suffering and animal cruelty were initially moral issues falling within "the purview of personal ethics",[11] they nevertheless came to be partially addressed at the national level, but not at all at the international level. In the latter case, animal suffering became conflated with disease prevention, so that in the same way animal cruelty was seen as a production issue at the national level, animal disease at the international level became a trade issue.

The reasons for these developments largely stem from the fact that both anti-cruelty regulation and animal welfare are based on utilitarianism, which balances animal interests against human interests. The practical outcome has been an inequitable bias in favour of human interests, where a pragmatic approach has led to farm animals being treated as commodities. This approach does not exclude farm animals from the sphere of human concern, but regulation focusses on issues important to production and trade, rather than mitigating animal suffering. Moreover, stakeholders understand animal cruelty in their own way, so that regulation invariably covers only a small part of the total problem.

In addition, consumer expectations regarding the supply and availability of animal products overlap with these issues, leading to two significant consequences. First, consumption is said to be the goal of production, so it is important to satisfy consumer demands,[12] and, second, consumer pressure also carries a political dimension capable of influencing law and policy.[13] The stumbling block, however, lies with the extent of commodification apparent in the sector, accompanied by the fact that by and large, government has relinquished regulation to market forces.[14] Consequently, improvements to animal wellbeing depend on the vagaries of the market, which are dominated by industry stakeholders, placing the burden on consumers or animal welfare groups to initiate reforms.[15] This is an onerous

[10]For the influence of English law and policy, David Favre and Vivien Tsang, "The Development of Anti-Cruelty Laws during the 1800's" (1993) Spring (1) *Detroit College of Law Review*; Lois Laimene Lelanchon "Detailed Discussion of Anti-Maltreatment Laws in France and Spain", (2013) *Animal Legal & Historical Center*, Michigan State University College of Law, available from https://www.animallaw.info/article/detailed-discussion-anti-maltreatment-laws-france-and-spain

[11]Bernard E Rollin, "Putting the Horse before the Descartes: My Life's Work on Behalf of Animals", Temple University Press, Philadelphia (2011), 31.

[12]Adam Smith, *An Inquiry into the Nature and Causes of the Wealth of Nations*, Digital Edition, ΜεταLibri (2007), 218, available from https://www.ibiblio.org/ml/libri/s/SmithA_WealthNations_p.pdf

[13]Matthew Hilton, *Consumerism in Twentieth-Century Britain: The Search for a Historical Movement*, Cambridge University Press (2003), 1.

[14]F Bailey Norwood and Jayson L Lusk, *Compassion, by the Pound: The Economics of Farm Animal Welfare*, Oxford University Press, New York (2011), 258.

[15]Ibid, 258.

expectation given that government determines law and policy, both of which already favour production.

This leads to the final point, a discussion on whether societal sway can counterbalance these forces. In his seminal tome, *The Civilizing Process*, Norbert Elias argues that over centuries society becomes less tolerant of violence and thus more civilised, a process that also applies to attitudes towards animals. The civilising process thus has the potential to reduce violence towards animals, as well as diminish their levels of commodification. Steps towards this process were established by at least the nineteenth century with the introduction of anti-cruelty regulation and the advent of animal protection organisations. However, civilising forces can be countermanded by de-civilising forces, which for farm animals primarily derive from economic imperatives of the sector. As with any industry, law and policy result from input by a range of stakeholders, with the present regime reflecting political struggles among producers, animal activists, consumers and government, to name a few. Depending on which group gains the ascendancy, regulation can either advance towards the kinder treatment of animals or entrench their treatment as commodities.[16] Currently, de-civilising forces in some jurisdictions are becoming stronger. Developments such as Ag-Gag laws, regulatory capture and the political power of the agricultural product sector aggravate power imbalances.[17] While government has the power to correct these imbalances, it may lack the political will to do so.

As already discussed in the last chapter of this book, this reinforces Robert Garner's remarks that the state is unlikely to create and enforce "moral obligations to protect animals. . . when they clash" with human demands.[18] Yet, this is precisely what needs to happen. Elias was perhaps more optimistic, identifying a process that would see humanity becoming ever more civilised so that 1 day humans might view farm animals as living beings worthy of protection because they exist within the human sphere, rather than seeing them as commodities to be exploited.

Bibliography

Curtis SE (1988) Animals in food production—American issues. Appl Anim Behav Sci 20(151): 153

Elias N 2000) The civilizing process, sociogenetic and psychogenetic investigations, revised edn. Blackwell Publishing, UK

Favre DS, Tsang V (1993) The development of anti-cruelty laws during the 1800s. Detroit College Law Rev 1:1

[16] S E Curtis, "Animals in Food Production – American Issues", (1988) 20 *Applied Animal Behaviour Science*, 151, 153.

[17] Pamela Fiber-Ostrow and Jarret S Lovell, "Behind a Veil of Secrecy: Animal Abuse, Factory Farms, and Ag-Gag Legislation", (2016) 19 (2) *Contemporary Justice Review*, 230, 232, DOI: https://doi.org/10.1080/10282580.2016.1168257

[18] Robert Garner, *A Theory of Justice for Animals: Animal Rights in a Nonideal World*, Oxford University Press, New York (2013), 61.

Fiber-Ostrow P, Lovell JS (2016) Behind a veil of secrecy: animal abuse, factory farms, and ag-gag legislation. Contemp Justice Rev 19(2):230. https://doi.org/10.1080/10282580.2016.1168257

Friedman H, McMichael P (1989) Agriculture and the state system, the rise and decline of National Agricultures, 1870 to the present. Sociol Rural XXIX-2:93

Garner R (2013) A theory of justice for animals: animal rights in a nonideal world. Oxford University Press, New York

Hilton M (2003) Consumerism in twentieth-century Britain: the search for a historical movement. Cambridge University Press

Hont I (2010) Jealousy of trade: international competition and the nation-state in historical perspective. Belknap Press of Harvard University Press, Harvard

Lelanchon LL (2013) Detailed discussion of anti-maltreatment laws in France and Spain. Animal Legal & Historical Center, Michigan State University College of Law. https://www.animallaw.info/article/detailed-discussion-anti-maltreatment-laws-france-and-spain

McEldowney, John, Grant, Wyn, Medley, Graham et al (2013) The regulation of animal health and welfare. Routledge, Taylor & Francis Group

Norwood FB, Lusk, JL (2011) Compassion, by the pound: the economics of farm animal welfare. Oxford University Press, New York

Rollin BE (2011) Putting the horse before the Descartes: my life's work on behalf of animals. Temple University Press, Philadelphia

Salt H (1922) Animals' rights: considered in relation to social progress (1894), revised edn. G Bell and Sons Ltd. https://ia600901.us.archive.org/32/items/cu31924030305332/cu31924030305332.pdf

Smith A (2007) An inquiry into the nature and causes of the wealth of Nations. Digital Edition, ΜεταLibri. https://www.ibiblio.org/ml/libri/s/SmithA_WealthNations_p.pdf

Swabe J (1999) Animals, disease and human society, human-animal relations and the rise of veterinary medicine. Routledge

Williams CD (2003) Liberating the enlightenment: how a transformed relationship with animals can help us transcend modernity. Relig Educ 29(1):95

Woods A (2004) The construction of and animal plague, foot and mouth disease in nineteenth-century Britain. Soc Hist Med 17(1):23

Woods A (2011) From cruelty to welfare: the emergence of farm animal welfare in Britain, 1964–71. Endeavour 36(1):14

Table of Cases

S. Riley, *The Commodification of Farm Animals*, Animal Welfare 21,
https://doi.org/10.1007/978-3-030-85870-4

Table of Statutes, Legislative Instruments and Debates

Australia

1901, *Inclosed Lands Protection Act 1901* (NSW), 7.4.1
1995, *Criminal Code Act 1995* (Cth), 7.4.1
2005, Australian Meat and Live-stock Industry (Standards) Order 2005 made under the Australian Meat and Live-stock Industry Act 1997, 6.3
2005, *Summary Offences Act 2005* (Qld), 7.4.1
2011, Australian Standards for the Export of Livestock (Version 2.3) 2011 and Australian Position Statement on the Export of Livestock, 6.3
2016, *Surveillance Devices Act 2016* (SA), 7.4.1
2019, *Criminal Code Amendment (Agricultural Protection) Act 2019* (Cth), 7.4.1

European Union

Council Directive 98/58/EC of 20 July 1998, Concerning the Protection of Animals Kept for Farming Purposes, 7.4.1
Council Directive 1999/74/EC of 19 July 1999, Laying down Minimum Standards for the Protection of Laying Hens, 7.3.1
Council Directive 2007/43/EC of 28 June 2007, laying down Minimum Rules for the Protection of Chickens Kept for Meat Production, 7.3.1

Ireland

1635, *Act Against Plowing by the Tayle, and pulling the Wooll off Living Sheep*, 3.4.1

S. Riley, *The Commodification of Farm Animals*, Animal Welfare 21,
https://doi.org/10.1007/978-3-030-85870-4

United Kingdom

1844, The Royal Charter of the Royal College of Veterinary Surgeons (UK), 4.2
1848, *An Act to prohibit the Importation of Sheep, Cattle, or other Animals, for the Purpose of Preventing the Introduction of Contagious or Infectious Disorder*, 2.6
1848, Privy Council, Orders in Council 4 September 1848, 2.6
1848, Privy Council, Orders in Council 22 September 1848, 2.6
1848, Privy Council, Orders in Council 3 November 1848, 2.6
1849, *An Act to amend the Laws in force for the Encouragement of British Shipping and Navigation*, 2.6
1849, *An Act for the more Effectual Prevention of Cruelty to Animals*, 3.4.2, 6.2, 6.3
1850, *An Act for the More Effectual Prevention of Cruelty to Animals in Scotland* (UK), 3.4.2, 6.2
1865, Consolidated Order of 25 September 1865, 4.4.1
1866, *An Act to Amend the Law Relating to Contagious or Infectious Diseases in Cattle and Other Animals* 29 & 30 Victoria c 2, 4.4.1
1866, A*n Act to amend the Law relating to Contagious Diseases amongst Cattle and other Animals in Ireland,* 29 & 30 Victoria c 4, 4.4.1
1866, *An Act to amend the Act of the Eleventh and Twelfth Years of Her present Majesty, Chapter One hundred and seven to prevent the spreading of contagious or infectious Disorders among Sheep, Cattle, and Other Animals,* 29 & 30 Victoria c 15, 4.4.1
1876, *Cruelty to Animals Act* (UK), 3.4.2, 6.2
1878, *An Act for Making Better Provision Respecting Contagious and Infectious Diseases of Cattle and Other Animals and for Other Purposes*, 2.6
1878, Privy Council, The Foreign Animals Order, 10 December, 1878, 2.6
1881, *Veterinary Surgeons Act* (UK), 4.1, 4.2
1894, *An Act to Consolidate the Contagious Diseases (Animals) Acts 1878 to 1893* (UK), 4.4.2
1896, *Diseases of Animals Act* (UK), 4.4.2
1911, *Protection of Animals Act* (UK), 3.4.2, 3.5, 6.2
1943, Winston Churchill, Agreement with Portugal, 12 October, 1943, House of Commons Debate, 5.3.1

United States of America

1641, *The Massachusetts Body of Liberties,* 3.4.1
1990, Kansas General Assembly, *Kansas Statutes*, Chapter 47, Livestock and Domestic Animals, § 47-1827, 7.4.1
2012, Montana General Assembly, *Title XXXVIII Crimes and Punishment*; Peace Officers And Public Defenders, Chapter 578, § 578.013, 7.4.1
2019, Montana General Assembly, *Montana Code Annotated 2019, Title 81 Livestock,* Chapter 30 Protection of Farm Animals and Research Facilities, § 81-30-103(2)(e) 7.4.1

Treaties and Other International Instruments

1373, *Governments of the Kingdom of England and the Kingdom of Portugal*, 5.3.1
1662, *Treaty of Peace and Commerce between Great Britain and Tunis*, 5.3.1, 5.3.2
1734, *Treaty of Commerce between Great Britain and Russia*, 5.3.1, 5.3.2
1815, *Convention of Commerce between the United Kingdom of Great Britain and Ireland and the United States of America, and Declaration by United Kingdom Regarding St Helena*, 5.3.1
1851, Ministère des Affaires Étrangères, *Procès-Verbeaux, de la Conférence Sanitaire Internationale*, 5.3.3
1852, *Sanitary Convention between France, Portugal, Sardinia, Tuscany and Turkey*, 5.3.3
1882, *Convention between France and Great Britain for the Regulation of Commercial and Maritime Relations and Declaration Prolonging the Treaties in Force*, 5.3.2
1887, *Convention Designed to Remove the Danger of Epizootic Diseases in the Territories of the Two Countries*, 5.3.2, 5.4, 5.5
1892, *International Sanitary Convention between Austria-Hungary, Belgium, Denmark, France, Germany, Great Britain, Italy, the Netherlands, Portugal, Russia, Spain, Sweden-Norway, and Turkey, with Annexes I to V*, 5.3.3
1922, *Commercial Treaty between the Esthonian Republic and the Kingdom of Hungary*, 5.3.2
1923, *Treaty of Commerce and Navigation made between Denmark and Finland*, 5.3.2
1923, *International Convention Relating to the Simplification of Customs Formalities*, 5.3.2, 5.3.3
1924, *International Agreement for the Creation at Paris of an International Office for Dealing with Contagious Diseases of Animals and Annex*, 5.1, 5.3.3, 5.4, 6.2, 7.4.1
1924, *Treaty of Commerce and Navigation Between Italy and Albania*, 5.3.2
1925, *Convention of Commerce and Navigation with Final Protocol Sweden and Czechoslovakia*, 5.3.2, 5.3.3
1927, *International Convention for the Abolition of Import and Export Prohibitions and Restrictions*, 5.3.2, 5.3.3

S. Riley, *The Commodification of Farm Animals*, Animal Welfare 21,
https://doi.org/10.1007/978-3-030-85870-4

1928, *Treaty of Commerce and Navigation between Bulgaria and Turkey*, 5.3.2, 5.3.3

1928, *Treaty of Commerce and Navigation with Final Protocol and Protocol concerning the Import and Export Régime in Italy Between Hungary and Italy*, 5.3.2, 5.3.3, 5.4

1928, *Treaty of Commerce and navigation with Protocol Between the Union of South Africa and Germany*, 5.3.2

1930, *Convention of Commerce and Navigation Between Poland and Roumania*, 5.3.2, 5.4, 5.5

1935, *International Convention for the Campaign Against Contagious Diseases of Animals*, 5.3.2, 5.3.3, 5.4

1935, *International Convention Concerning the Transit of Animals, Meat and Other Products of Animal Origin*, 5.3.2, 5.3.3, 5.4, 5.5, 7.4.1

1935, *International Convention Concerning the Export and Import of Animals, Meat and Other Products of Animal Origin*, 5.3.2, 5.3.3, 5.4

1945, Food and Agriculture Organization of the United Nations 1.1

1946, *Constitution of the World Health Organization*, 5.1

1951, *International Plant Protection Convention*, 5.4

1959, *Agreement Concerning Co-Operation in the Field of Veterinary Science*, 5.3.2, 5.3.3, 5.4, 5.5

1975, *Convention on the Elimination of all Forms of Racial Discrimination,* 5.3.1

1952, *Agreement Concerning Epizootic Diseases Between the Kingdom of Greece and The Federal people's Republic of Yugoslavia*, 5.3.2, 5.4, 5.5

1982, *The United Nations Law of the Sea Convention,* 5.3.1

1987, *European Convention for the Protection of Pet Animals*, 3.1

1992, *Treaty on European Union,* 7.4.1

1994, *Marrakesh Agreement Establishing the World Trade Organization,* 5.3.2

1994, *Agreement on the Application of Sanitary and Phytosanitary Measures*, 5.3.2, 5.3.3

1994, *General Agreement on Tariffs and Trade* (GATT) 1994, 5.3.2

1997, *International Plant Protection Convention 1997*, 5.4

1997, *Treaty of Amsterdam amending the Treaty on European Union, the Treaties Establishing the European Communities and Certain Related Acts*, 7.4.1